IET MATERIALS, CIRCUITS AND DEVICES SERIES 68

High Quality Liquid Crystal Displays and Smart Devices

Other volumes in this series:

Volume 2	**Analogue IC Design: The current-mode approach** C. Toumazou, F.J. Lidgey and D.G. Haigh (Editors)
Volume 3	**Analogue–Digital ASICs: Circuit techniques, design tools and applications** R.S. Soin, F. Maloberti and J. France (Editors)
Volume 4	**Algorithmic and Knowledge-based CAD for VLSI** G.E. Taylor and G. Russell (Editors)
Volume 5	**Switched Currents: An analogue technique for digital technology** C. Toumazou, J.B.C. Hughes and N.C. Battersby (Editors)
Volume 6	**High-frequency Circuit Engineering** F. Nibler *et al.*
Volume 8	**Low-power High-frequency Microelectronics: A unified approach** G. Machado (Editor)
Volume 9	**VLSI Testing: Digital and mixed analogue/digital techniques** S.L. Hurst
Volume 10	**Distributed Feedback Semiconductor Lasers** J.E. Carroll, J.E.A. Whiteaway and R.G.S. Plumb
Volume 11	**Selected Topics in Advanced Solid State and Fibre Optic Sensors** S.M. Vaezi-Nejad (Editor)
Volume 12	**Strained Silicon Heterostructures: Materials and devices** C.K. Maiti, N.B. Chakrabarti and S.K. Ray
Volume 13	**RFIC and MMIC Design and Technology** I.D. Robertson and S. Lucyzyn (Editors)
Volume 14	**Design of High Frequency Integrated Analogue Filters** Y. Sun (Editor)
Volume 15	**Foundations of Digital Signal Processing: Theory, algorithms and hardware design** P. Gaydecki
Volume 16	**Wireless Communications Circuits and Systems** Y. Sun (Editor)
Volume 17	**The Switching Function: Analysis of power electronic circuits** C. Marouchos
Volume 18	**System on Chip: Next generation electronics** B. Al-Hashimi (Editor)
Volume 19	**Test and Diagnosis of Analogue, Mixed-signal and RF Integrated Circuits: The system on chip approach** Y. Sun (Editor)
Volume 20	**Low Power and Low Voltage Circuit Design with the FGMOS Transistor** E. Rodriguez-Villegas
Volume 21	**Technology Computer Aided Design for Si, SiGe and GaAs Integrated Circuits** C.K. Maiti and G.A. Armstrong
Volume 22	**Nanotechnologies** M. Wautelet *et al.*
Volume 23	**Understandable Electric Circuits** M. Wang
Volume 24	**Fundamentals of Electromagnetic Levitation: Engineering sustainability through efficiency** A.J. Sangster
Volume 25	**Optical MEMS for Chemical Analysis and Biomedicine** H. Jiang (Editor)
Volume 26	**High Speed Data Converters** Ahmed M.A. Ali
Volume 27	**Nano-scaled Semiconductor Devices** E.A. Gutiérrez-D (Editor)
Volume 29	**Nano-CMOS and Post- CMOS Electronics: Devices and modelling** Saraju P. Mohanty and Ashok Srivastava
Volume 30	**Nano-CMOS and Post-CMOS Electronics: Circuits and design** Saraju P. Mohanty and Ashok Srivastava
Volume 32	**Oscillator Circuits: Frontiers in design, analysis and applications** Y. Nishio (Editor)
Volume 33	**High Frequency MOSFET Gate Drivers** Z. Zhang and Y. Liu
Volume 38	**System Design with Memristor Technologies** L. Guckert and E.E. Swartzlander Jr.
Volume 39	**Functionality-enhanced Devices: An alternative to Moore's law** P.-E. Gaillardon (Editor)
Volume 43	**Negative Group Delay Devices: From concepts to applications** B. Ravelo (Editor)
Volume 47	**Understandable Electric Circuits: Key concepts, 2nd Edition** M. Wang
Volume 60	**IP Core Protection and Hardware-assisted Security for Consumer Electronics** A. Sengupta and S. Mohanty

High Quality Liquid Crystal Displays and Smart Devices

Volume 2: Surface alignment, new technologies and smart device applications

Edited by
Shoichi Ishihara, Shunsuke Kobayashi and Yasuhiro Ukai

The Institution of Engineering and Technology

Published by The Institution of Engineering and Technology, London, United Kingdom

The Institution of Engineering and Technology is registered as a Charity in England & Wales (no. 211014) and Scotland (no. SC038698).

First published 2019

The Institution of Engineering and Technology
Michael Faraday House
Six Hills Way, Stevenage
Herts, SG1 2AY, United Kingdom

www.theiet.org

British Library Cataloguing in Publication Data
A catalogue record for this product is available from the British Library

ISBN 978-1-78561-923-6 (Hardback Volume 2)
ISBN 978-1-78561-924-3 (PDF Volume 2)
ISBN 978-1-78561-925-0 (Hardback Volume 1)
ISBN 978-1-78561-926-7 (PDF Volume 1)
ISBN 978-1-78561-939-7 (Hardback Volumes 1 and 2)

Typeset in India by MPS Limited

Contents

About the editors

Editor (1)
Dr. Shoichi Ishihara, Professor
Graduate School of Engineering, Osaka Institute of Technology
5-16-1, Ohmiya, Asahi-ku, Osaka, 535-8585 Japan
Phone +81-6-6954-0970
E-mail: shoichi.ishihara@oit.ac.jp

Dr. Shoichi Ishihara, Ph.D. in engineering, has been a professor at the Osaka Institute of Technology since 2010. He is a member of the Japanese Liquid Crystal Society, and the Society for Information Display. He got his B.S. in synthetic chemistry at the Nagoya Institute of Technology in 1973 and received his Ph.D. in physics and electronics at Osaka Prefecture University in 2000. He joined Matsushita Electric Industrial Co., Ltd., currently, Panasonic Corporation, in 1973. During 1978–1979, he had been as a fellow at MIT, Boston, USA. From 1996 to 2002, his activities have focused on the alignment control of LCs and the optical design of retardation films. There he had developed optically compensated birefringence (OCB) mode LCD-TVs and put them into the practical use for the first time. From 2003 to 2010, he worked at Sharp Corporation where he put the photo-aligned LCD panels (UV2A mode panels) and transverse bend alignment (TBA) mode LCDs to practical use first in the world.

Dr. Shoichi Ishihara's special field is display mode and interaction between LC molecules and alignment layers. Also, optical design of LCD is his strong area. His current work is on developing alignment-free film substrates and chemical gas sensors using cholesteric LCs. He is now the president of the Japanese Liquid Crystal Society and serves as an associate editor of Journal of the Society for Information Display. He holds about 130 Japanese Patents and 42 US Patents and has published 31 original papers.

Editor (2)
Dr. Shunsuke Kobayashi, Professor Emeritus
Tokyo University of Science, Yamaguchi
Phone +81-3-5387-7081
E-mail: lcd.kobayashi@gmail.com

Dr. Shunsuke Kobayashi is currently Professor Emeritus at the Tokyo University of Agriculture and Technology (TUAT) and Professor Emeritus and Founding Director of liquid Crystal Institute at Tokyo University of Science, Yamaguchi.

He obtained his Ph.D. degree in Electronic Engineering from the University of Tokyo in 1964. From 1964 to 1973, he was a Member of Staff at the Institute of Physical and Chemistry Research (RIKEN) in Wako-city, Japan; he conducted research on far-infrared detectors and he made a demonstration of defect free TN-LCD in1972. He was an associate and full professor at the TUAT from 1973 to 1996 in Electronic Engineering. His major research on LCDs focuses on the realization of defect free TN, STN, FLC-LCD, defect-free photo-alignment and practical field-sequential colour LCDs. He was Professor Emeritus at the TUAT since 1996. And then he became the professor at the Tokyo University of Science, Yamaguchi, and he has conducted a national project of nanoparticles as the Director of Liquid Crystal Institute at TUS-Y from 1997-2014. He has served as an associate editor of the MCLC since 1985.

He was a recipient of various awards: SID Jan Rajchman Prize in 1996, The Tokyo Metropolitan Governor Prize in1996, The Japanese Ministry of Education and Science Award in 2011, Honoured Member of the International Liquid Crystal Society 2012 for his research of defect-free colour LCDs and services as the vice president of ILCS and SID. His research is recorded in a book *Progress in Liquid Crystal Science and Technology*, in honour of Shunsuke Kobayashi's 80th birthday Edit. H.-S. Kwok, S. Naemura, and H.L.Ong World Scientific (2013). He has published 270 original papers and holds 60 patents.

Editor (3)
Dr. Yasuhiro Ukai, Technical Consultant
Ukai Display Device Institute
Phone +81-78-754-8839
E-mail: yasuhiro.ukai@spice.ocn.ne.jp

Dr. Yasuhiro Ukai is presently a technical consultant at the UDDI (Ukai Display Device Institute), and a senior member of the Society for Information Display and a life member of the Japan Applied Physics Society. He graduated from the Department of the Electrical Engineering, Osaka University, in 1968, and joined Hosiden Electronics Co., Ltd.

Since 1979 he has engaged mainly in R & D and commercialization of top-gate type a-Si TFT-LCD. He participated in producing the world's first 10-inch monochrome reflective a-Si TFT-LCD for Apple Macintosh portable in 1989 as a chief engineer. In 1994, he contributed as a project leader to the mass production of TFT-LCD mounted on the Boeing 777 as the world's first cockpit display. In 1999, he received his Ph.D. degree from the Tokyo Institute of Technology in Japan.

His doctoral thesis discussed the technology that contributed to the practical application of a-Si-TFT-LCD such as top gate structure by plasma doping, wide viewing angle by halftone grey scale, low reflection by Mo/MoO. In 1999, he joined Sony Corp. and worked at STLCD (a joint venture company of Sony and Toyota), where he contributed in establishing the mass-production technology in an area of low-temperature poly-Si (LTPS) TFT-LCDs, especially practical application of system-on-panel as technical manager.

He was a chief distinguished engineer and deputy general manager of Corporate R&D Display Device Development, working on the technology strategy and development of TFT-LCDs, this technology was applied to produce the product for in-cell touch panel. He retired from Sony in 2008. He has received Japan Display '86 Best Paper Award, Japan Display'89 Outstanding Paper Award, and SID'96 Best Application Paper Award, for his more useful and practical work.

He served as a co-editor of the *Journal of Display Technology* (A joint IEEE/OSA Publication) from 2008 to 2011.

He has published 50 original papers and holds 40 USPs.

Preface

This book celebrates 50 years since liquid crystal display (LCD) technologies were first demonstrated. LCDs have grown in that short time from a laboratory curiosity to a US$100 billion industry and are now indispensable in our daily life and careers – an impressive example of successful innovation and commercialisation.

Features of LCDs are excellent image quality, long life-time, low-cost and low power consumption. Early problems such as restricted viewing angle and slow response speed have now been overcome and field-effect LCDs are omnipresent. Research departments in the industry and universities cooperate to further advance the performance of LCDs and liquid crystal materials, by integrating chemistry, physics, optics, electrical and electronic engineering, biology and mechanical engineering specialities such as robotics and ergonomics. Currently the LCD industry is working on the development of high resolution 8k displays and arbitrarily shaped LCD-dashboards for cars, VR/AR, and 3D applications, while active research topics include the alignment mechanisms of liquid crystal molecules on surfaces and human perception of images.

LCDs are today the most important flat panel display and their development shows no sign of slow-down. In this book, we aim to describe the trends, challenges and potential solutions of this exciting technology to young scientists and engineers working on liquid crystal displays and in its numerous related fields. The book targets not only liquid crystal experts but also newcomers. The extensive list of references and a discussion of the latest developments and challenges serve this aim. Many recent state-of-the-art technologies are introduced in this book. Special emphasis is put on display quality, such as image sticking, contrast ratio and colour hue which have not been sufficiently dealt with in other books. Moreover, future trends and applications of liquid crystals materials and LC-devices following historical development and how they affect society are discussed. Of particular interest, emerging applications such as gas sensors and smart windows are covered.

The book includes 41 chapters, with contributions from 70 authors from all over the world.

Chapters 1–7 of Volume 1 cover the development of LCDs. Insight into the technical progress of LCDs is important for understanding their strong influence on social trends. Within the past 20 years LCD research has progressed from high-definition LCDs to large size displays with wide viewing angles and fast response times. Mobile phones, for example, have become an ubiquitous everyday product of LCD technology. The history of LCD developments is described in these chapters by showing milestone products with their photos and describing how researchers who participated in developing new LCD products solved problems for crossing over the so-called Darwin Sea. The trend of next-generation LCDs is discussed, starting in chronological order with the historical development of display operating modes, such as TN (twisted nematic), MVA (multi-domain vertical alignment), IPS (in-plane switching), FFS (fringe field switching), with proper reference to the inventors. Recent development trends are large-sized, high-definition LCDs and the enhancement of their colour gamut. In these chapters, high image quality and large dynamic-range technologies are discussed. Finally, the competitive advantage of LCDs versus organic light emitting diodes (OLEDs) and LCDs with arbitrary shape are discussed.

In Chapters 8–13 (Volume 1), current applications such as automotive, flexible displays, signage, etc. are presented. In particular vehicle displays have recently made remarkable progress. Their requirements from the point of view of manufacturing and user are outlined. The flexible display section addresses substrate and panel manufacturing processing. As an example for the latest technology, the advanced design of large, high-definition, curved LCDs with enhanced sense of immersion is described.

Today's diverse LCDs owe their advancement to untiring research and development efforts, mainly by research and development groups in industry working in functional materials and components. Recent developments of LC materials, quantum dots (QDs), backlight optics, and thin-film transistors, including oxide semiconductors and liquid crystalline organic semiconductors, are discussed in Chapters 14–20 (Volume 1). Transistors that are indispensable for driving LCDs are diversified into a-Si, high-temperature poly-Si (HTPS), low-temperature poly-Si (LTPS), oxide semiconductors (OSs) and organic thin film transistors (TFTs). TFT-LCDs are widely spread and these technologies are also applied to the active matrix OLED (AMOLED) backplane. Si LSIs are called microelectronics and it is important to understand how to miniaturize these elements. LCD is a human-machine interface where we can touch and obtain information, and its size and resolution are important factors. Along with the establishment of TFT-LCD technology, TFTs are now being fabricated on a 3 m square glass substrate with an accuracy of a few microns or less by micro-fabrication technology. Fabrication technology of these TFTs which contributed greatly to the establishment of giant microelectronics and their physical properties is discussed in detail. Quantum dots are essential for enhancing the colour gamut of LCDs. The current QD technology trends and its applications are discussed in detail.

In Chapters 1–9 of Volume 2, the technology of surface alignment of liquid crystal molecules is discussed. The quality and resolution of LCDs depends essentially on the surface alignment in an LCD. In these chapters, several alignment processes and planar alignment materials are discussed with a focus on non-mechanical photo-alignment. Moreover, newly developed technologies not requiring surface alignment are discussed. For investigating and evaluating LC-alignment and the interaction between molecules and substrates, new measuring equipment has been developed. Image sticking and flickering which are serious problems related to LC-alignment are investigated with such instruments.

In Chapters 10–16 of Volume 2, cutting-edge LCD technologies such as reflection mode LCDs, field sequential ferroelectric LCDs and LC lenses are discussed. In particular, the optical design of LC lenses which is essential for VR/AR displays and glass-free 3D-LCD-TVs is described. The basic principles of field-sequential-colour LCDs which are getting attention for low power consumption applications are discussed.

In Chapters 17–21 of Volume 2, applications of liquid crystals for uses other than displays are discussed. This covers applications such as gas sensors, smart windows, antennas for satellite communication, security devices and actuators; their design concepts and material requirements are discussed. Several examples for potential new liquid crystal applications are presented.

Liquid crystal science involves a wide range of disciplines and technical fields such as physics, electronics engineering, chemistry, biology, optics, ergonomics, sociology, and economics. Since these fields strongly cooperate, we trust that new liquid crystal devices apart from LCDs will come true.

We would like to offer our heartfelt thanks to the chapter authors. We especially thank Dr. Martin Schadt for his encouragement and advice. We are also grateful to the staff of the IET, Mr. Paul Deards, Ms. Olivia Wilkins, and Ms. Sarah Lynch, for giving us the opportunity to write an excellent book, and for their interest and coordination in the publishing of the book.

Shoichi Ishihara
Shunsuke Kobayashi
Yasuhiro Ukai

Chapter 1
Evolution of alignment material technology

Hiroyoshi Fukuro[1], Hideyuki Endo[1], and Go Ono[1]

The alignment layer is one of the important components in LCDs because the film makes an impact for the display quality by controlling the initial alignment of liquid crystals. It also affects the electrical properties of panels which decide the reliability of LCDs.

In this chapter, the design of alignment materials responded to mass fabrication process, such as the printing, the rubbing and so on, is described. In addition, the required properties of the alignment materials to generate and keep high display quality in LCDs are reviewed from the viewpoint of the alignment of liquid crystals and electrical properties of the film and LC cells. Finally we introduce alignment materials for new technology using the irradiation of UV, the photo-alignment and PSA technology. Photo-alignment technology is just applying FFS LCDs for smartphones and laptop computers with high density. PSA technology is widely applied to TV applications. The design of alignment materials corresponding to the each panel and fabrication process is also explained.

1.1 Introduction

Initially, liquid crystal displays (LCDs) were practically used in the first half of the 1970s through applications of segment-type LCDs to electronic calculators and clocks and so on. Subsequently, LCDs evolved to matrix-type displays, which have greater capacity, and numerous LCD modes have been developed. At first, the twisted nematic (TN) and super twisted nematic (STN) mode LCDs were developed using the passive matrix driving. Next, the active matrix driving using thin film transistor (TFT) became mainstream, and LCD modes such as in-plane switching (IPS), fringe field switching (FFS), multi-domain vertical alignment (MVA), and polymer-sustained vertical alignment (PSVA) were developed. As a result, LCD applications, including laptop computers, monitors, smartphones, and televisions, have diversified significantly.

LCDs display its image by altering the orientation of liquid crystals by applying the appropriate voltage to liquid crystal between electrodes. Thus, a

[1]Nissan Chemical Corporation, Japan

precise control of the liquid crystal orientation, especially the initial orientation control when voltage is not applied, is important, and it is achieved by using liquid crystal alignment layers coated on the surface of substrates.

For the liquid crystal alignment materials, chemically and thermally stable polyimides have been industrially used. Through the progress in LCDs, numerous materials based on polyimides have been developed and practically used for each LCD mode. And they have been improved in various ways from the perspectives of alignment control in accordance with LCD modes, accommodating the evolution of manufacturing processes and improving the electrical characteristics of LCD panels. In this chapter, we describe the history of development of liquid crystal alignment materials, their required properties, recent progress and the future view, including our research results.

1.2 Alignment materials from the aspect of LCD fabrication process

1.2.1 Composition of materials

In general, alignment materials based on polyimides comprise polymers, organic solvents and additives. As the components of polymers, polyamic acid obtained by polymerising diamines and tetracarboxylic dianhydrides in organic solvents such as *N*-methyl-2-pyrrolidone (NMP), *N,N*-dimethylformamide (DMF), *N,N*-dimethylacetamide (DMAc), or gamma-butyrolactone (GBL), or polyimides obtained by chemical imidization reaction of the polyamic acids are used. Typical examples of tetracarboxylic dianhydrides and diamines are indicated in Table 1.1.

For example, pyromellitic dianhydride (PMDA) produces a polyimide alignment layer with high glass transition temperature and high LC alignment stability. However, the low solubility of polyamic acids made from PMDA limits the solvents and diamines with which it can be applied. To improve these issues, the combination with alicyclic dianhydrides is used [1].

After coating alignment materials onto the substrate by appropriate printing methods, they are baked at a temperature of at least 200 °C to form approximately

Table 1.1 Typical monomer structures for polyimide alignment layers

	Dianhydrides	Diamine
Aromatic	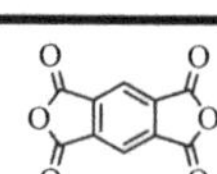 **PMDA**	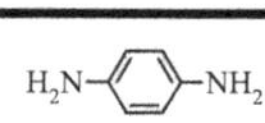 **PDA** 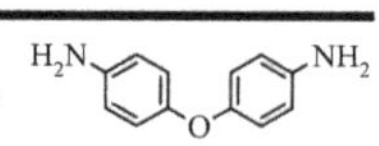 **DDE** 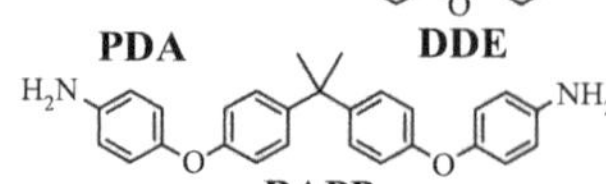 **BAPP** 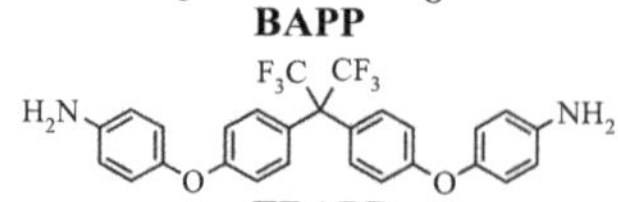 **FBAPP**
Alicyclic	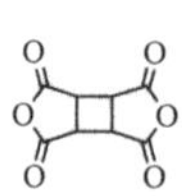 **CBDA** 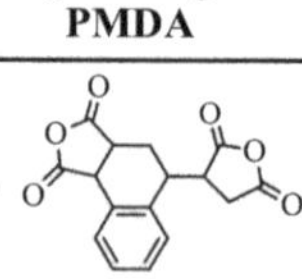 **TDA** 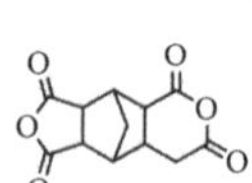 **NDA**	

100 nm thickness polyimide film. If the polymers are polyamic acids, baking at 200 °C or above is necessary for the conversion to polyimide, but in case of soluble polyimides, baking at lower temperature is also possible. To prepare such polyimides dissolved in organic solvents, diamines or acid dianhydride with flexible structure or a bulky substituent (for example, TDA or NDA in Table 1.1) are chosen.

Based on the above, to prepare polyamic acids or polyimides used for liquid crystal alignment layer, suitable acid dianhydrides and diamines must be selected with consideration of solubility in organic solvent as well as various properties described below.

1.2.2 Printing process

Dipping, spin-coating, flexo-printing and inkjet (IJ) printing are well known as coating method and flexo-printing and IJ printing are adapted in LCD mass production. Flexo printing needs highly viscous solution and the alignment material coated on the substrate cannot spread easily, and thus, it is difficult to obtain uniformly flat film. Therefore, it is important to optimise the molecular weight of polymer and solvent composition in alignment materials solution. It is common to use solvents such as NMP or GBL as main solvent that can dissolve the polymer, into which solvent such as ethylene glycol monobutyl ether or dipropylene glycol monomethyl ether with low surface tension is added in order to improve wetting properties to the substrate. IJ printing is useful coating method at factories that use large substrates, since it is difficult to prepare flexo-printing plates of a sufficient size. Since highly viscous varnish cannot be ejected using IJ printing, the varnish with viscosity of no greater than 10 mPas and solid content of 1–3 weight per cent is used. Liquid droplets ejected from IJ head do not form film immediately after adhering to the substrate surface, but spread out over time, contact each other and form uniform film through levelling process (Figure 1.1).

In the case of IJ printing, uniformity of film thickness and edge precision are very important. Recently, LC panels have become higher definition, and so the pixel size continues to become ever smaller, making it even more difficult to secure uniform film thickness of the alignment layer due to effects from stepped features at the substrate surface. For example, array substrates have deep contact holes, and these stepped features make it easier for unevenness in the film thickness to arise. Ma *et al.* evaluated film thickness unevenness after varying dot pitch of the adhered

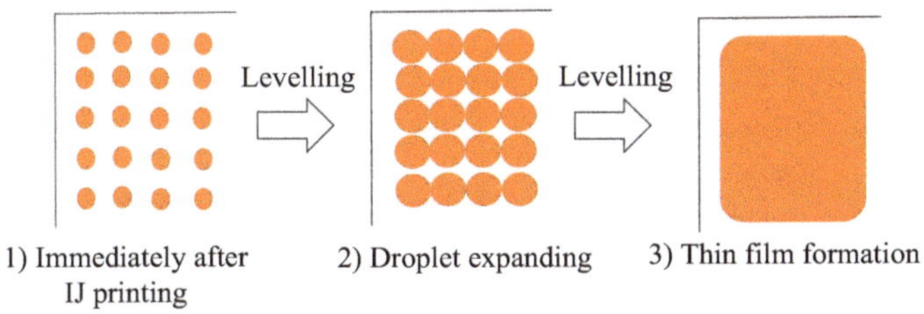

Figure 1.1 Film formation process by IJ printing

droplets and density and depth of the contact holes [2]. As an improvement approach based on the alignment material, there is a usual manoeuvre (which was also used for flexo-printing) of optimising the molecular weight of the resin component and solvent composition used. However, caution is needed as there are IJ heads that do not exhibit solvent resistance to NMP, and so there are limitations to its amount that can be used. LC panels are expected to continue to become higher definition in the future, and so cooperation among panel manufacturers, equipment manufacturers, and material manufacturers is desired to further improve coating properties.

1.2.3 Rubbing process

Rubbing method, which began with research by Maugin [3], is a simple method with inexpensive equipment, and so it is one of alignment methods still in use today. Since the rubbing method involves directly rubbing the surface of the alignment layer with cloth, this results in (1) contamination of the surface, (2) particles from the alignment layer and the rubbing cloth, (3) damage to the alignment layer surface and (4) static electricity, and these factors reduce the LCD quality and yield. It is the responsibility of the alignment layer to solve these issues.

To remove surface contamination or adhered particles, the substrate is washed using pure water or a mixed solvent of water and isopropyl alcohol (IPA) after rubbing process. Thus, the alignment layer must sufficiently retain alignment ability to liquid crystals, or in other words, the anisotropy of the film, after the washing process. To reduce the issues of particles and damage caused by delamination or scraping of the liquid crystal alignment layer, it is effective to add a few weight per cent of additive to the polymer component contained in the varnish. Silane coupling agent can improve the adhesion property to the substrates and it is common to use cross-linking agent (e.g., an epoxy cross-linking agent) that can react with the carboxylic acid group in order to improve film hardness. However, if too much of such additives are included, anisotropy of the film by rubbing diminishes, and LC alignment ability reduces. To simultaneously achieve the necessary LC alignment and film strength, it is necessary to optimise polymer component in addition to additives. Particularly for IPS/FFS, the effects in the rubbing process are the main issues of image sticking and lower contrast ratio, and so various studies have been conducted on factors such as material design for alignment material [4], selection of rubbing cloth [5], and evaluation of surface anisotropy resulting process after rubbing [6].

1.2.4 ODF and sealing process

Compared to the conventional vacuum injection method, the one drop fill (ODF) process has such benefits as using smaller amount of LC and shortening time it takes to fill with LC. On the other hand, while in the vacuum injection method, LC contacts completely cured sealing material, in the ODF method, LC contacts the sealing material while it is still uncured or in a partially cured state, and so contamination from the sealing material is easy to occur [7]. In addition, since multiple

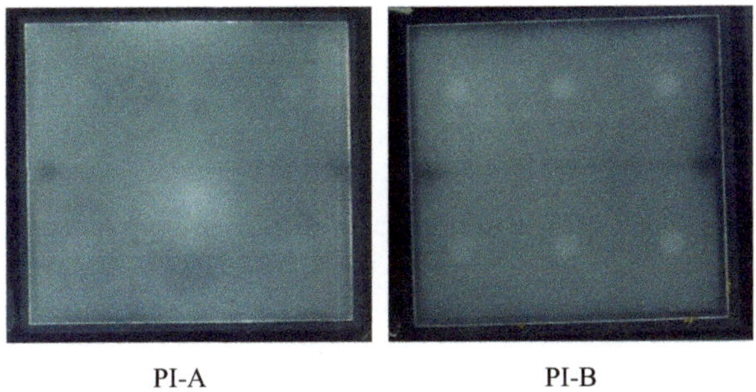

Figure 1.2 *Example of ODF mura observed in LC cells (lattice-shaped and dot-shaped non-uniformity)*

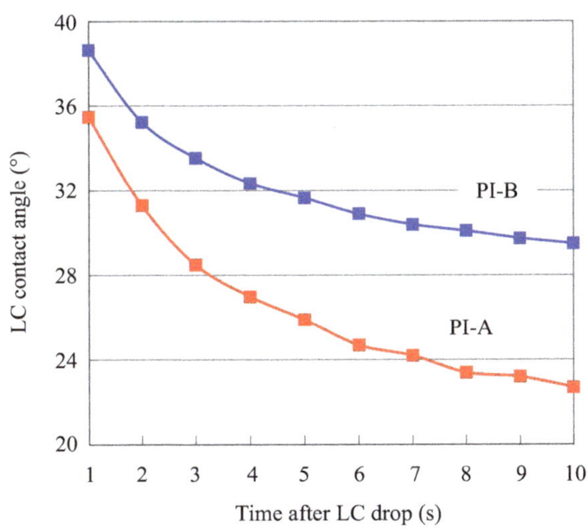

Figure 1.3 *LC contact angles of two alignment materials with different ODF mura*

droplets are dropped on the display pixel area before the substrates are adhered together, dot-shaped non-uniformity or lattice-shaped non-uniformity in areas where droplets collide with each other is easy to generate [8,9]. Figure 1.2 shows examples of dot-shaped non-uniformity and lattice-shaped non-uniformity caused when using two types of vertical alignment materials that differ in LC wettability.

ODF non-uniformity appears more prominently at PI-B, which has lower wettability than PI-A, as shown in Figure 1.3. To improve ODF non-uniformity due to alignment materials, it is effective to (1) increase anchoring strength, (2) prevent locally adsorption of impurities within LC and (3) increase LC wettability. Particularly with PSVA, which uses LC with reactive monomer, non-uniformity derived from uneven adsorption of the reactive monomer to the alignment layer surface can occur. Thus, ODF non-uniformity is a serious issue, as detailed in Section 1.4.2.

Recently, the development of flexible LCD is accelerated as one of the targets in next-generation LCDs. Therefore, higher adaptability for these fabrication processes is likely to be needed for alignment materials.

In particular, curing with lower temperature, improvement of printing ability and higher adhesive property for the flexible substrate will be required for it. To develop the alignment materials for flexible LCDs, the design of suitable polymer structures, and the selection of additives or solvents is of increasing significance.

1.3 Alignment materials from the aspect of the function

1.3.1 Display quality

The most important function of an alignment layer for LCD is to arrange the liquid crystal molecules in a panel in a desired orientation, such as horizontally, vertically or in a twisted manner. To develop a high-quality LCD, it is necessary to achieve a uniform director distribution across the entire display region of the LCD.

When the azimuthal angle is in-plane with the substrate surface, the angle of the liquid crystal director with respect to the polar angle is called the pre-tilt angle. There is considered to be an appropriate value for the pre-tilt angle depending on the LCD modes. For example, for the IPS and FFS LCDs, the pre-tilt angle closer to 0° is better based on viewing angle considerations, while for the TN mode, an angle of approximately 3°–5° is preferable from the perspective of preventing a reverse tilt. As methods for obtaining a high pre-tilt angle using the alignment materials, a commonly known method is to introduce fluorine groups or side chains such as long-chain alkyl groups to the polyimide [10–12].

Uniaxial orientation of liquid crystals in the azimuthal direction is formed when the liquid crystal molecules are in contact with the anisotropic surface of alignment layer through rubbing process. Thus, to achieve a uniform orientation, it is extremely important to investigate the relation between the alignment state of liquid crystals and the alignment layer surface after rubbing process, and numerous research results have been reported on this subject to date [13,14]. The mechanism by which the alignment layer which consists of polyimide achieves uniaxial orientation of the liquid crystals as a result of the rubbing process is considered to result from a combination of steric effects generated from the micro-grooves formed at the surface and intermolecular interactions between the liquid crystal molecules and the uniaxially oriented polyimide chains [15,16]. Kim *et al.* analysed the contribution of both factors on liquid crystal alignment using X-ray diffraction and NEXAFS for alignment layers for the IPS mode LCD, finding that the latter is predominant [17].

Suppressing the deterioration of the display quality as a result of driving the LCD panels can also be said to be an important role of the alignment layer. Specifically, it is necessary to suppress a falloff in the display quality such as display unevenness which is caused by the backlight irradiation, contamination of impurities from the sealant and infiltration of moisture from outside of the cells.

It is common for a decrease in the voltage holding ratio (VHR) to be related to that in the display quality such as display unevenness. Regarding the correlation between the molecular structure of polyimide and VHR, Nishikawa *et al.* investigated the effect of polarisation of the diamine constituting the polyimide, and reported that low-polarisation structure of polymer exhibits high VHR properties [18]. Sato *et al.* evaluated VHR using the alignment layer having double layer structure in which a soluble polyimide is coated on polyamic acid films as top layer, discovering that the component on the top side of the alignment layer, which is in contact with the liquid crystals, has a dominant effect on VHR [1].

1.3.2 Image retention and image sticking

An image sticking in LCD panel is a phenomenon in which after continuing to display a fixed pattern, this pattern remains visible even after changing the display to a separate pattern. It arises due to the changes in the voltage-transmittance characteristics of the LCD panel as a result of driving it. Avoiding image sticking is an extremely important aspect of display quality when used by the end user. In this section, the term 'image sticking' will be consistently to indicate this phenomenon. Regarding the image sticking in LCDs, refer to Chapter 7 in which the phenomenon and mechanism are described in detail.

LCD image sticking can be broadly divided into 'area image sticking' and 'line image sticking', as shown in Figure 1.4. Causes of area image sticking include (1) charge accumulation due to the polarisation induced within the liquid crystal cells through driving and (2) deviation of the liquid crystal director caused by insufficient anchoring energy of the alignment layer to the liquid crystal. On the other hand, line image sticking is considered to be caused by the localisation of ion impurities in the liquid crystal layer.

Generally, LCDs are driven by alternating voltage, but for various reasons, a DC component is superimposed in the voltage waveform applied to the liquid crystal cells via TFT. This DC component depends on the driving voltage waveform, and when the quantity of this DC component differs by the display pattern region, it causes image sticking. Applying DC induces internal polarisation in the liquid crystal cells, changing the voltage-transmittance characteristics and causing image sticking. It has been reported that when DC is applied to liquid crystal cells,

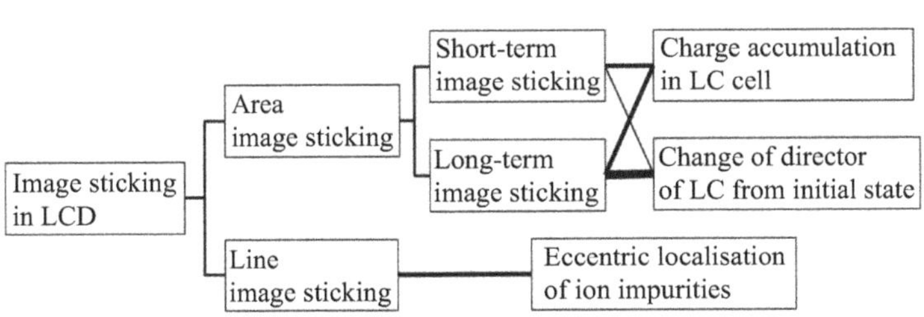

Figure 1.4 Classification and root causes of image sticking in LCD. The strength of correlation of each items are shown as the thickness of line

ion impurities contained in the liquid crystal form an electric double layer at the liquid crystal–alignment layer interface, and the accumulation as a residual DC voltage in the liquid crystal cells results in image sticking [19]. Mada *et al.* explained that the abnormal absorption current arising when DC voltage is applied to an LCD is caused by the electric double layer formed by the ions in the liquid crystal layer [20,21].

Image sticking caused by the accumulation of the DC component (residual DC voltage) is an issue that can generally occur in TN, VA, as well as IPS/FFS LCD, and is closely related to the electrical conductive behaviour of the alignment layer and the dielectric properties of the liquid crystal cell.

Nakanowatari *et al.* approximated a liquid crystal cell as a stack of dielectric components as an equivalent circuit, and reported that a residual DC voltage arises from the difference in the time constant of the relaxation process between the alignment layer and the liquid crystal [22]. On the other hand, Mizusaki *et al.* showed that the residual DC voltage can be theoretically explained by the adsorption/desorption of ionic impurities present in the liquid crystal to/from the polyimide alignment layer interface [23,24].

Based on the above, attempts have been made to quickly eliminate the residual DC voltage accumulated in liquid crystal cells by reducing the resistance of the alignment layer. For example, PI-C, into which a diamine component with electron donor properties has been introduced, exhibits relatively low volume resistivity ($\rho = 8.0 \times 10^{12}$ Ωcm) and faster attenuation of residual DC voltage accumulated in liquid crystal cells with respect to PI-D ($\rho = 1.4 \times 10^{15}$ Ωcm), which has the same tetracarboxylic dianhydride component, as shown in Figure 1.5.

It is known that for image sticking caused by deviation of the liquid crystal director is particularly significant in the in-plane switching mode of LCD, such as in IPS and FFS. Thus, a large anchoring energy for azimuthal orientation can be considered the most important property of alignment layers for IPS and FFS panels. Examples of evaluation using a torque balancing method have been reported for the

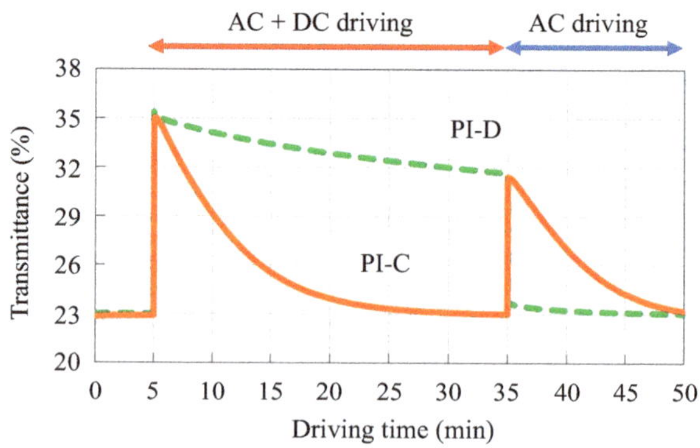

Figure 1.5 Changes in transmittance of LC test cells with applying and removing process of DC voltage

azimuthal anchoring energy in the IPS and FFS modes [25,26]. However, this method requires accurately measuring an extremely small twist angle in the strong anchoring region of 10^{-3} J/m^2 or higher, which required for mass-produced alignment layers. Thus, it is difficult to practically evaluate the anchoring of practical materials by this method. As a result, a simpler and more practical evaluation has been proposed to use IPS cells with herringbone comb electrodes to perform long-term AC driving, and to measure the change in the liquid crystal director angle (Δ_{angle}) in the azimuthal direction both before and after [27]. Momoi *et al.* used this evaluation method to report that the phenomenon of liquid crystal director deviation is correlated with the rheology at the surface of alignment layer [28].

1.4 Progress in alignment materials

1.4.1 Photo-alignment materials for IPS mode

Photo-alignment technologies are non-contact methods to achieve uniform LC alignment by irradiating a polymer thin film with linearly polarised ultraviolet light (LPUV). Thus, they are widely investigated because of having a large potential to resolve issues of the rubbing method, such as rubbing scratches, reduced contrast ratio due to non-uniformity of the rubbing, and breakage of TFT elements due to static electricity. Generally, photoreactive groups used for a photo-alignment layer can be categorised as (1) those that cause axially selective photoisomerisation, such as azobenzene [29,30]; (2) those that cause photodimerisation, such as cinnamic acid [31]; and (3) those that use axially selective photodegradation through the irradiation of deep ultraviolet light onto a polyimide [32]. Till now, various polymers using such photoreactive groups have been studied as photo-alignment layers, but due to their superior thermal and mechanical stability, as well as having the electrical characteristics required of a liquid crystal alignment film, polyimides are selected for practical applications.

In case of applying photo-alignment to IPS LCDs, since there is no deformation to the polymer chain arrangement by mechanical stress, a highly uniform LC alignment is obtained with a pre-tilt angle of approximately 0°. As a result, changes in black luminance due to the pre-tilt angle formation and the viewing angle properties are improved [33,34]. However, conventional photo-alignment layers have the image sticking issue (called AC image sticking) caused by weak anchoring due to low anisotropy induced at the thin film surface after LPUV exposure. In IPS LCDs, the liquid crystal director rotates in-plane with the substrate under applied voltage. AC image sticking is a phenomenon in which this rotated liquid crystal director does not return to the pre-determined position even after the voltage is released. It is considered that AC image sticking is prone to occur when LC on the alignment layer surface align with a lower alignment order and when the alignment layer surface is easily deformed by rotational torque of the liquid crystal [34,35]. Anchoring energy has been reported to be correlated with the magnitude of anisotropy induced in thin films [36]. As a result, polyimides that can induce large anisotropy by LPUV exposure, which are chemically, thermally, and mechanically stable, have been developed. Faetti *et al.* reported that by irradiating LPUV light at 365 nm onto an alignment layer made from a polyimide (Azo-PI) contained in the main chain of an azobenzene group, larger anisotropy is obtained than that induced by rubbing [37]; moreover, this alignment layer results in very strong anchoring. In addition, Usami *et al.* reported that by optimising the mobility of the polyimide main chain, uniform liquid crystal alignment can be obtained even in region of smaller exposure energy [38].

Meanwhile, there have also been studies on the photo-alignment of liquid crystals using a polyimide consisting of cyclobutane tetracarboxylic dianhydride (CBDA) (Figure 1.6). We evaluated the correlation between AC image sticking and anisotropy induced in a thin film by LPUV irradiation using CBDA polyimides. Table 1.2 summarises the characteristics of the polyimides used in this investigation. For comparison, AC image-sticking property of the cell fabricated by rubbing method is also listed.

AC image sticking is evaluated by measuring the change in azimuthal angle ($\Delta\theta$) of the LC director from the pre-determined position before and after AC driving in a liquid crystal cell with comb-like electrodes as described in literatures [27,35]. Figure 1.7 shows the magnitude of the anisotropy induced in various polyimide films when exposed to an optimum irradiation dose. PI-E and PI-F, which exhibit high crystallinity, induce larger anisotropy, while PI-G, which has a flexible structure, has smaller anisotropy. It has been found that the anisotropy induced by irradiation with LPUV shows the same tendency as that of rubbing process. However, although the

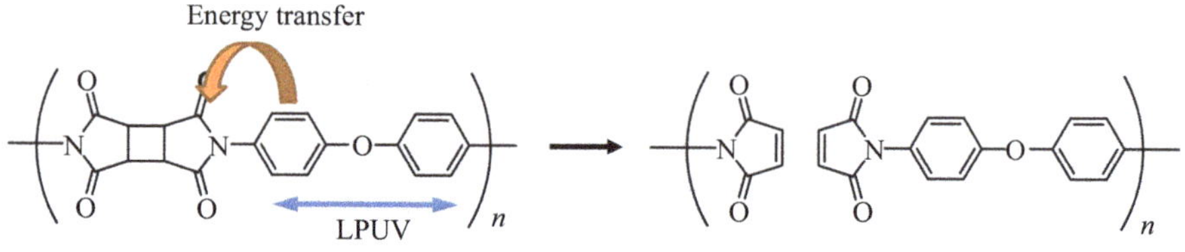

Figure 1.6 Photo-reaction mechanism of polyimide comprising cyclobutane units

Table 1.2 Characteristics of polyimide samples and their AC image-sticking property of the LC cell fabricated by the rubbing method

Samples	**Characteristics**	**AC image-sticking property of the LC cell fabricated by rubbing method**
PI-1	– Partially flexible and longer structure – High crystallinity	Excellent
PI-2	– Partially flexible structure – High crystallinity	Good
PI-3	– Flexible structure	Poor

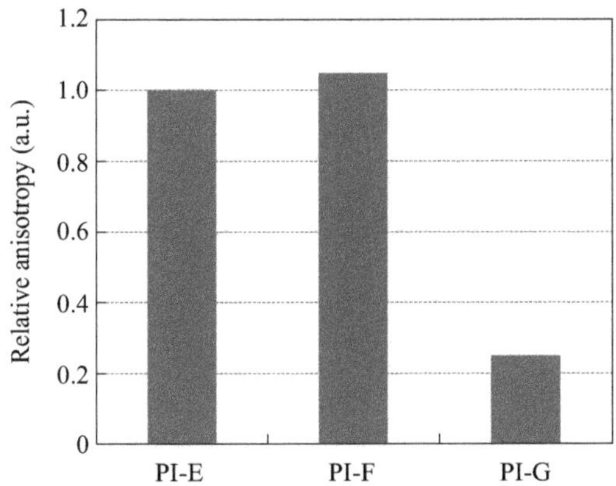

Figure 1.7 Relative anisotropy of PI-E, PI-F and PI-G films irradiated by appropriate dose with 254 nm LPUV light

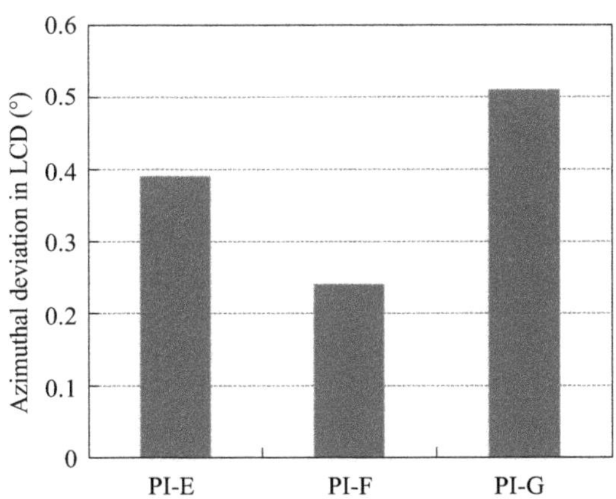

Figure 1.8 Azimuthal deviation of LC director from predetermined position on PI-E, PI-F and PI-G films irradiated by appropriate dose with 254 nm LPUV light. Test cell type: FFS mode

anisotropy values for PI-E and PI-F are almost equal, PI-F has superior AC image-sticking performance as shown in Figure 1.8. In this way, a different material design is clearly needed for the photo-alignment due to differences in performance as an alignment layer depending on the alignment process.

A cleavage reaction of cyclobutane ring occurs as a result of the light energy absorbed at the diamine unit and transferred to the acid dianhydride side. Therefore, it is important to select the irradiation wavelength which the diamine group can absorb UV light in axis selectivity. Table 1.3 shows the anisotropy inducing in a PI-H film irradiated by LPUV light at different wavelengths, while Figure 1.9 shows the absorption spectrum of this film. When irradiated with LPUV light at

Table 1.3 Anisotropy change generated in PI-H film irradiated with 254 nm and 313 nm LPUV light as a function of exposure energy

	Irradiation condition (J/cm²)				
	254 nm		**313 nm**		**Rubbing**
	1.0	**1.8**	**1.0**	**1.8**	
Anisotropy (a.u.)	0.49	0.72	0.12	0.25	1.0

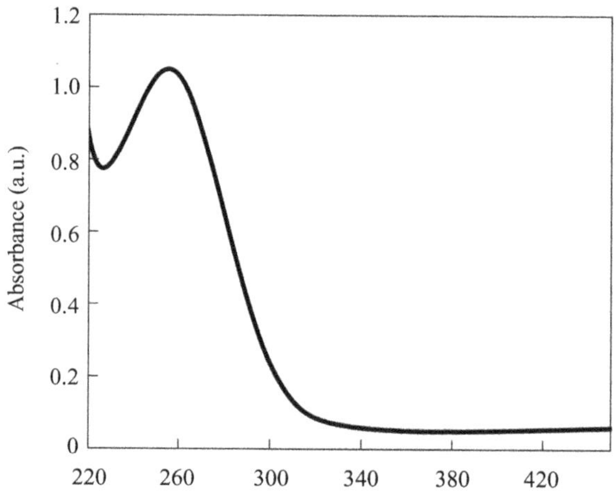

Figure 1.9 UV absorption spectra of PI-H

254 nm, the small anisotropy is induced after 1.0 J/cm^2 exposure and increasing the exposure dose increases the anisotropy. On the other hand, the anisotropy induced when irradiated by 313 nm LPUV light is extremely smaller than that irradiated by 254 nm light, and thus, it can be seen that the photoreaction does not proceed easily.

Since the photo-alignment of the polyimide having a cyclobutane ring is based on photo-degradation of the polymer main chain which is not so high efficiency, there are issues in throughput. With respect to these issues, modifications to the polyimide structure result in significant improvements in photo-reaction efficiency. Thus, studies have been continued from perspectives on improving both throughput and AC image sticking, which have been issues in the practical application of photo alignment technology for IPS LCDs. As the result, we have developed photo-alignment materials which exhibit sufficient performance for mass production.

1.4.2 Polymer-sustained vertical alignment technology

Recently, there have been attempts to use photo-reactions for liquid crystal alignment control as an approach different from the photo-alignment methods using

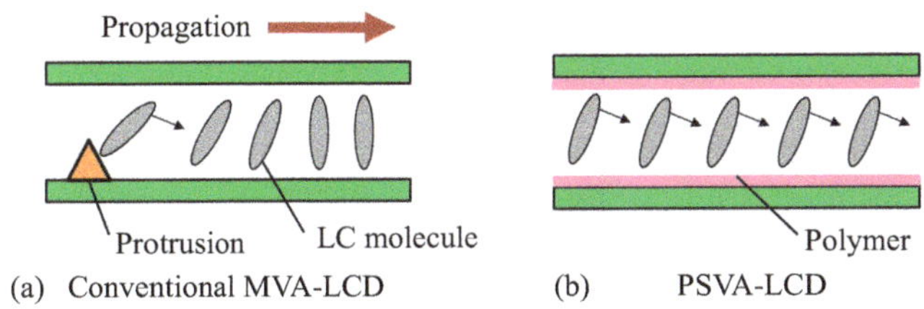

Figure 1.10 The switching mechanism comparison between conventional MVA-LCD and PSVA-LCD

LPUV irradiation. By combining ITO electrode patterning with photo-polymerisation technology, PSVA, which was developed as an advanced version of MVA, succeeded in eliminating ITO slits and protrusions, which were the cause of reduced transmittance even though they played the role of controlling the orientation of liquid crystals in the MVA mode [39]. Figure 1.10 shows a comparison of MVA and PSVA. In PSVA, a reactive monomer (RM) such as biphenyl diacrylate is added to the liquid crystal and photo-polymerisation is performed through ultraviolet irradiation in the state of tilting the liquid crystals by applying voltage. Consequently, a polymer layer is formed at the alignment layer interface and pre-tilt angle is generated by the liquid crystals being immobilised at a slight tilt. As a result of this polymer layer formation, PSVA achieves faster response time, improved contrast ratio and transmittance compared to MVA, and exhibits less image sticking caused by residual DC [39,40]. In PSVA, liquid crystals align with pre-tilt angle of 88°–89° to the substrate. Therefore, the deviation of pre-tilt angle tend to be recognise in visual as a transmittance change compared to MVA whose pre-tilt angle is 90°. From this reason, the changes in pre-tilt angle after driving in PSVA become visible as image sticking. Nakanishi *et al.* studied the effect of RM structure on image sticking caused by pre-tilt change. It has been reported that the pre-tilt changes are correlated with image sticking and more rigid RM makes the pre-tilt change smaller, which means image sticking does not occur easily.

As described above, PSVA has superior display performance compared to conventional MVA. However, the generation of ODF non-uniformity called 'ODF mura' centred on the liquid crystal dropdown point is a major issue when filling a liquid crystal with RM using the ODF method. Figure 1.10 shows the location dependence of the pre-tilt angle in liquid crystal cells in which ODF mura is strongly observed and in which it is mostly not observed. The area observed as ODF mura, which are more white, have a larger pre-tilt angle than the other area. From the factor investigation, the amount of RM in the ODF mura area is found to be greater than that in the other area. Therefore, alignment layers that can help a uniform distribution of RM in the liquid crystal cells are studied. Figure 1.11 shows the relation between ODF mura and surface energy of the alignment layer, and it is found that reducing the surface energy of the alignment layer improves ODF mura.

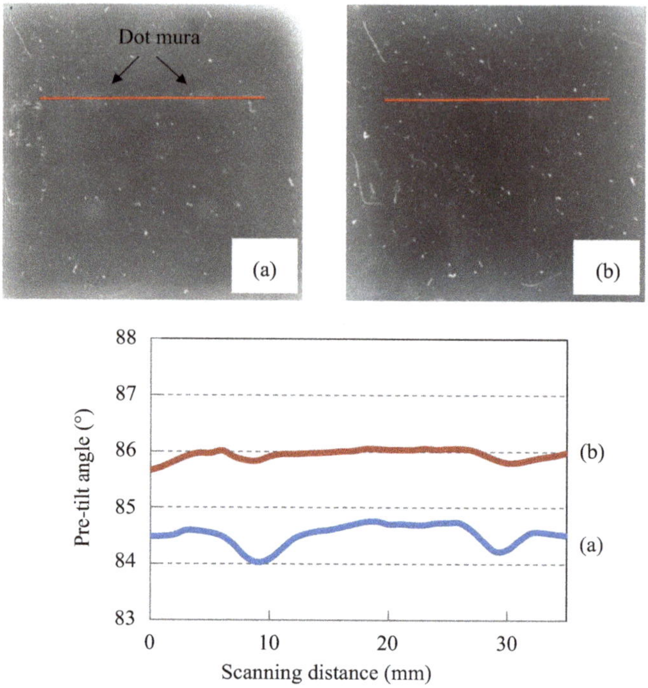

Figure 1.11 Location dependence of pre-tilt angle in (a) LC cell which dot mura is observed and (b) LC cell which has less dot mura, when pre-tilt angle is measured along the red line

1.5 Summary

As described above, liquid crystal alignment materials are indispensable for the production of high-quality LCDs, and have greatly advanced together with LCDs. To liquid crystal alignment materials, not only bulk properties based on their molecular structure but also surface properties should be taken into account. Furthermore, mutual interaction with liquid crystal and peripheral materials such as sealant, in addition various properties that conform with manufacturing processes are important. However, their systematic study cannot yet be considered sufficient.

On the other hand, new alternative technologies such as OLED and μLED are being developed as display devices. In particular, OLEDs have acquired a certain share in the smartphone market, and have begun gaining market share for TV applications. To compete with such display devices, LCD technologies need to be improved continuously. In addition to improvements regarding conventional issues of such as image sticking and non-uniformity in display, further advances in manufacturing processes are needed to obtain higher yield in high image quality LCDs. In particular, major future themes of liquid crystal alignment materials are to be improving the response speed, the light transmittance and reliability of LCD panels. In addition, new alignment techniques may be discovered in the near future. In such circumstances as well, the previous knowledge about alignment techniques and the interactions between alignment layer and the liquid crystals or the peripheral materials will be very useful.

Furthermore, some new devices and applications are being developed, such as flexible LCDs, dimming devices, high frequency devices, such as liquid crystal antennas, and thus, new properties of liquid crystal alignment materials are being demanded. The liquid crystal alignment layer plays an important role in these devices as well, and the development of liquid crystal alignment materials conforming to the required properties is underway. To achieve success in these areas, it is believed necessary to construct new platform technologies.

We have assisted in the evolution of LCDs through the development of liquid crystal alignment materials. We continue to believe that liquid crystal alignment material technology will contribute to the future evolution of LCDs.

References

[1] Fukuro H., Sawahata K., Sato T. and Endo H. 'Optimal alignment materials and technologies for various LCDs'. SID International Symposium Digest of Technical Papers, Long Beach, US, May 2000. **31**, 434–37.

[2] Ma G., Wang Y.S.D, Ren J., *et al.* 'Analysis of ink-jet-printing mura in the PI film on TFT-LCDs'. SID International Symposium Digest of Technical Papers, Los Angeles, US, May 2017. **48**, 1466–68.

[3] Mauguin M.C. 'Sur les cristaux liquides de Lehmann'. *Bull. Soc. Fr. Mineral. Crystallogr.* 1911; **34**, 71–117.

[4] Tsutsui K., Sakai T., Goto K., Sawahata K., Ishikawa M. and Fukuro H. 'New rubbed alignment material with no image sticking for IPS-LCD'. SID International Symposium Digest of Technical Papers, Baltimore, US, May 2003. **34**, 1166–69.

[5] Ishihara S., Hoshino Y., Horikawa R., Maeda T. and Yoshida H. 'Novel rubbing cloth providing an alignment layer with low pretilt angle and large azimuthal anchoring energy'. SID International Symposium Digest of Technical Papers, Boston, US, June 2012. **43**, 1418–21.

[6] Zhang W., Song S.H., Li J., Xiao A., Chen X. and Min T.Y. 'Analysis of rubbing mura in fringe field switching liquid crystal display'. SID International Symposium Digest of Technical Papers, Boston, US, June 2012. **43**, 1243–46.

[7] Kamiya H., Tajima K., Toriumi K., *et al.* 'Development of one drop fill technology for AM-LCDs'. SID International Symposium Digest of Technical Papers, San Jose, US, June 2001. **32**, 1354–57.

[8] Lee H., Song S., Lee J.P., Xiao A., Min T. and Su S. 'Analysis of liquid crystal drop mura in high resolution mobile thin film transistor liquid crystal displays'. SID International Symposium Digest of Technical Papers, San Diego, US, June 2014. **45**, 97–99.

[9] Woo B.C., Kim K.H., Kim J.S., Kim W.H. and Han S.Y. 'Mechanism for lattice-shape mura in TFT-LCD'. Proceedings of the 17th International Display Workshops, Fukuoka, Japan, December 2010. 109–12.

[10] Seo D.S., Kobayashi S. and Nishikawa M. 'Study of the pretilt angle for 5CB on rubbed polyimide films containing trifluoromethyl moiety and analysis of the surface atomic concentration of F/C(%) with an electron spectroscope for chemical analysis'. *Appl. Phys. Lett.* 1992; **61**(20), 2392–94.

[11] Fukuro H. and Kobayashi S. 'Newly synthesized polyimide for aligning nematic liquid crystals accompanying high pretilt angles'. *Molecular Crystals and Liquid Crystals Incorporating Nonlinear Optics.* 1988; **163**(1), 157–62.

[12] Nishikawa M., Matsuki Y., Bessho N., Iimura Y. and Kobayashi S. 'Characteristics of polyimide liquid crystal alignment films for active matrix-LCD use'. *J. Photopolym. Sci. Technol.* 1995; **8**(2), 233–40.

[13] Stöhr J., Samant M. G., Cossy-Favre A., *et al.* 'Microscopic origin of liquid crystal alignment on rubbed polymer'. *Macromolecules*, 1998, **31**(6), 1942–46.

[14] Toney M., Russell T., Logan J., Kikuchi H., Sands J. and Kumar S. 'Near-surface alignment of polymers in rubbed films'. *Nature* 1995; **374** (6524), 709–11.

[15] Pidduck A. J., Bryan-Brown G. P., Haslam S., *et al.* 'Atomic force microscopy studies of rubbed polyimide surfaces used for liquid crystal alignment'. *J. Vac. Sci. Technol., A* 1996; **14**(3), 1723–28.

[16] Ito T., Nakanishi K., Nishikawa M., Yokoyama Y. and Takeuchi Y. 'Regularity narrowness of the intervals of the microgrooves on the rubbed polymer surfaces for liquid crystal alignment'. *Polym. J.* 1995; **28**(3), 240–46.

[17] Kim, D-G., Lee W. K., Ahn H. J., *et al.* 'Effect of the surface anisotropy of a rubbed polyimide film on liquid crystal alignment properties'. SID International Symposium Digest of Technical Papers, San Antonio, US, June 2009. **40**(3), 1641–43.

[18] Nishikawa M., Suganuma T., Tsuda Y., Bessho N., Iimura Y. and Kobayashi S. 'Properties of voltage holding ratios of liquid crystal cells using organic-solvent-soluble polyimide alignment films'. *Jpn. J. Appl. Phys., Part 2*, 1994; **33**(8A), L1113–L1116.

[19] Yasuda Y., Naito H., Okuda M. and Sugimura A. 'Observation of adsorption and desorption processes of impurity ions in nematic liquid crystal cells'. *Mol. Cryst. Liq. Cryst.*, 1995; **263**(1), 559–65.

[20] Mada H. and Suzuki H. 'Reverse hysteresis loop of nematic liquid crystals in *C-V* characteristic due to static electric field'. *Jpn. J. Appl. Phys., Part 2, Letters*, 1987, **26**(7), L1092–94.

[21] Mada H. 'Increasing phenomenon of abnormal absorption current in liquid crystal cells'. *Jpn. J. Appl. Phys., Part 2: Letters* 1990; **29**(1), L123–25.

[22] Nakanowatari J. and Ono K., 'Effect of the thickness of alignment layer on the low-frequency dielectric dispersion of a nematic liquid crystal'. *Jpn. J. Appl. Phys., Part 2, Letters*, 1996, **35**(1B), L111–13.

[23] Mizusaki M., Yoshimura Y., Yamada Y. and Okamoto K. 'Analysis of ion behavior affecting voltage holding property of liquid crystal displays'. *Jpn. J. Appl. Phys.*, 2012. **51**(1), 014102/1–014102/6.

[24] Mizusaki M., Miyashita T. and Uchida T. 'Kinetic analysis of image sticking with adsorption and desorption of ions to a surface of an alignment layer'. *J. Appl. Phys.*, 2012, **112**(4), 044510/1–044510/9.

[25] Uchida T., Hirano M. and Sakai H. 'Director orientation of a ferroelectric liquid crystal on substrates with rubbing treatment: The effect of surface anchoring strength'. *Liq. Cryst.* 1989; **5**(4), 1127–33.

[26] Han K-Y., Vetter P. and Uchida T. 'Determination of molecular inclination in rubbed polymer for liquid crystal alignment by measuring retardation'. *Jpn. J. Appl. Phys., Part 2, Letters*, 1993; **33**(9A), L1242–44.

[27] Suzuki T., Matsushima J., Sasaki Y., *et al.* 'High contrast ratio in-plane-switching TFT-LCD with ion beam irradiated polyimide film as LC alignment layer'. Proceedings of the 12th International Display Workshops in conjunction with Asia Display, Takamatsu, Japan, December 2005. 57–60.

[28] Momoi Y., Sato O., Koda T., Nishioka A., Haba O. and Yonetake K. 'Surface rheology of rubbed polyimide film in liquid crystal display'. *Opt Mater Exp.,* 2014; **4**(5), 1057–66.

[29] Ichimura K. 'Photoalignment of liquid-crystal systems'. *Chem. Rev.*, 2000, **100**(5), 1847–74.

[30] Gibbons M.W., Shannon S.T., Sun S.T. and Swetlin B.J. 'Surface-mediated alignment of nematic liquid crystals with polarized laser light'. *Nature*, 1991, **351**(6321), 49–50.

[31] Schadt M., Schmitt K., Kozinkov V. and Chigrinov V. 'Surface-induced parallel alignment of liquid crystals by linearly polymerized photopolymers'. *Jpn. J. Appl. Phys.*, 1992, **31**(7), 2155–64.

[32] Hasegawa M. and Taira Y. 'Nematic homogeneous photo alignment by polyimide exposure to linearly polarized UV'. *J. Photopolym. Sci. Technol.* 1995; **8**(2), 241–48.

[33] Matsukawa H., Shimano Y., Suefuji M., Umeda Y. and Mimura K. 'The world's first photo-aligned IPS-LCDs for a TV use'. Proceedings of the 19th International Display Workshops in conjunction with Asia Display, Kyoto, Japan, December 2012. 1581–84.

[34] Kunimatsu N., Sonoda H., Hyodo Y. and Tomioka Y. 'Photoalignment technology for high performance IPS-LCDs: IPS-NEO Technology'. SID International Symposium Digest of Technical Papers, San Diego, US, June 2014. **45**, 1406–1409.

[35] Momoi Y., Tamai K., Furuta K., *et al.* 'Mechanism of image sticking after long-term AC field driving of IPS mode'. *J. Soc. Inf. Disp.*, 2010, **18**(2), 134–40.

[36] Oka S., Mitsumoto T., Kimura M. and Akahane T. 'Relationship between surface order and surface azimuthal anchoring strength of nematic liquid crystals'. *Phys. Rev. E.*, 2004, **69**(6 Pt1), 061711.

[37] Faetti S., Sakamoto K. and Usami K. 'Very strong azimuthal anchoring of nematic liquid crystals on uv-aligned polyimide layers'. *Phys. Rev. E.*, 2007, **75**(5 Pt1), 051704.

[38] Usami K., Sakamoto K., Tamura N. and Sugimura A. 'Improvement in photo-alignment efficiency of azobenzene-containing polyimide films'. *Thin Solid Films.*, 2009, **518**(2), 729–34.

[39] Hanaoka K., Nakanishi Y., Inoue Y., Tanuma S., Koike Y. and Okamoto K. 'A new MVA-LCD by polymer sustained alignment technology'. SID International Symposium Digest of Technical Papers, Seattle, US, May 2004. **35**(1), 1200–3.

[40] Nakanishi Y., Hanaoka K., Shibasaki M. and Okamoto K. 'Relation between monomer structure and image sticking phenomenon of polymer-sustained-alignment liquid crystal displays'. *Jpn. J. Appl. Phys.*, 2011, **50**(5R), 051702.

Chapter 2

Photo-alignment technology for vertically aligned LCDs

Manuel Koechlin[1], Yuichiro Yamada[1], and Qian Tang[1]

Rolic® light-controlled molecular orientation (LCMO) is a state-of-the-art photo-alignment technology used in today's mass production for vertically aligned (VA) liquid crystal display (LCD), where high-resolution LC director patterning with well-defined azimuthal and pretilt angles is required.

We provide a description of the LCMO mechanism and show the positive impact of the lower pretilt angle on the intrinsic transmittance of VA-LCDs. Novel LCMO-VA materials for ultra-high definition LCDs are also presented.

2.1 Introduction

Homogeneous and stable alignment of liquid crystals (LCs) with well-defined bias tilt angles over large areas is a prerequisite for large-scale manufacturing of liquid crystal displays as well as other LC based optical devices. Conventional alignment methods such as the rubbing of polyimide layers or geometrical patterning (such as topologies or protrusion) via photo lithographic processing were the methods of choice for large scale production of LCDs. The rubbing of polyimide generates dust, static electricity and mechanical damage which lower the yield and performances of LCDs. Also, the micro-topology reduces LCD performances such as light transmission and response time [1]. Moreover, complicated and expensive processing is required to realize multi-domains with different LC director orientation, in order to get wide viewing angle. Therefore, more reliable production methods are needed to reduce costs of large flat panel displays and improve their performances. One of the most attractive manufacturing methods is the photo-alignment.

In the 1980s, first attempts were made to induce molecular reorientation in guest host materials, comprising mainly azo guest molecules, by exposing to ultra violet (UV) light [2–4]. In the early 1990s, two research groups showed polarized UV exposure to induce an in-plane anisotropy in a polymer doped with an azo dye [5] and in polyvinylcinnamate (PVCin) polymers based on photo-cross-linkable

[1]Rolic Technologies Ltd., Switzerland

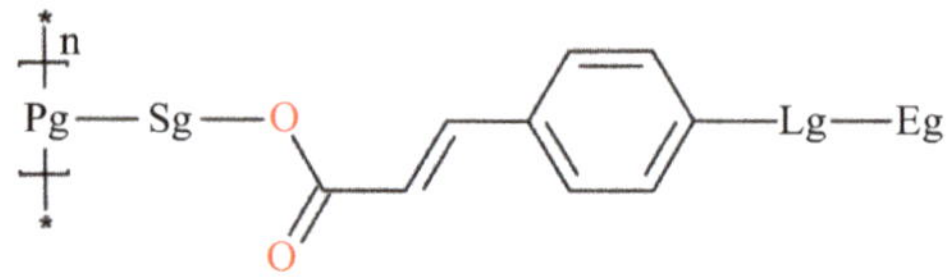

Figure 2.1 Base molecular structure of typical LCMO-VA materials used in mass production of VA-LCDs

cinnamate chromophores [6]. Meanwhile, other materials and photo-alignment processes have been proposed, such as photo degradation with UV-C light [7]. Therefore, mainly three fundamentally different aligning strategies by linearly polarized light were pursued, namely (i) optically induced cis–trans isomerization of single chromophore [5], (ii) photo degradation or photo rubbing of polyimide layers [7], and (iii) photo-crosslinking [6].

Because of in-situ photo-crosslinking during processing [6,8], Rolic proprietary LCMO photo alignment technology provides possibilities to obtain inherently thermally and optically stable alignment films. This technology allows an easy achievement of high resolution azimuthal LC-director patterns with defined bias angles, from homogeneous planar to homeotropic orientation, depending on the target application [6,8–13]. This renders LCMO the fundamental photo-alignment technology for VA-LCDs. Furthermore, LCMO-VA photo-aligned multi-domain VA-LCDs have been demonstrated by Rolic already in 1999 [14]. Commercial introduction of this technology has been achieved in 2008 thanks to the enormous development progress done at Rolic together with an efficient collaboration with panel manufacturers [1,15].

It is obvious that the photo-reaction involved as well as the related LCD performance depend on the chemical structure of the alignment material. Figure 2.1 shows the basic molecular structure of a typical state of the art LCMO-VA material: A polymer comprising a side chain cinnamate photo-reactive group linked on one side, via a linker (L_g), to an end group (E_g) and on the other side to a moiety bearing polymerizable groups (P_g), via a spacer (S_g). Typical LCMO-VA materials used in this study are polyamic acids with a side chain terminated by a fluorinated end group, $E_g = -C_nF_{2n+1}$.

Based on Rolic long-time expertise and design capabilities, the different groups can be designed to control alignment, pretilt angle and their stability as well as other required electrical- and electro-optical performances such as voltage holding ratio (VHR), residual DC (RDC) and image sticking. In this chapter, we will discuss the intrinsic photo-crosslinking mechanism of LCMO-VA and its role in enhancing stability of LCMO-VA alignment layers. Furthermore, we show the future potential of LCMO-VA technology with improved transmittance.

2.2 LCMO mechanism [16]

This section summarizes the investigation of photo-reactions of typical LCMO-VA materials exposed to linearly polarized UV (LPUV) light. The investigation includes chemical identification of photo-reaction products and spectroscopic investigation of LCMO-VA films.

2.2.1 Method development with model compounds

Ethyl 4-hydroxycinnamate is used as model compound for the method development. Photo-reactions of this model compound result in three main products (Figure 2.2): no-reaction (Trans-1), cis-isomerization (Cis-1), and cyclo-addition or dimerization (Dimer-1). The Cis isomer (Cis-1) and the cyclo-dimer (Dimer-1) are also synthesized as reference compounds.

Figure 2.3 shows the comparison of isotropic UV–Vis absorption spectra of trans-1, cis-1, and dimer-1. It clearly indicates that the positions of the absorption bands, as well as their intensity, are strongly affected by the presence and the configuration of the vinyl double bond. As a general trend, absorption spectra of cinnamate containing molecules, i.e., trans-1 and cis-1, exhibit two characteristic absorption peaks in two distinctive ranges. The first is located below 230 nm, whereas the second one appearing as broad and intense peak with the maximum absorption at approx. 290 nm. The origin of this second absorption band for trans-1 and cis-1 was attributed to the π–π* transition of the vinyl double bond and the aromatic ring being in conjugation. However, the UV absorption spectrum of dimer-1 shows the presence of a strong absorption at approx. 195 nm, attributed to the cyclobutane group.

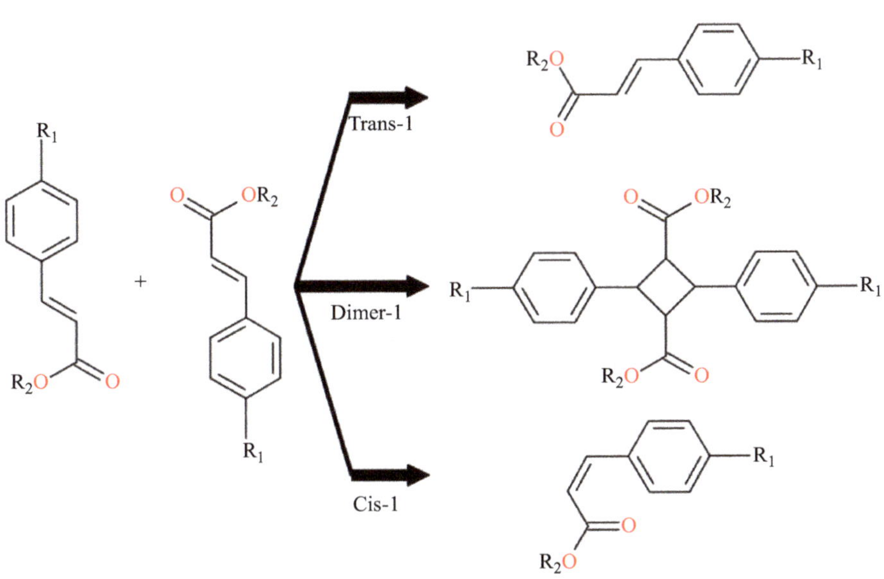

Figure 2.2 Main possible photo-reaction products after exposure to UV light; Model compound with $R_1 = OH$ and $R_2 = C_2H_5$

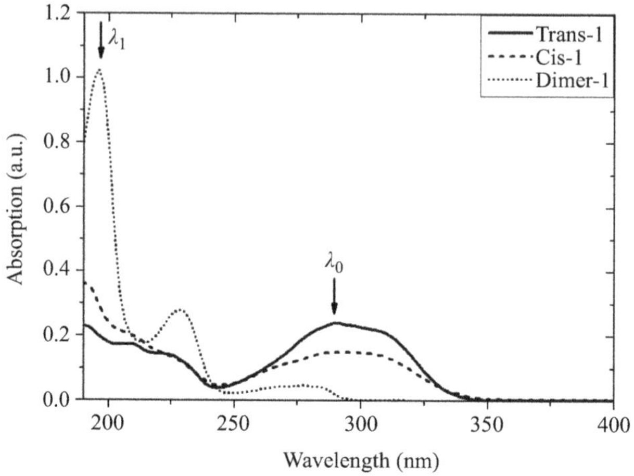

Figure 2.3 Isotropic absorption spectra of cinnamate model compounds in acetonitrile solutions ($C = 5.6 \times 10^{-6}$ mol · L^{-1})

2.2.2 *Chemical identification of photo-reaction products in LCMO-VA thin films after LPUV exposure*

Approximately 90-nm-thick films of typical LCMO-VA are obtained by spin-coating on glass substrates, thermally dried and baked and finally exposed to LPUV (UVB) at different doses between 20 and 3,000 mJ/cm^2.

The exposed films are insoluble in common solvents, even after exposure with low doses, which is already a clear indication that crosslinking of the LCMO-VA materials occurs.

For the analysis of the reaction products to identify [2+2] cyclo-addition dimers, the above exposed films are scratched from the substrates and hydrolyzed (degradation) in an acidic medium. The resulting mixtures, corresponding to different exposure doses, are stirred under reflux until total dissolution of the scratched solid material. Figure 2.4 shows some possible dimers that might result from hydrolysis. A fine-tuned LC/MS (liquid chromatography coupled with electrospray mass spectroscopy) method is used for the analysis of degradation products. Chemically stable dimers such as A, G, and H are also synthesized as reference compounds.

The retention time and the corresponding mass experimentally obtained with LC/MS are shown in Table 2.1.

After that, the solutions resulting from hydrolysis are analyzed with the same method as for the reference compounds. The obtained results show the existence of peaks corresponding to the retention time and the mass of dimer G and dimer D at 3,000, 200 as well as at low LPUV doses of 20 mJ/cm^2, confirming that dimerization of LCMO-VA effectively takes place during LPUV exposure already at low LPUV doses.

2.2.3 *Spectroscopic investigation of LCMO thin films*

The spectroscopic investigation is related to the occurrence of anisotropy in LPUV exposed LCMO-VA polymer thin films. Thin films (150 nm) of typical LCMO-VA

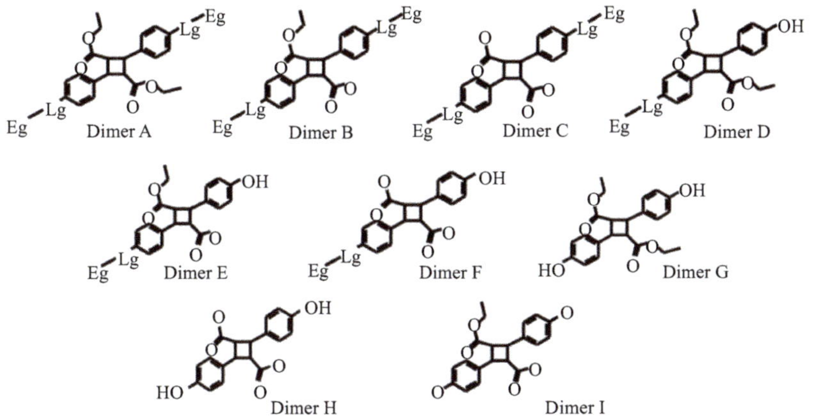

Figure 2.4 Some possible dimers that can result from acidic degradation of LPUV exposed typical LCMO-VA thin films

Table 2.1 Retention times (min) and mass of reference dimers

Compound	LC (min)	*m*/*z* (DA)
Dimer G	9.89; 10.32	384.8
Dimer E	17.42; 17.61	586.6
Dimer D	20.80; 21.08; 21.44	614.6
Dimer B	28.858	816.4
Dimer A	32.23; 32.63; 32.84; 3	844.4

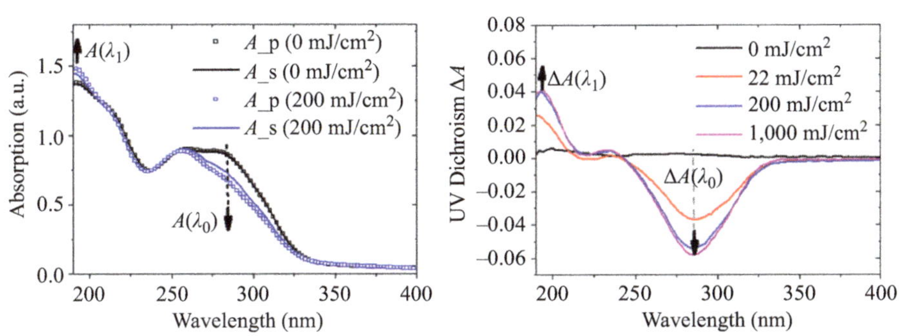

Figure 2.5 Left: Anisotropic UV-absorption spectra of LCMO-VA film. Right: Dichroism $\Delta A = (A_p - A_s)$ of a 150 nm LCMO-VA thin film after exposure to LPUV

are obtained by spin-coating on quartz substrates, thermally dried and baked. Polarized UV-absorption spectrums of the thin films with polarization direction in parallel (A_p) and perpendicular (A_s) to the LPUV direction are measured, before and after LPUV exposure with different doses between 20 and 1,000 mJ/cm^2 (Figure 2.5 left).

Unexposed LCMO-VA films exhibit a large and strong absorption at $\lambda_0 =$ 284 nm, corresponding to the π–π^* transition of the vinyl double bond and the aromatic ring being in conjugation. After LPUV induced polymerization, double bonds are consumed, leading to the diminution of this peak. Simultaneously the second short wavelength absorption peak at $\lambda_1 = 192$ nm, corresponding to the absorption of cyclobutane group, increases. This shift $\lambda_0 \rightarrow \lambda_1$ results from the reduction of π-conjugation of the higher conjugated cinnamic side chains into cyclobutane upon cross-linking, in line with the UV–Vis absorption results on model compounds and with LC/MS results which demonstrate the occurrence of dimers in thin films of LCMO-VA after exposure to LPUV.

The dichroic UV-absorption $\Delta A = (A_p - A_s)$ before and after LPUV exposure (Figure 2.5 right) shows the optical anisotropy in LCMO-VA films after LPUV exposure. As expected, the LCMO-VA thin film without exposure to LPUV exhibits practically no dichroism. However, after exposure to LPUV, the initially isotropic absorption of the non-irradiated film becomes strongly negatively dichroic at $\lambda_0 \sim 284$ nm in the range of cinnamate absorption and positively dichroic at $\lambda_1 \sim 192$ nm, i.e., in the cyclobutane absorption range.

2.2.4 LCMO-VA alignment mechanism and performances

As a result of these investigations, we have shown that crosslinking of cinnamate groups occurs to form cyclobutane groups in LCMO-VA thin films, starting from low LPUV exposure doses. We have also shown that this linear photo-polymerization of LCMO-VA layers is accompanied with pronounced optical changes and anisotropies which are known to force liquid crystals to align homogeneously. Based on these results, we proposed the following mechanism which may individually or collectively lead to the uniform LC photo-alignment.

The results on photo-crosslinking of LCMO-VA films can be interpreted at the mesoscopic level as follows: (i) cinnamate chains oriented in parallel to the polarization direction of the LPUV light have a larger absorption (and thus photo-reactivity) than those oriented perpendicularly, which leads to (ii) a preferred depletion of the parallel chains by photo-reaction and finally (iii) to a preferred creation of cyclobutane photo-products with their long axis parallel to the polarization direction. According to this model, the negative dichroism at 284 nm can be explained according to (ii) and the positive dichroism at 192 nm is related to the alignment and organization of the resulting cyclobutane along polarization direction. It is clear that cyclobutane subunit has an isotropic distribution and cannot contribute to the anisotropy, however, the resulting photo-reaction products bearing aromatic moiety, are anisotropic. Therefore one can clearly conclude that for such typical LCMO-VA, cyclobutane groups align parallel to the polarization direction of the LPUV light, which in turn induces alignment of liquid crystals parallel to the polarization direction.

On the microscopic level, we are investigating and probing cinnamate and phenyl groups orientation of LCMO-VA alignment layers using other analytical tools such as NEXAFS (Near-edge X-ray absorption fine structure). At this stage, the results indicate that, prior to LPUV exposure, side-chain chromophores of LCMO-VA are aligned perpendicular to (x,y) plane with phenyl groups randomly aligned around the

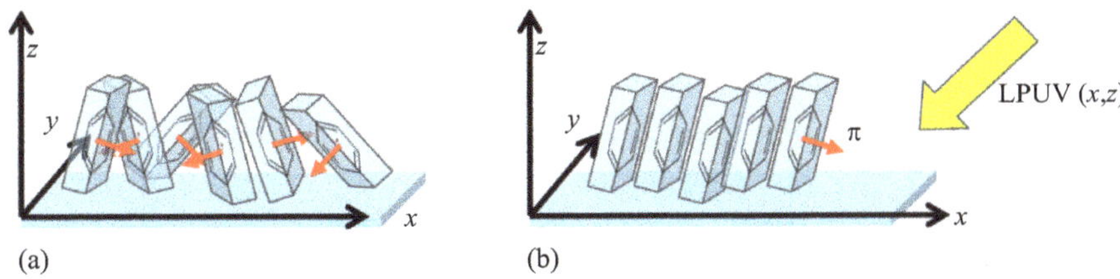

Figure 2.6 Schematic illustration of phenyl rings of side-chain chromophores before (a) and after (b) LPUV exposure in (x, z) plane, with red arrows indicating the direction of π-orbitals of phenyl groups

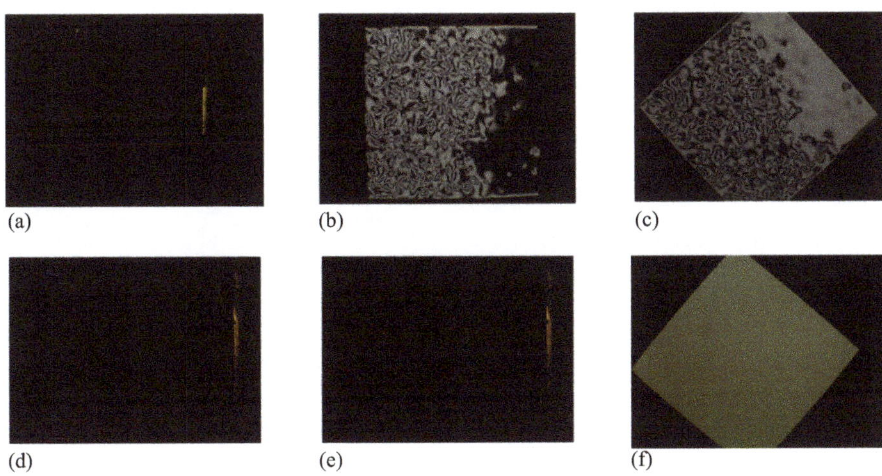

Figure 2.7 Comparison of alignment homogeneity (between cross-polarizers): The upper row shows a sample without LPUV exposure with (a) no voltage applied; (b) 5 V applied; and (c) 5 V applied and rotated by 45°. The lower row shows a sample with LPUV exposure with (d) no voltage applied; (e) 5 V applied; and (f) 5 V applied and rotated by 45°

normal as schematically shown in Figure 2.6(a). However, after the LPUV exposure, microscopic rearrangement occurs and, at least some of, phenyl groups are reoriented in the direction perpendicular to LPUV, i.e., to plane (x,z). It is obvious that Figure 2.6(b) is just a simplified scheme for illustration purpose since, it is not necessary that all side chain chromophores rearrange, only part of them can be sufficient to induce the required alignment and pretilt homogeneity. More detailed investigations are still in progress to clarify the microscopic picture.

Figure 2.6(a) shows that the LC is vertically oriented (homeotrope) even without LPUV exposure. The measured pretilt angle of 90° is in line with the normal and random in-plane orientation of chromophores as explained in Figure 2.6(a). However, application of an external voltage induces defects as shown in Figure 2.7(b) and (c), hence, this alignment is not uniform and not stable. After LPUV exposure and, hence, crosslinking and subsequent rearrangement of side chain chromophores, liquid

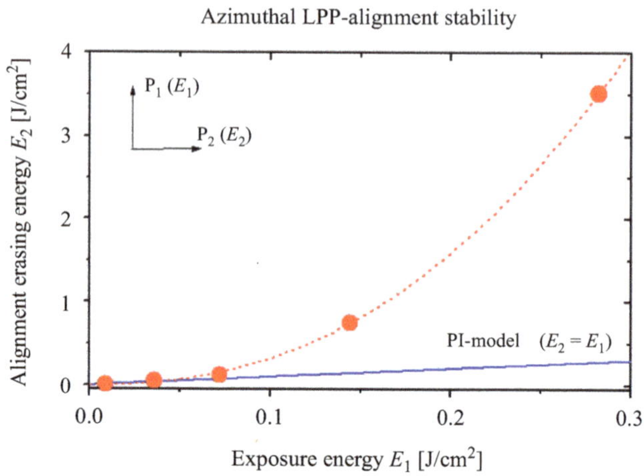

Figure 2.8 Azimuthal LCMO alignment stability, with model calculated curve (solid) and experimental curve (dashed)

crystals are aligned vertically (Figure 2.7(d)) with a tuned pretilt angle in the range of 87°–89° as required for VA-LCDs. Because of the high sensitivity and crosslinkability of Rolic LCMO-VA materials, very uniform and stable alignment is achieved already at low LPUV exposure doses. External changes such as voltage (Figure 2.7(e–f)) or temperature have no substantial effect on the alignment quality and pretilt angles.

Because of the generation of uniform and stable pretilt angles without fluctuations over large areas, LCMO-VA technology has been successfully applied to Gen 10 factory as the basis technology for the state-of-the-art ultraviolet-induced multi-domain vertical alignment (UV^{2}A) production technology [1,15]. LCMO-VA allows manufacturing of advanced new generation LCD-TV panels with reduced manufacturing costs and low energy consumption [1,15]. The transmittance, contrast ratio, and response time of LCD-TV are drastically improved compared to the conventional MVA [1].

Another experiment to demonstrate the azimuthal stabilization through photo-crosslinking is the following. LCMO thin films with different LPUV dose are made. The oriented films are then exposed to an "alignment erasing" LPUV light with the polarization direction perpendicular to the previous one. Figure 2.8 shows the erasing energy needed to destroy the photo-alignment in function of LPUV energy. The model calculated curve is a linear while the experimental one is exponential, showing that much higher energy is actually needed to erase the original photo-alignment, because of stabilization through crosslinking.

2.3 Multi-domain VA-LCD and further development

2.3.1 *Multi-domain VA-LCD*

The main advantage of a multi-domain VA-LCD is the increased viewing angle, since the overall birefringence shows much less dependence on light passing the LCD at oblique viewing angles.

Figure 2.9 shows the LC configuration, the fabrication principle and the picture of a multi-domain VA-LCD as demo-sample from Rolic [9,14]: LCMO-VA thin film coated on Substrate 1 (S1) is exposed in four different azimuthal directions using a photomask to select and cover the respective areas (domains) with approx. 100 × 100 μm size; LCMO thin film on Substrate 2 (S2) is exposed perpendicularly. As a result the LC directors in the four domains form the same pretilt angles θ but azimuthal angles differing by 90° on S1 and a pretilt angle of 90° on S2.

In 2008, Sharp UV^{2}A display mode [1,17] using Rolic® LCMO technology was applied to manufacture LC-TVs (liquid crystal television) because of its excellent properties such as high contrast ratio, high transmittance, fast response, and high production yield. After that, the UV^{2}A display mode has been widely used for several applications.

The UV^{2}A configuration is shown in Figure 2.10. The method is slightly different to the process described above, as individual exposures are done along both a long-side direction on the TFT substrate and a short-side direction on the CF substrate, respectively, as shown in Figure 2.10(a) and (b) (arrows denote the projected k-vector of the incident LPUV light). After assembly of the two substrates, pixels in reverse TN mode comprising four divided multi-domain structures are obtained. When applying an electric field, liquid crystal molecules incline from vertical to

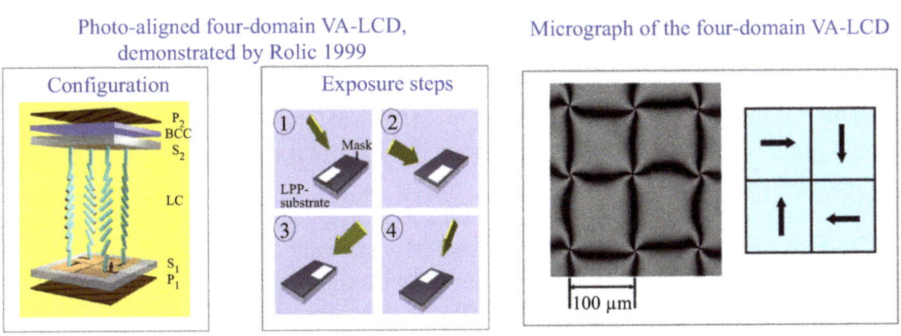

Figure 2.9 Left: Concept of photo-aligned four-domain VA-LCD. Right: Picture of a multi-domain VA-LCD demo-sample

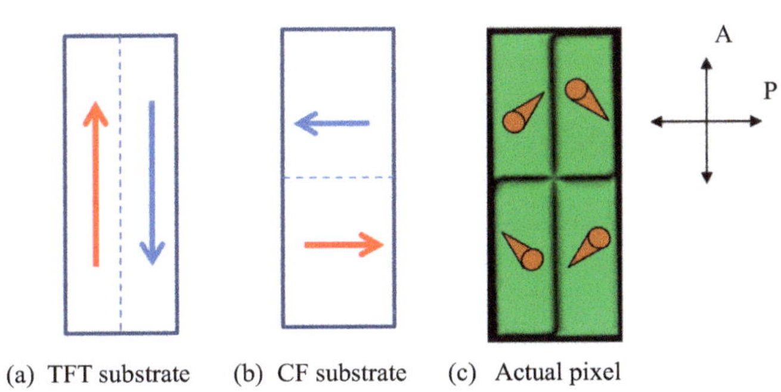

Figure 2.10 UV^{2}A configuration

homogeneous alignment. The optical axis of each domain moves toward 45° direction as illustrated by cones in Figure 2.10(c). Between the domains the cross shaped dark lines emerge.

Transmittance, which is one of the most important properties for LCD panel architecture, decreases when LC-TV moves to higher definition such as 4K or 8K ultra-high definition, because the dark line width is independent of the pixel size and therefore the aperture ratio decreases.

In order to improve the transmittance of UV^2A display mode, we have been focusing on the pretilt angle, which is induced by LCMO photo alignment film. This part of work is presented in Section 2.3.2.

Nevertheless, other properties such as contrast ratio and switching time will also be directly influenced by the pretilt angle, the results of the related investigations is briefly presented below.

2.3.2 Influence of the pretilt angle on contrast ratio and switching time

One of the key advantages of the VA-LCD versus other display modes is the high contrast ratio, which is defined as

$$C = \frac{T_{ON}}{T_{OFF}} \tag{2.1}$$

It is usually governed by T_{OFF} (transmitted light intensity in the off-state). For simple VA-LCD (anti-parallel alignment without twist and only one domain) T_{OFF} depends mainly on the pretilt angle. On the other hand, T_{ON} (transmitted light intensity in the on-state) is rather depending on the pixel design, which is not taken into account in these measurements.

For the measurement simple VA cells with pretilt angles between 85.5° and 90° were placed between crossed polarizers at an azimuthal angle of 45°. Indeed we observed that T_{ON} is almost independent of the pretilt angle and that T_{OFF} increases with decreasing pretilt angle. The contrast values are shown in Figure 2.11 (red squares).

The theoretical contrast ratio C_{th} is also plotted (Figure 2.11, blue squares):

$$C_{th} = \left[k_1 \sin\left(\frac{\pi}{\lambda}\Delta n\, d\right) + \frac{1}{k_2}\right]^{-1} \tag{2.2}$$

using $\lambda = 500$ nm as the wavelength, $d = 4.2$ µm as the cell gap, $\Delta n = n_o - n'_e(\theta)$. $n_o = 1.483$ is the ordinary refractive index of the LC, $n_e = 1.582$ the extraordinary refractive index, and $n'_e(\theta)$ the refractive index depending on the pretilt angle θ:

$$n'_e(\theta) = \frac{(\cos\theta)^2}{n_o{}^2} + \frac{(\sin\theta)^2}{n_e{}^2} \tag{2.3}$$

The constants $k_1 = 0.35$ and $k_2 = 21{,}000$ are determined empirically, representing the normalization factor and account for the system limitation of the contrast measurement device respectively.

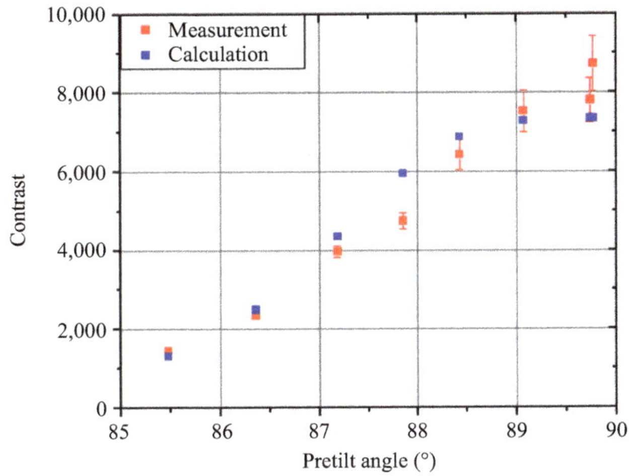

Figure 2.11 Contrast ratio dependence on pretilt angle

Table 2.2 Calculated g(θ)

Pretilt angle	$\mathbf{g(\theta)}$
89.99°	1.00
89°	1.10
88°	1.16
87°	1.21
85°	1.30

Nevertheless, the influence of the pretilt angle on the contrast ratio obtained in this investigation is incomplete to evaluate the case of a real panel, because T_{OFF} depends on the alignment (with or without twist) and T_{ON} is also influenced by the dark line width, which will be discussed in Section 2.4.

The theoretical formula for the switching time of a VA-LCD can be found in the literature ([18] pp. 41–43). For instance the decay time t_{off} may be expressed as

$$t_{off} = \frac{\tau_0}{2} f(\delta) g(\theta) \tag{2.4}$$

$$\tau_0 = \frac{\gamma d^2}{K_{33} \pi^2} \tag{2.5}$$

where τ_0 is the LC director reorientation time, γ the rotational viscosity of the LC, d the cell gap, K_{33} the elastic constant of LC. The function $f(\delta)$ is a factor depending on the phase retardation induced when switching the cell from on- to off-state and $g(\theta)$ a factor accounting for the pretilt angle: the slightly weaker restoring elastic torque due to the decreased pretilt angle leads to slight increase of t_{off}. $g(\theta)$ can be calculated by the Erickson–Leslie equation without using the small angle approximation. Some values are shown in Table 2.2 ([18] page 54).

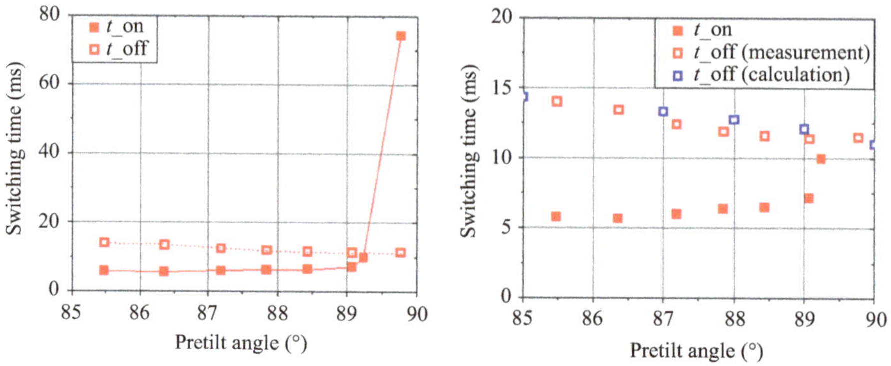

Figure 2.12 Switching time dependence on pretilt angle

A similar expression exists for the rise time (t_{on}) which is however more complicated and depends on the applied voltage.

Figure 2.12 shows the switching times (t_{on} and t_{off}) for different pretilt angles. t_{on} is measured by applying a sharp voltage step from 0 Volts to V_{max}, which corresponds to the voltage where the display shows the maximum transmittance (T_{max}). Similarly, t_{off} is measured by applying a voltage step from V_{max} to 0 V. The switching time t_{on} is then defined as the time within which the transmittance changes from 10 percent to 90 percent of T_{max} (and vice versa for t_{off}). The rotational viscosity γ of the LC used is 148 mPas. For pretilt angles lower than 87.5°, t_{off} increases whereas t_{on} decreases slightly, leading to a larger and larger discrepancy between the two values, which is not ideal for a high-end display. The calculated t_{off} values shown in Figure 2.12 are according to (2.4) and (2.5) using $K_{33} = 16.5$ pN and $d = 4.3$ μm, leading to $\tau_0 = 16.8$ ms and $f(\delta) = 1.19$ ([18] page 43).

2.4 Low pretilt LCMO materials for high transmittance [19,20]

Before any development of new LCMO materials, the first work to do is the simulation of the dark line width as well as the transmittance as a function of the pretilt angle, in order to predict the theoretical achievable improvement level. During the development work then, it is preferable to investigate the possible transmittance improvement at a laboratory scale, to minimize the number of tedious and expensive trials on a mass production line.

2.4.1 Simulation results

2.4.1.1 Two-dimensional simulation

A two-dimensional simulation was carried out to study the impact of the pretilt angle on the dark line width. A Cartesian coordinate *xyz* is defined as shown in Figure 2.13, where the *y*-axis is perpendicular to the paper plane and coincides with the dark line. The alignment of liquid crystals under application of an electric field is calculated by means of simulation tool of LCD-master provided by Shintech,

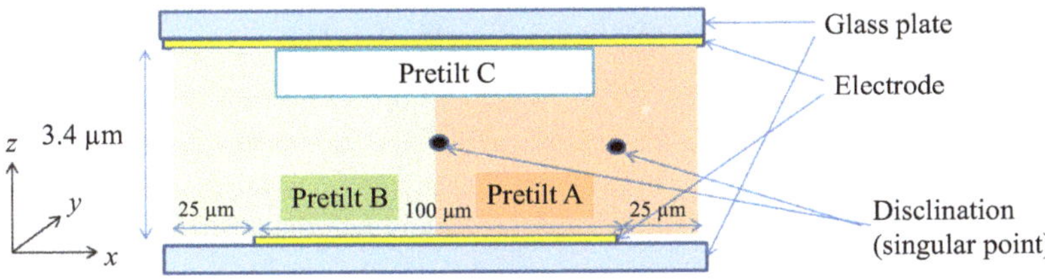

Figure 2.13 2D Simulation model: At the bottom surface the pretilt direction of pretilt A and B is opposite with respect to the y-axis

Table 2.3 LC material and cell parameters

LC parameters		Cell parameters	
K_{11}	14.6 pN	Cell gap	3.4 μm
K_{22}	10.7 pN	Applied voltage	±7.5 V
K_{33}	16.5 pN		
$\Delta\varepsilon$	−3.1		
Δn (589 nm)	0.0996		

where the distortion-free energy of the Oseen–Frank form and the electrostatic free energy are considered. In Figure 2.13, pretilt A and B on the bottom substrate are both along the y-axis, but toward the opposite direction of each other, and pretilt C on the top surface is along the x-axis as shown in Figure 2.10(a) and (b).

Liquid crystal materials and LCD cell parameters used in the simulation are listed in Table 2.3. LC parameters of a common liquid crystal mixture with negative dielectric anisotropy are used.

Figure 2.14 shows the typical simulation result of both liquid crystal alignment and transmittance under an electric voltage of 7.5 volts. Two disclinations emerge, i.e., one is at the center and the other at the right edge of the electrode, respectively. At the vicinity of the disclination, distortion of alignment takes place, so that the transmittance at the center of disclination has a minimum value. The dark line width of the disclination line is plotted as a function of the pretilt angle in Figure 2.15. As is clear from Figure 2.15, the dark line width becomes thinner in accordance with decrease of the pretilt angle, namely, the dark line width at the pretilt angle of 87.5° is about 18 percent thinner than the one at the pretilt of 88.5°.

2.4.1.2 Three-dimensional simulation

Next, we move our attention forward to the actual pixel design to confirm the quantitative value in the transmittance improvement. Recently customers prefer to watch large size higher definition televisions such as 65″ diagonal with 4K and 8K ultra-high definition. Assuming a display size of 65″ diagonal and 8K definition, the pixel size is 62.3 × 189.9 μm. Since a TFT-array, gate and source lines and a storage capacitor where the light cannot pass through are designed in the pixel, the aperture

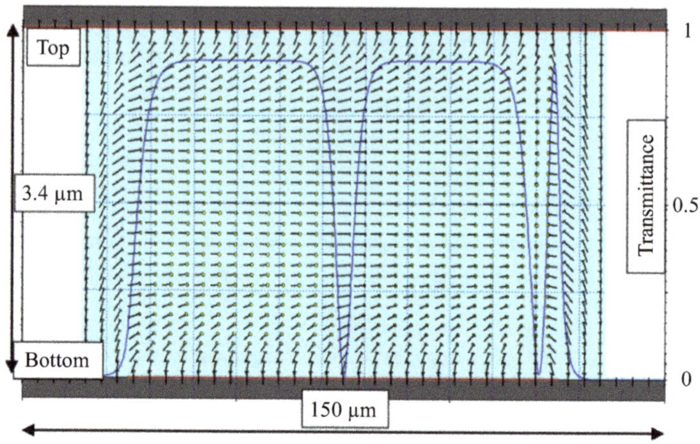

Figure 2.14 2D Simulation results: nails and solid line show LC director and transmittance, respectively, the electrodes on the substrates are drawn as rectangles

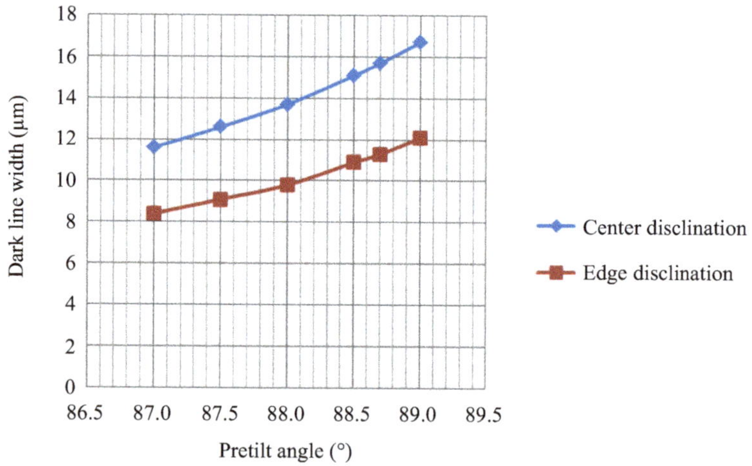

Figure 2.15 Pretilt angle vs. dark line width: calculation was performed by 2D simulation

ratio is estimated up to 42 percent. This means that 58 percent of backlight intensity is wasted, therefore, we have to focus on using the remaining 42 percent effectively.

Figure 2.16 shows the 3D simulated design of 4 domain (4D) and 8 domain (8D) structure. The reason why we provide 8D structure is that MPGC (multi-pixel gamma control) method [21], which improves the viewing angle property, requires 8D structure. The exposure direction of each domain is set as same as that shown in Figure 2.10. Simulation is carried out using 3D LCD master provided by Shintech.

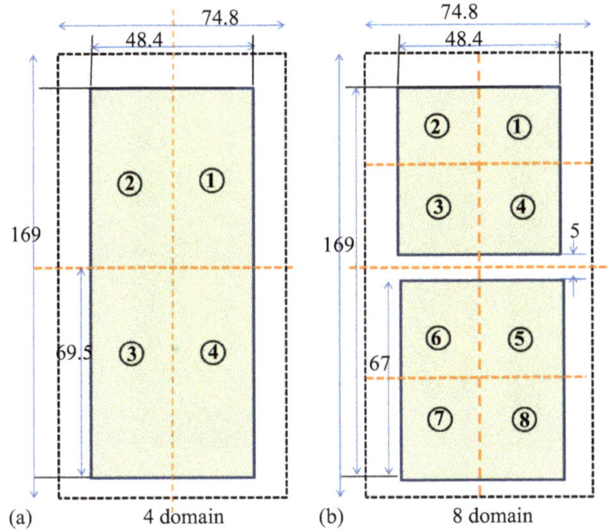

Figure 2.16 *Pixel design for 3D simulation (dimensions in μm): the colored area is an electrode on TFT substrate, on CF substrate the electrode (dotted line) is larger by 5 μm at each side than that on TFT substrate simulation*

The typical simulation result of transmittance is plotted as two-dimensional contour map in Figure 2.17. Comparing Figure 2.17(a) with Figure 2.10(c), the cross-shaped dark lines are reproduced well by the simulation.

The transmittance per pixel is calculated by summing the transmittance of each lattice in the pixel. As shown in Figure 2.18, the transmittance increases, when the pretilt angle becomes smaller. The cases plotted in Figure 2.18 use the so-called straight source line structure, where the source line is wired straightly along the long side of the rectangular pixel.

For the 8D case, 8–10 percent increase of transmittance for LCDs using the new Rolic® LCMO material is calculated.

2.4.1.3 Source line design

The transmittance using straight source line structure is insufficient since the dark line at the edge of electrode occupies a considerable area of the pixel aperture. To prevent the loss of transmittance caused by the edge dark line, an overlapping zigzag source design is suitable. When SHA (Super High Aperture) technology [22] is applied to TFT array design, it is not difficult to wire a zigzag source line. Figure 2.19 shows the zigzag source line design and simulation result of pretilt angle of 87.5°. Comparing with Figure 2.17, the dark line is hidden by the source line and the transparent area, which was hidden by the straight source line, appears and contributes to enhance the transmittance. The further improvement ratio of the zigzag source line against the straight source line is about 3.3 percent.

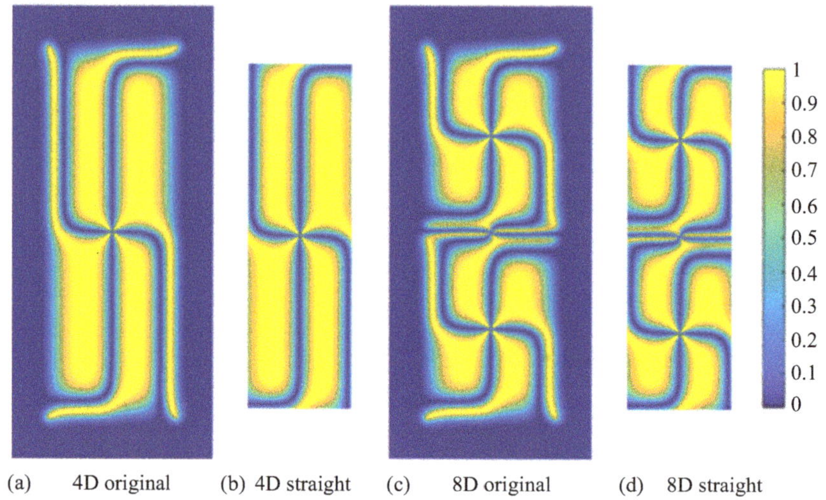

Figure 2.17 Simulation results of 4D and 8D cases: the color from dark to light gray represents transmittance: (a) and (c) use no mask, (b) and (d) assume a straight source line structure

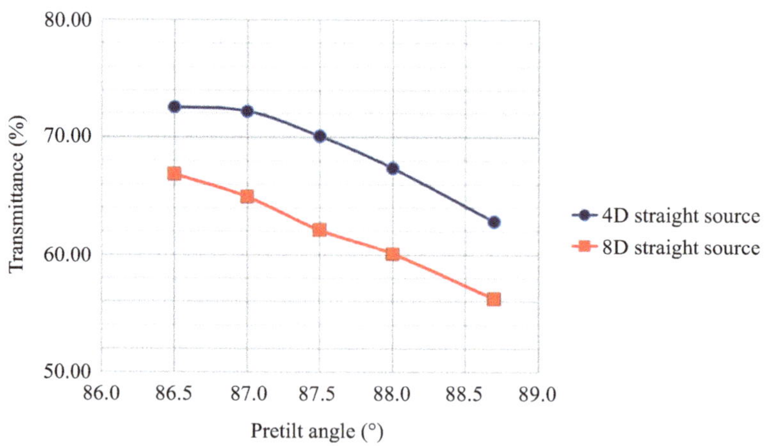

Figure 2.18 Transmittance vs. pretilt angle: calculation was done by 3D simulation

2.4.2 Experimental investigation of dark line width and transmittance improvement

2.4.2.1 Experimental approach

For the investigation of the impact of the pretilt angle on the dark line width, the biggest issue is how to eliminate the experimental error related to the positioning of the photomask. Rolic has set up the following method to overcome this problem.

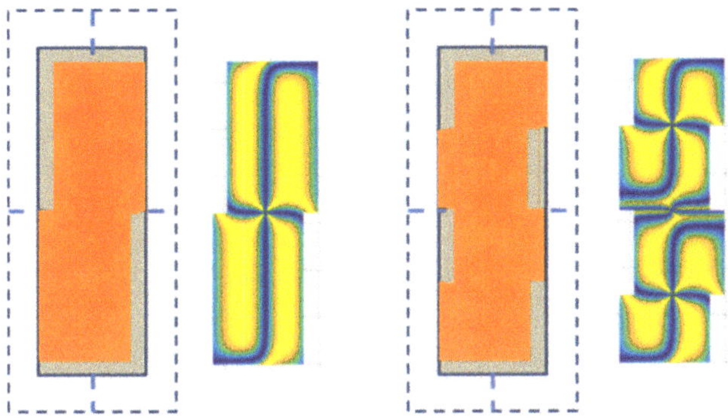

(a) Straight source line (b) Zig-zag source line

Figure 2.19 Source line design: simulation result of the straight and zigzag source line

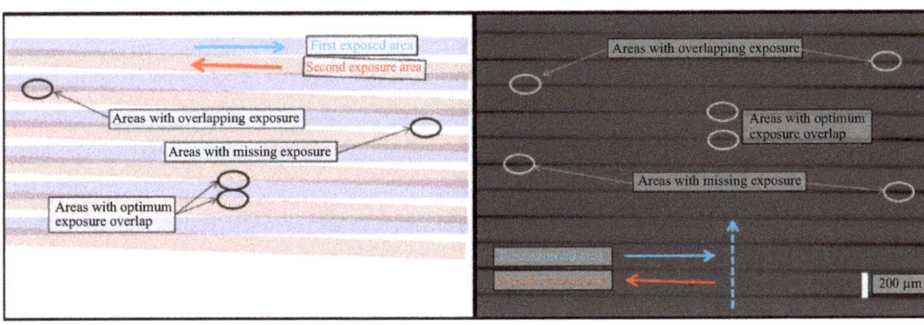

Figure 2.20 Left: photo mask positioning for the 2-step UV exposure. Right: photograph of experimental cells under the electric field

Liquid crystal cells with UV^{2}A configuration similar to Figure 2.10 are manufactured using a chromium photo mask. The photo mask comprises 200 μm wide alternating chromium and transparent stripes which block and let pass the UV-light during the exposure step. After coating a pair of substrates with a photo alignment film, two exposures are performed on the bottom substrate, while for the second exposure the photo mask is slightly rotated by 1°–2°. This approach ensures the occurrence of an area with an optimal overlap, therefore a minimal dark line width (between the overlapped- and the non-exposed areas), which is independent to the process condition. Figure 2.20 (left) shows schematically the positioning of the photo mask during the first and second exposures, light and dark grey stripes indicate the two exposure areas and the non-exposed areas are drawn in white. The solid and dashed arrows indicate the two opposite inclined exposure directions on the bottom and the single exposure on the top substrate, respectively.

The top substrate is single exposed without photo mask and assembled with the bottom substrate to get a cell exhibiting 200 μm wide domains. Figure 2.20 (right) shows the picture of a typical cell obtained by polarized optical microscopy while applying a voltage to the cell. The solid arrows depict the exposure directions of the bottom substrate and the dotted arrow depicts the exposure direction of the top substrate. The dark lines are visible between the strip-like domains with variable width and a minimal width in the middle.

2.4.2.2 Experimental results

By taking photographs of the test cell between crossed polarizers, the transmittance of the dark area is plotted in the direction perpendicular to the line and on the minimal width position. Figure 2.21 shows typical transmittance plot of two cells with pretilt angle of 88.6° and of 87.3°, respectively. Usually, the dark line width is defined at 90 percent transmittance. The thinner dark line can be clearly observed for the cell with 87.3° pretilt angle. This implies that the lower pretilt angle alignment can improve the transmittance of UV^2A. Nonetheless, we prefer to use the integration of the non-transmissive area for better quantification of the "dark line width."

Applying the same method, the first Rolic sample for mass production trial at CEC Panda LCD is investigated. With the pretilt angle of 87.7°, the integration of the plotted dark area is 10 units, while the standard pretilt angle of 88.5° gives 15 units, corresponding to a theoretical transmittance improvement of 7–8 percent for 4 domains pixel design.

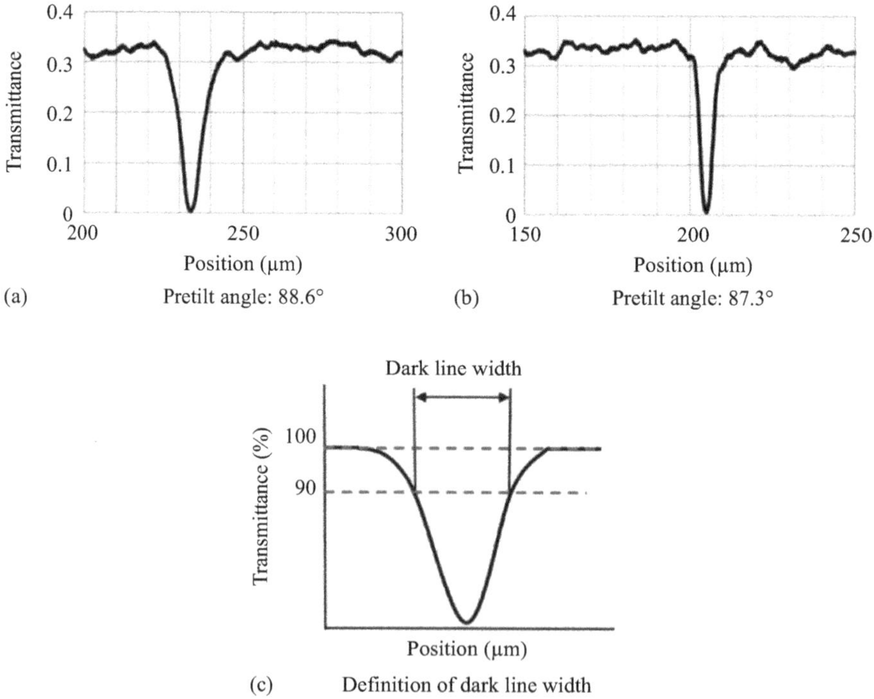

Figure 2.21 Dark line width: dark line cross section and line width definition

Table 2.4 Transmittance and contrast of panels processed with low pretilt photoalignment material

Alignment material	**Pretilt angle**	**Contrast ratio**	**Transmittance increase**
Standard	88.5°	3,769	Reference
Low pretilt	87.7°	3,833	8%

2.4.2.3 Investigation of low pretilt material in mass production

The Rolic sample for mass production trial, which is mentioned before, was investigated by CEC Panda. Test results are summarized in Table 2.4.

By using the Rolic low pretilt material, the overall transmittance is increased by approx. 8 percent compared to the reference, which match pretty well with the experimental prediction.

With lower pretilt angle it was expected to have light leakage as shown in Section 2.3.2, but the increase of transmittance due to the reduction of dark line width compensates this loss. Therefore, the contrast ratio remains unchanged.

2.5 Conclusion

UV^{2}A display mode using Rolic® LCMO technology is a well-established manufacturing technology for LC-TV panels. We demonstrated that the mechanism is by photo-crosslinking of cinnamate group through [2+2] cycloaddition.

For the further development related to the improvement of overall transmittance for ultra-high resolution display panel, we showed the dependence of the dark line width upon the pretilt angle by means of simulation. We described also an easy experimental method for preliminary evaluations of new low pretilt material candidates. This may allow panel manufacturers, to reduce the number of tedious and expensive tests on their mass production line.

Finally, we confirmed the prediction of transmittance improvement with the first ready-to-use low pretilt formulation on a mass production line.

References

[1] Miyachi K., Kobayashi K., Yamada Y. and Mizushima S. 'The world's first photo alignment LCD technology applied to generation ten factory'. SID Symposium Digest of Technical Papers 2010, vol. 41, pp. 579–582.

[2] Eich M. and Wendorff J. H. 'Erasable holograms in polymeric liquid crystals'. *Makromol. Chem., Rapid Commun.* 1987, vol. 8, pp. 467–471.

[3] Todorov T., Nikolova L. and Tomova N. 'A new high-efficiency organic material with reversible photoinduced birefringence'. *Appl. Optics.* 1984, vol. 23, pp. 4309–4312.

[4] Ichimura K., Suzuki Y., Seki T., Hosoko A. and Aoki K. 'Reversible change in alignment mode of nematic liquid crystals regulated photochemically by command surfaces modified with an azobenzene monolayer'. *Langmuir.* 1988, vol. 4, pp. 1214–1216.

[5] Gibbons W. M., Shannon P. J., Sun S. T. and Swetlin B. J. 'Surface-mediated alignment of nematic liquid crystals with polarized laser light'. *Nature.* 1991, vol. 351, pp. 49–50.

[6] Schadt M., Schmitt K., Kozinkov V. and Chigrinov V. 'Surface-induced parallel alignment of liquid crystals by linearly polymerized photopolymers'. *Jpn. J. Appl. Phys.* 1992, vol. 31, pp. 2155–2164.

[7] Hasegawa M. and Taira Y. 'Nematic homogeneous photo alignment by polyimide exposure to linearly polarized UV'. *J. Photopolym. Sci. Technol.* 1995, vol. 8, pp. 241–248.

[8] Schadt M. and Seiberle H. 'Optical patterning of multidomain LCDs'. *J. SID* 1997, vol. 5, pp. 367–370.

[9] Schadt M., Seiberle H. and Schuster A. 'Optical patterning of multi-domain liquid-crystal displays with wide viewing angles'. *Nature.* 1996, vol. 381, pp. 212–215.

[10] Schadt M., Seiberle H., Schuster A. and Kelly S. M. 'Photo-generation of linearly polymerized liquid crystal aligning layers comprising novel, integrated optically patterned retarders and color filters'. *Jpn. J. Appl. Phys.* 1995, vol. 34, pp. 3240–3249.

[11] Seiberle H. and Schadt M. 'Photoalignment and photo-patterning of planar and homeotropic liquid-crystal-display configurations'. *J. SID.* 2000, vol. 8, pp. 67–71.

[12] Ibn-Elhaj M. and Schadt M. 'Optical polymer thin films with isotropic and anisotropic nano-corrugated surface topologies'. *Nature.* 2001, vol. 410, pp. 796–99.

[13] Seiberle H., Bachels T., Benecke C. and Ibn-Elhaj M. 'Volume Photo-Aligned Retarders'. *IEICE Trans. Electron.* 2007, vol. E90-C, pp. 2088–93.

[14] Seiberle H., Schmitt K. and Schadt M. 'Multidomain LCDs and complex optical retarders generated by photo-alignment'. Eurodisplay '99 Proceedings; Berlin, Germany, Sep. 1999. VDE-Verlag; 1999. pp. 121–25.

[15] Yamada Y., Sakurai T. and Miyachi K. 'Advanced technologies for vertical alignment. liquid crystal display'. Key Note, SID-ME Spring Meeting 2011.

[16] Ibn-Elhaj M., Chappellet S. and Lincker F. 'Rolic® LCMO photo alignment technology: Mechanism and application to large LCD panels'. *Sol. State Phenom.* 2012, vol. 181–182, pp. 3–13.

[17] Miyachi K., Kimura N., Yamada Y. and Mizushima S. 'The UV2A technology for large size LCD-TV panels'. Proceedings of the 17th International Display Workshop (IDW'10), Fukuoka, Japan, December 2010, LCT1-1, pp. 13–16.

[18] Wang H. 'Studies of liquid crystal response time'. Doctoral dissertation, University of Central Florida, USA, 2005.

[19] Yamada Y., Tang Q., Koechlin M. and Yamamoto Y. 'Investigation of transmittance dependence upon pre-tilt angle in photo aligned vertical LCDs'. SID 2017, digest; Los Angeles, USA, May 2017, pp. 708–11.

[20] Tang Q., Koechlin M., Seiberle H., *et al.* 'Novel photo-alignment material for ultra-high definition VA LCDs'. ICDT 2018 Proceedings; Guangzhou, China, April 2018, pp. 526–31.

[21] Ishihara S. 'The latest display technology in LCD-TV "AQUOS"'. *Sharp Technical Journal*, no. 28, vol. 96, pp. 15–20, November 2007.

[22] Washizuka I. and Kubo M. 'Development of a 20.1-inch diagonal SXGA TFT-LCD for replacement of CRT monitors'. Proceeding of AM-LCD, Amp1-1, pp. 99–102, 1999.

Chapter 3
Azo-dye photoalignment materials

Vladimir Chigrinov[1,2]*, Abhishek Srivastava*[2]*, and Hoi-Sing Kwok*[2]

The new azo-dye photoaligning films are robust and possess very good aligning properties for liquid crystal (LC) cells such as anchoring energies and voltage holding ratios as well as a high thermo and UV stability both in glass and plastic substrates with very high resolution on submicron level. The most important characteristics of azo-dye photoaligning films will be compared with those of rubbed polyimide films usually used for LC alignment. In this chapter, we will also describe the new applications of azo-dye photoalignment: (i) LC display devices, such as ferroelectric LCD for IPS, field sequential color and microdisplay applications; (ii) LC photonics devices, such as LC lenses; (iii) LC E-paper devices, such as electrically and optically rewritable LC E-paper; (iv) nanoscopic optically engineered polymeric thin film polarizers.

3.1 Introduction

In this chapter, we concentrate on one aspect of LCD manufacturing, that of the alignment surface. In particular, we present a review of azo-dye photoalignment materials and technologies. Photoalignment has been proposed and studied for a long time [1–6]. It was soon shown that these materials could provide high-quality alignment of LC molecules in a cell. Over the past 20 years, a lot of improvements and variations have been made for photoalignment. Commercial photoalignment materials are now readily available. Many new applications, in addition to the alignment of LCD, have been proposed and demonstrated. In particular, the application of photoalignment to active optical elements in optical signal processing and communications is currently a hot topic in photonics research [7].

[1]Foshan University, P.R. China

[2]Department of Electronic and Computer Engineering, Hong Kong University of Science and Technology, Hong Kong

Photoalignment possesses obvious advantages in comparison with the usually "rubbing" treatment of the substrates of liquid crystal display (LCD) cells. Possible benefits for using this technique include:

(i) Elimination of electrostatic charges and impurities as well as mechanical damage of the surface;
(ii) A controllable pretilt angle and anchoring energy of the LC cell, as well as its high thermo and UV stability and ionic purity;
(iii) Possibility to produce the structures with the required LC director alignment within the selected areas of the cell, thus allowing pixel dividing to enable new special LC device configurations for transflective, multi-domain, 3D and other new display types;
(iv) Potential increase of the manufacturing yield, especially in LCDs with active matrix addressing, where fine tiny pixels of a high resolution LCD screen are driven by thin film transistors on a silicon substrate;
(v) New advanced applications of LC in fiber communications, optical data processing, holography and other fields, where the traditional rubbing LC alignment is not possible due to the sophisticated geometry of LC cell and/or high spatial resolution of the processing system;
(vi) Ability for efficient LC alignment on curved and flexible substrates;
(vii) Manufacturing of new optical elements for LC technology, such as patterned polarizers and phase retarders, tunable optical filters, polarization non-sensitive optical lenses, with voltage-controllable focal distance.

This chapter presents the status of the current research in LCD photoalignment and photo-patterning based on azo-dyes. In this chapter, we shall concentrate on a recent approach of ours, which is rather promising, namely, the photo-induced reorientation of azo-dyes [4]. This technique of photoalignment does not involve any photochemical or structural transformations of the molecules. As well, the new photoaligning films are robust and possess a good aligning properties such as anchoring energies and uniformity. They can be very useful for the new generation of the LCDs as well as in new photovoltaic, optoelectronic and photonic devices based on highly ordered thin organic layers. Examples of such applications are light emitted diodes (OLED), solar cells, optical data storage and holographic memory devices. The novel and highly ordered layer structures of organic molecules may exhibit certain physical properties, which are similar to the aligned LC layers.

The principle aims of the chapter are:

1. To describe the physical mechanisms of LC azo-dye photoalignment with a special emphasis to the most useful photosensitive materials and preparation procedures suitable for the purpose;
2. To summarize LC surface interaction in photoaligned LC cell to produce the required LC pretilt angles, anchoring energy, ionic purity, IR and UV stability and sensitivity to the activating light exposure;
3. To show how to produce perfect vertical, twisted, rewritable and other LC photoalignment in nematic, ferroelectric, lyotropic and discotic materials on glass and plastic substrates and special LC cell configurations (Si waveguide and 3D surfaces, superthin tubes, etc.);

4. To compare various applications of photoalignment technology for in-cell patterned polarizers and phase retarders, transflective and micro-displays, security and other LC devices;
5. To present the recent results in applications of azo-dye photoaligning and photo-patterning technology in LCDs.

3.2 LC surface interaction in azo-dye photoaligned cell

When the azo-dye molecules are optically pumped by a polarized light beam, the probability for the absorption is proportional to $\cos^2 \theta$, where θ is the angle

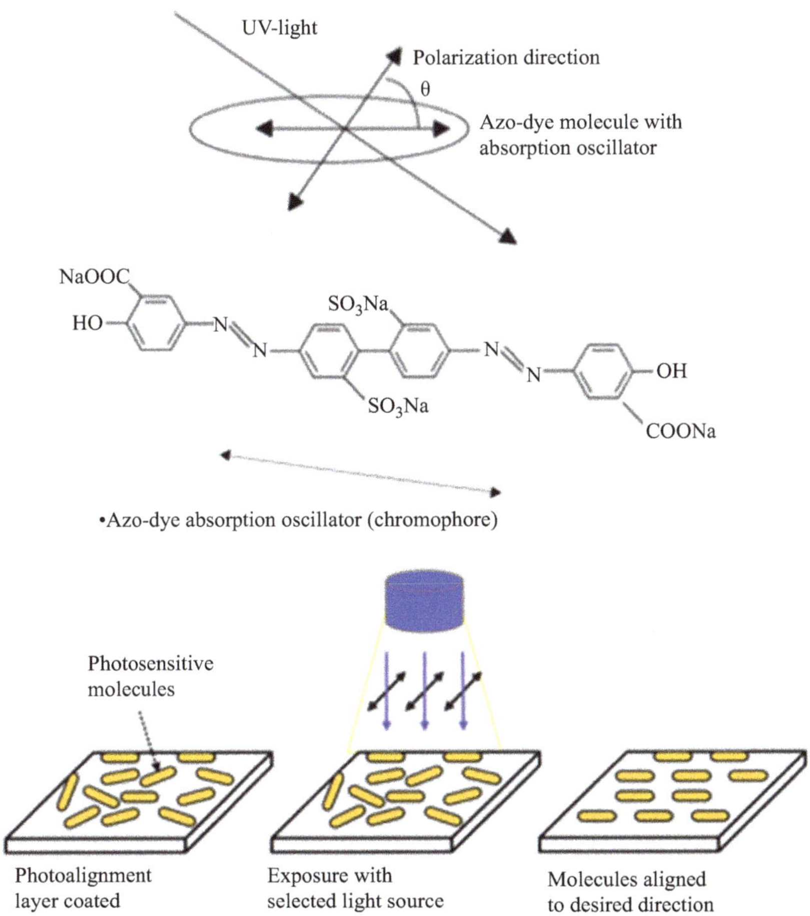

Figure 3.1 The photo-induced order in photochemical stable azo-dye films: (a) the geometry of the effect and (b) azo-dye molecule, having the absorption oscillator (chromophore) perpendicular to the long molecular axis; below nanolayer of azo-dye (2–15 nm) [4,5]

between the absorption oscillator of the azo-dye molecules and the polarization azimuth of the irradiating light (Figure 3.1).

Therefore, the azo-dye molecules, which have their absorption oscillators (chromophores) parallel to the light polarization azimuth, get excess of energy and reorient to minimize the energy. This results in an excess of chromophores in a direction at which the absorption oscillator is perpendicular to the polarization azimuth of the light. The chromophore is parallel to the long molecular axis of the azo-dye (Figure 3.1), i.e., azo-dye molecules are tending to align their long axes perpendicular to the UV-light polarization. Actually, the thickness of the layer in in the nanoscale range about 2–15 nm is quite enough to align LC in the bulk. Figure 3.1 shows a molecule of azo-dye SD1 (sulfonic dye 1), which is a very efficient dye for the photoalignment [5].

3.2.1 Pretilt angle generation in azo-dye photoaligning materials

To measure the pretilt angle, the LC cells were assembled with substrates antiparallel to UV irradiation directions and were filled with LC mixture. The crystal rotation method was used to measure their pretilt angles [5]. Pretilt angle of 0.7°, 0.1°, 5.3° and 3.8° is obtained with the exposure method (a), (b), (c) and (d), respectively (Figure 3.2). The oblique irradiation of non-polarized light proved to be the most efficient in order to obtain the larger pretilt angles. The dependence of the pretilt angle on the exposure energy in the method (c) is shown in Figure 3.3. The pretilt angle was temperature stable and does not change after heating the sample up to 100 °C for 10 minutes. Recently similar results for LC tilt angle using SD1 azo-dye were obtained also in [9].

Photoaligned vertically aligned nematic (VAN) mode was developed using the commercially available and UV-light-stable-aligning polyimide (PI) material [10].

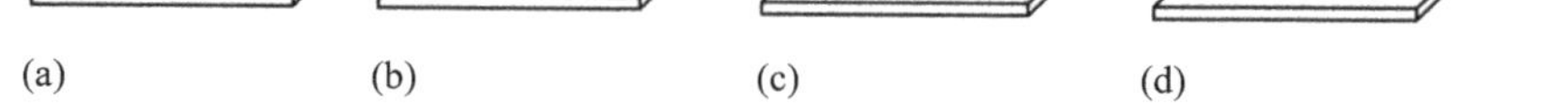

*Figure 3.2 Various methods of UV light irradiation to attain the pretilt angle of LC layer on the photoaligned film. Case (a): two-step exposure by **y**-polarized light vertical to the plane of the azo dye film (**xy** plane) and oblique p-polarized light; (b): case (a) with **s**-polarized light used at the second step; (c): case (a) with non-polarized light at the second step; (d): one-step exposure with oblique non-polarized light. The activated UV light propagates in the **xz** plane at an oblique angle with the substrate $\theta = 45°$ [8]*

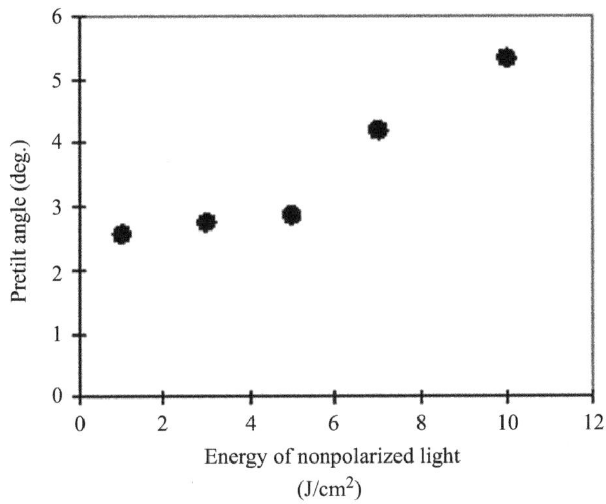

Figure 3.3 Dependence of the pretilt angle on the energy density of the obliquely irradiated nonpolarized light after the irradiation of vertical polarized light (Figure 3.2(c)) with the energy density of 10 J/cm^2 [12]

To provide the photoaligning of LC, the polyimide was illuminated by oblique non-polarized UV light. It has been shown that doping a PI with water-soluble sulfuric azo-dye (SD1, Figure 3.1) yields a reliable and perfect electrooptical performance of the VAN mode. This PI/SD1 mixture, used for the preparation of the photoaligning layer, provides the same parameters of VAN mode response as the conventional rubbing technique. The electrooptic response of the photoaligned VAN mode and VAN mode obtained by rubbing was very similar despite a certain difference in the LC pretilt angles on the substrate.

3.2.2 Anchoring energy in azo-dye photoalignment materials

The anchoring energy is one of the most important parameters, which characterizes the LC alignment quality [5]. It usually consists of the two parts, azimuthal and polar, that describe how easy to change the alignment of the LC director at the surface. The anchoring energy is a part of the LC free energy, needed to realign the LC director from the preferred orientation at the substrate, e.g., z-axis ($\theta = \varphi = 0$, Figure 3.4) up to the angles θ and φ in polar and azimuthal direction [5]:

$$W(\theta, \varphi) = 1/2 W_\theta \sin^2\theta + 1/2 W_\varphi \sin^2\varphi \approx {}^1\!/\!_2 W_\theta \theta^2 + {}^1\!/\!_2 W_\varphi \varphi^2 \quad (3.1)$$

where W_θ and W_φ are the corresponding polar and azimuthal anchoring energy. The value of W_θ and W_φ is a very important characteristic of LC-surface interaction. The polar and azimuthal LC anchoring energy should be sufficiently high to provide a high LC optical quality and fast switching.

A thickness of the photoaligned azo-dye layer strongly affects the LC alignment quality. The azo-dye can be used to align the LC even at very small thickness.

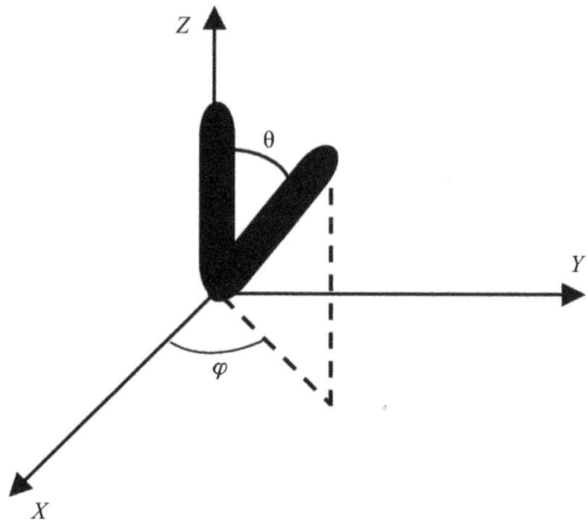

Figure 3.4 *The anchoring energy is a part of the LC free energy, needed to realign the LC director from the preferred orientation at the substrate, such as the z-axis ($\theta = \varphi = 0$) up to the angles θ and φ in polar and azimuthal direction: $W(\theta, \varphi) = W_\theta \theta^2 + W_\varphi \varphi^2$, where W_θ and W_φ are the corresponding polar and azimuthal anchoring energy [5]*

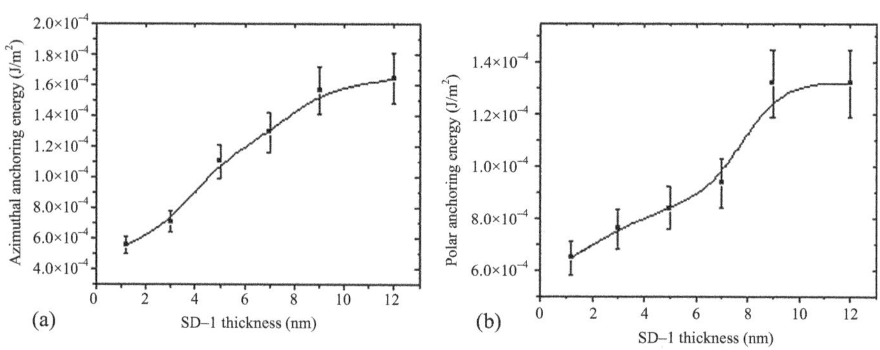

Figure 3.5 *LC polar and azimuthal anchoring energy as a function of the thickness of SD-1 photoaligning layer [5]*

Figure 3.5 shows the measured anchoring energy of the photoalignment layer as a function of the layer thickness [11]. It was found that even a 1 nm thick (almost a monolayer) alignment layer could produce uniform LC alignment. However, the anchoring energy was reduced by a factor of two.

The results can be understood by noting that the film with a thickness of 1 nm may not be continuous. In fact, it should be in the form of islands. Thus the coverage of the substrate is not perfect leading to weaker alignment effects. However, alignment of LC does not require a continuous film due to the tendency of the LC molecules to align themselves to minimize elastic deformation energy, and therefore, the LC molecules

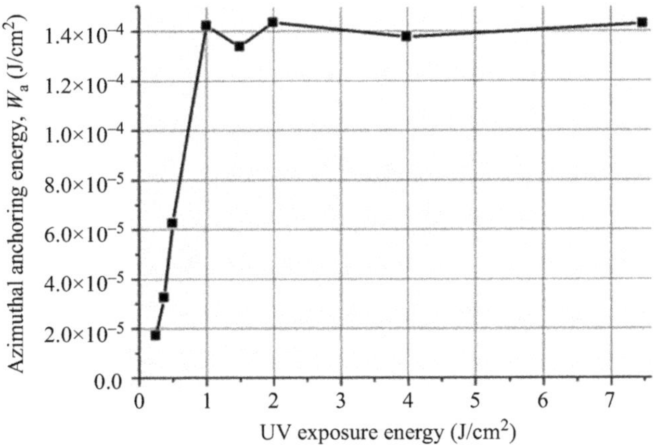

Figure 3.6 The azimuthal LC anchoring energy of the azo-dye layers as a function of the exposure time [5,12]

align by the discontinuous layer but the anchoring energy is small. Figure 3.5 points to the fact that the alignment is a surface effect. Once a continuous film is obtained, LC alignment can be achieved with good anchoring properties. This is another supporting fact that the azo-dyes are good agents for photoalignment [11].

Another factor, which determines the anchoring energy of the photoaligned layers is the exposure time. In particular, the azimuthal LC anchoring energy of the azo-dye layers can be substantially varied changing the exposure time (Figure 3.6) [12].

A super-thin photoaligned azo-dye layer can be also successfully used for LC photoalignment [13]. This new method includes the formation of a very neat "textile knitwear" by a super-thin azo-dye SD1 layer and allows to avoid the spin-coating procedure. Moreover, the photosensitivity of azo-dye after photoalignment can be further reduced and "island" azo-dye structure onto the rough ITO surface can be prevented due to better adhesion of SD1 molecules. Using this super-thin SD1 layer as an alignment agent, the sufficiently high polar and azimuthal anchoring energy and a perfect LC alignment can be obtained [13]. The method allows to have a perfect LC photoalignment in large or curved cells and is very attractive for mass production. The quality of a photoaligned TN cell is shown in Figure 3.7 [5].

3.2.3 Characteristics of azo-dye photoalignment materials: voltage holding ratio (VHR) and residual DC voltage (RDC): image sticking

VHR is characterized by the amount of ion dopants in LC bulk, while RDC stated the amount of the ions on the surface of LC cell [5]. The acceptable values of VHR should be more than 99 percent even at high temperature $T = 80$ °C, while RDC should be as small as possible at least less than 50 mV [5]. VHR and RDC for the azo-dye photoaligned LC cell is >99 percent (at 60 °C) and <50 mV, respectively, which is comparable to PI (see Table 3.1). This implies that the azo-dyes studied

Figure 3.7 Polarized microscope photograph of LC alignment on photoaligned polymerized azo-dye film [5]. The size of the picture is 1 mm^2. Both homogeneous and twist LC alignment looks perfect between crossed polarizers as black and yellow regions

Table 3.1 Comparison of azo-dye photoalignment quality with rubbed PI

Parameter	**Pure Azo-dye**	**Stack layer**	**Photo polymerization**	**Thermal polymerization**	**PI**
RDC (mV)	80	10	8	5	33
VHR (%)	<90	99.1	99.05	99.13	99.87
Polar Anchoring energy (J/m^2)	10^{-4} to 10^{-3}	10^{-4} to 10^{-3}	10^{-4} to 10^{-3}	10^{-4} to 10^{-3}	3.76×10^{-3}
Image Sticking parameter ratio	1.13	1.028	1.025	1.02	1.03
Azimuthal anchoring energy (J/m^2)	$\sim 5 \times 10^{-5}$	1×10^{-4}	1×10^{-5}	1×10^{-5}	4.7×10^{-4}

*The stabilization of azo-dye alignment (2–4 columns) will be explained below.

can be applied as aligning layers in thin film transistors liquid crystal displays (TFT-LCDs), as their purity is high enough and there is no additional ionic impurity from the aligning layers both in the volume and the surface of LC cell.

Experimentally, it was found that indeed even with a light dosage of 50 mJ/cm^2, perfect alignment can be achieved in the LC cell by azo-dyes. Certain azo-dye derivatives were synthesized, which considerably improve the sensitivity of azo-dye SD1, so a very low dosage of UV-light is sufficient for a perfect uniform alignment of LC cell: 150 mJ/cm^2 for a non-polarized light and 20 mJ/cm^2 for a polarized light [14].

Image-sticking parameter is another very important parameter for the active matrix display panel, which is characterized by the change of LC transmission voltage curve after certain continuous switching [15]. The AC image-sticking parameter ISP of photoaligning materials in IPS-cells is defined by:

$$\mathrm{ISP} = T_2/T_1, \tag{3.2}$$

where T_1 is the initial LCD transmittance at room temperature and 4 V driving voltage; T_2 is the room temperature transmittance at 4 V determined after the application of 10 V for 64 h at 60 °C. The LCD in plane switching (IPS) voltage-transmission graphs in Figure 3.8 illustrate the definitions of T_1 and T_2. Decreasing image sticking of LC-alignment corresponds to smaller ISP-values; where ideal alignment/surface treatment results in no image sticking; i.e., ISP = 1. As a reference for virtually ideal alignment/surface treatment we used brushed IPS-LCDs with polyimide (PI)- alignment yielding ISP ~1.00–1.02. Image sticking of a pure azo-dye SD1 is rather high (1.13), but after stabilization, it can become even better than for the rubbed PI layer (see Table 3.1).

3.2.4 Super high resolution of azo-dye photoaligning materials

The maximum possible resolution (i.e., minimum alignment domain size) for the photoalignment of LCs looks better than 100 nm [16]. The minimum domain size of less than 100 nm was achieved by photoalignment of the sulfonic dye SD1. A two-step alignment involving Lloyd's mirror interference technology for the second step has been used to achieve this high resolution (Figure 3.9). The theoretical limit for the minimum period of the interference pattern is $\lambda/2$ and the 2-step

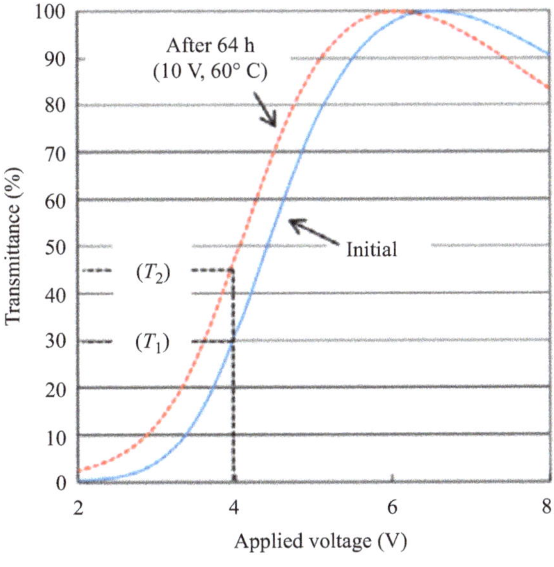

Figure 3.8 TcT₁ and T₂ for image-sticking parameter [15]

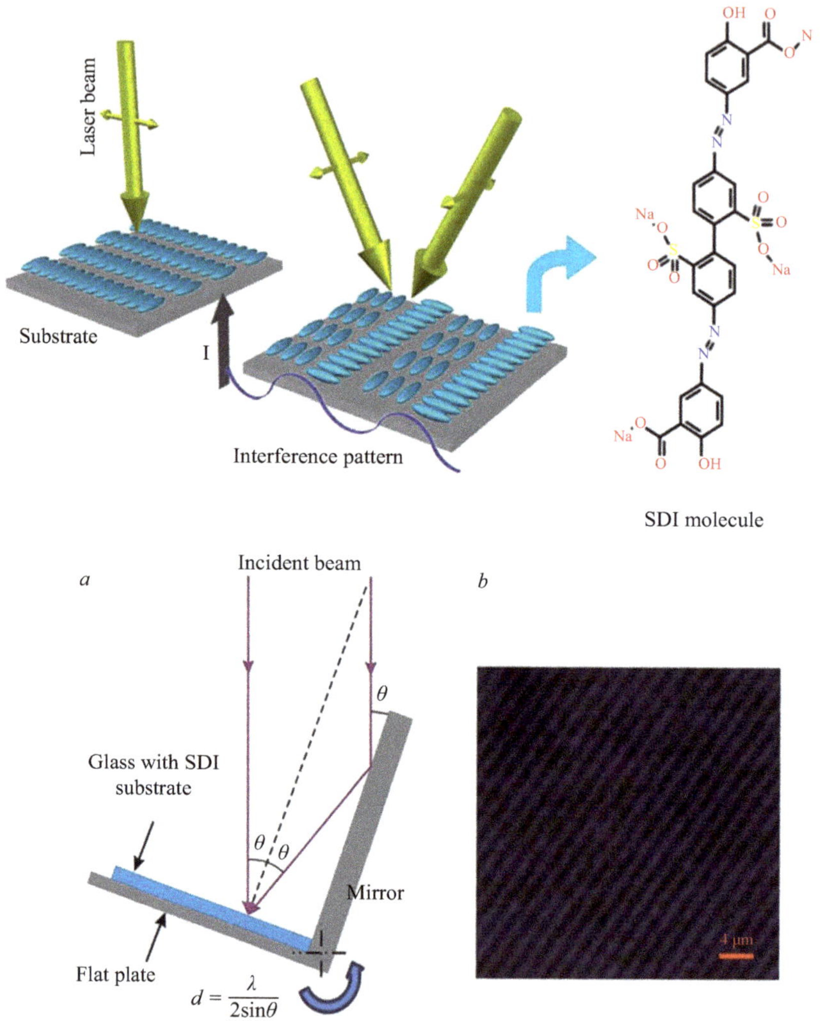

Figure 3.9 Two-stage exposure procedure for the fabrication of the gratings with submicron pitch. The right-hand insertion represents the molecular structure of the sulfonic azo dye SD1 [16]. The grating pattern formation by interference pattern is shown below

alignment provides the opportunity to achieve a domain size equal to $\lambda/4$. The wavelength of the exposing light $\lambda = 405$ nm, in the present case it is 405 nm, thus the theoretical limit of the system was about 100 nm [16].

3.2.5 LC photoalignment on plastic substrates

The LC alignment on plastic substrates is really an issue, as usual, rubbed polyimide (PI) requires a very high imidization temperature (> 250 °C), which is incompatible with usual plastic substrates that cannot tolerate such a high

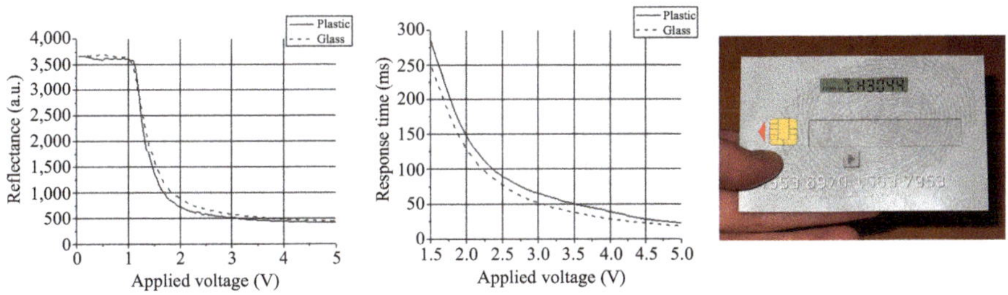

Figure 3.10 Comparison of the LC electrooptical performance in photoaligned TN LC cells on glass and plastic substrates. Reflectance (left) and response time (right) versus applied voltage. Reflective flexible photoaligned TN-LCD mounted in smart card prototype is shown below [18]

temperature [17]. At the same time, azo-dye photoalignment does not require a high temperature, and therefore, it is highly suitable for all plastic substrates [5].

Excellent LC photoalignment on plastic substrates with a high anchoring energy was achieved using the exposure energy less than 1.0 J/cm^2, which corresponds to the azimuthal anchoring energy > 10^{-4} J/m^2 [18]. A mixture of azo-dye SD1 with a mentioned above thermo-polymerized azo-dye SDA2 was used for the photoaligned LC cell fabrication. The LC pretilt angle of about 5° on the plastic substrate was made by a double exposure method (Figure 3.2). The reflectance as a function of applied voltage and response time was measured for both the plastic cell and glass reference cell (Figure 3.10).

A reflective contrast of 8:1 at normal incidence was demonstrated. A perfect uniform LC alignment was achieved on both plastic and glass substrates [18]. To demonstrate the alignment quality on plastic substrates a 9-digit reflective passive matrix TN-LCD mounted in a smart card was fabricated (Figure 3.10).

To maintain a uniform cell gap of 8 μm (second Mauguin minimum of MLC-6809-000 Merck LC mixture), semi-dry adhesive spacers were chosen. The electrooptical performance of the photoaligned plastic display was very similar to common TN-LCD fabricated, for comparison, by usual rubbing method on glass substrate.

A plastic color-twisted nematic liquid crystal display (TN-LCD) using an azo-dye photoalignment method was successfully developed [19]. It was confirmed that LC alignment on a plastic substrate is possible under continuous roll-to-roll processes. The irradiation energies in a double exposure by a polarized and oblique non-polarized UV light were 1 and 0.5 J/cm^2, respectively. The TN-LCD on a plastic substrate was 1/6 thinner and 1/10 lighter than TN-LCD using a glass substrate. A photoaligned plastic LCD plastic prepared by a continuous roll-to-roll coating and orientation process is shown in Figure 3.11.

3.2.6 Effect of humidity on the azo-dye photoalignment

A strong dependence of photoalignment quality of the azobenzenesulfonic dye on relative humidity has been observed [20]. The photoinduced phase retardation,

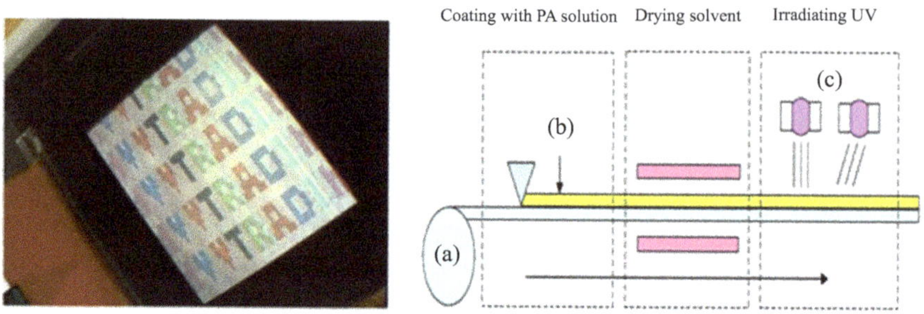

Figure 3.11 Left: Plastic color LCD photoaligned by azo-dyes; Right: A continuous roll-to-roll process of coating and orientation treatment by azo-dyes [19]

reorientation speed, order parameter of the alignment layer, and its alignment quality for LC under different relative humidities show dramatically different behaviors. At low humidity, the alignment ordering is low (Figure 3.12). As relative humidity increases, the film can be photoaligned better until about 60 percent RH, where the order parameter reaches above 0.8. As the humidity further increases, the alignment quality gets worse. According to [20], the best relative humidity range for the sulfonic dye SD1 photoalignment process is between 50 percent RH and 70 percent RH, which is recommended as a working window for the application of this photoalignment material for both photonic devices and displays. The humidity working window is also recommended for similar hydrophilic azobenzene photoalignment materials, where the specific value may vary according to different materials. Besides writing dynamics, the film alignment after excitation is also explored. A phase retardation increment after exposure has been observed at humidity above 30 percent RH. This increment is enhanced if the prealigned film gets more humidified after linearly polarized ultraviolet (LPUV) excitation: the film ordering increases substantially by about 4 times if the relative humidity changes from 20 percent RH to 80 percent RH, suggesting that a more ordered structure is preferred due to the insertion of water molecules (Figure 3.12). These phenomena are well explained by the semicrystalline property of the azobenzenesulfonic dye with water molecules, which is proposed based on X-ray diffraction (XRD) and depolarized optical microscope observations. The self-assembly of the photoaligned molecules with the insertion of water molecules contributes to the high ordering ($S > 0.8$). The transition from amorphous solid at low relative humidity to semicrystal at high relative humidity provides a new perspective of understanding the hydrophilic photoalignment materials [20].

3.2.7 Stability of photo-aligning materials: sensitivity to UV light: strengthening of liquid crystal photoalignment on azo-dye films: passivation by reactive mesogens

The realignment of the azo-dye is not desirable in most of the display applications, so the alignment stabilization is highly desirable [21,22]. Several efforts have already been made to stabilize the azo-dye photoalignment with liquid crystal polymer (LCP),

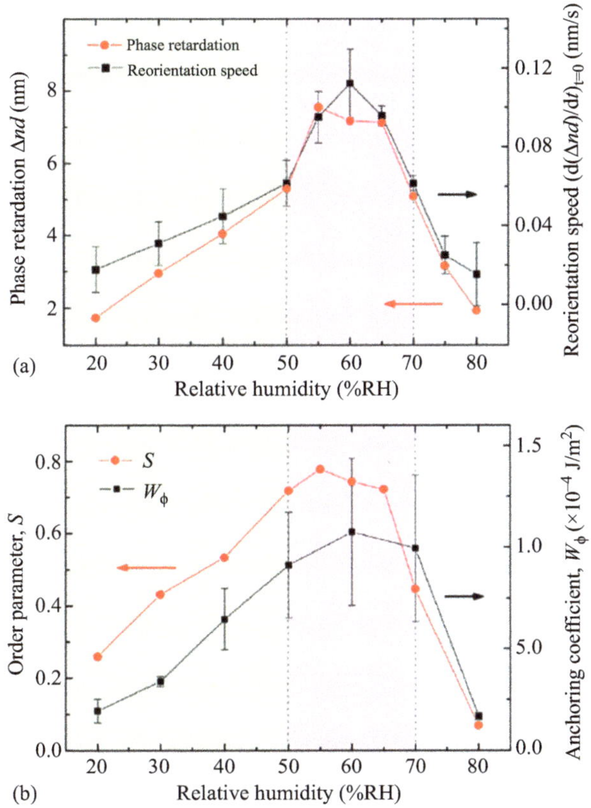

Figure 3.12 Ordering comparison of the photoalignment films induced by 1 J/cm² 365 nm LPUV at different relative humidity. (a) Phase retardation (red circles) and reorientation speed (black squares). (b) Order parameter (red circles) and azimuthal anchoring coefficient (black squares). The LPUV power is 5 mW/cm². Different background colors indicate different humidity regions for photoalignment: the green region indicates low humidity, which gives low ordering; the pink region indicates moderate humidity for best alignment quality; the blue region indicates high humidity which also gives low ordering

including polymer azo-dye composite, LCP layer on the top of the azo-dye, and introduction of the polarizable moiety in the azo-dye itself. For the polymer composite azo-dye alignment layer, it has been proposed to utilize azo-dye and LCP mixture with a proper ratio to form a photo-stable photoalignment, but VHR and RDC do not meet the industrial standards. The LCP layer on the top of azo-dye, an additional LCP layer offers strong constraint on the thickness of LCP. Moreover, the double-layer structure increases the manufacturing cost. On the other hand, the introduction of the polymer unit in azo-dye leads to poor alignment characteristics.

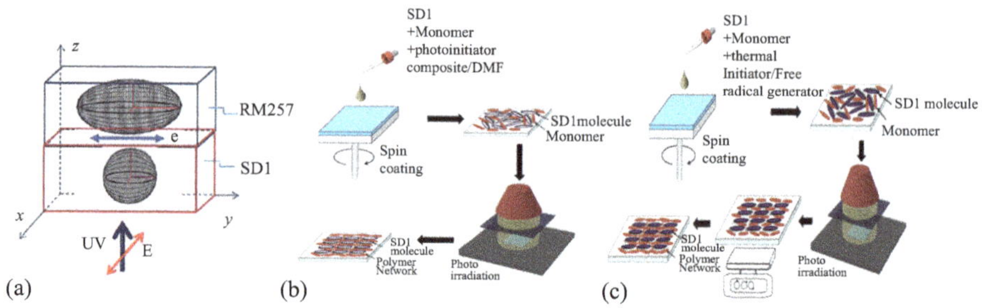

Figure 3.13 Stabilization of azo-dye by double (stack) layer (above) [21] and photo and thermo stabilization (below) [22]. The characteristics of the stabilized azo-dye layer are given in Table 3.1

Photo and thermal stabilization demonstrated a method for stabilizing and optimizing the sulfonic azo-dye SD1 alignment by using LCP composite for the better parameters suitable for the industry (Table 3.1, Figure 3.13). The process contains two-step irradiation: first for the SD1 alignment and second for the polymerization of LCP to build a robust network and then fix SD1 molecules one by one. Several experiments have been made to confirm the high thermal-tolerance and photo-stability. Additionally, the display-related parameters, i.e., VHR, RDC and image-sticking parameter, were measured and meet the industrial standard. Furthermore, the proposed photoalignment layer is capable to offer multiple-domain alignment with high optical quality. Thus, this composite photoalignment layer has immense potential to be applied in various LC displays.

3.3 Application of azo-dye photoalignment

Almost all the criteria of perfect LC photoalignment are met in case of azo-dye layers. Nowadays, azo-dye alignment materials can be used in LCD manufacturing, e.g., for the alignment of monomers in liquid crystal polymer (LCP) films for new generations of photonics and optics devices. Recently, a new application of photoaligned technology for the tunable LC lenses with a variable focal distance was proposed. New optically rewritable (ORW) LCD and photonics devices with a light controllable structure may include LC E-paper screens, LC lenses with a variable focal distance etc.

Fast ferroelectric liquid crystal devices (FLCD) are achieved through the application of nano-scale photo aligning (PA) layers in FLC cells. The novel photoaligned FLC devices may include field sequential color (FSC) FLC with a high resolution, high brightness, low power consumption and extended color gamut to be used for PCs, PDAs, switchable goggles, and new generation of switchable 2D/3D LCD TVs, as well as photonics elements.

Photoalignment possesses obvious advantages in comparison with the usually "rubbing" treatment of the substrates of LCD cells. Possible benefits for using these techniques include new advanced applications in displays and photonics, fiber communications, optical data processing, holography and other fields, where the traditional rubbing LC alignment is not possible due to the sophisticated geometry of LC cell and/or high spatial resolution of the processing system. New optical elements for LC technology, such as highly efficient polarizers based on polarization grating and patterned phase retarders, optical lenses with voltage controllable focal distance, optically rewritable devices, etc., can be also successfully made by photoalignment and photopatterning technology.

We will demonstrate below several applications of LCD photoalignment and photopatterning technology, such as LC E-paper, optical lenses, highly efficient polarizers, fast ferroelectric devices.

3.3.1 Optically-rewritable LCD

Optical rewritable technology (ORW) [5,7] is a modified method of azo-dye photoalignment that possesses traditional high azimuthal anchoring energy, up to 2×10^{-4} J/m^2, and has a unique feature of reversible in-plane aligning direction reorientation, i.e., rotation perpendicular to the polarization of an incident light. An ORW LC cell consists of two substrates with different aligning materials (Figure 3.14).

One aligning material is optically passive and keeps aligning direction on one substrate. The other aligning material is optically active and can change its alignment direction on being exposed with polarized light through the substrate. In comparison with electrically controlled plastic display, ORW can be significantly thinner and requires no ITO photolithography and etching on plastic substrate because no electrodes are needed. Switching and continuous gray scale of 5 bits or more are achieved by control of aligning direction of photoaligning azo-dye layer, which is insoluble in LC (Figure 3.14) [3].

Several methods to generate grayscale images on ORW e-paper have been proposed: (i) exposition of the photo-alignment layer with different amount of time or exposure energy, which strongly depends on the thermal treatment of the cell;

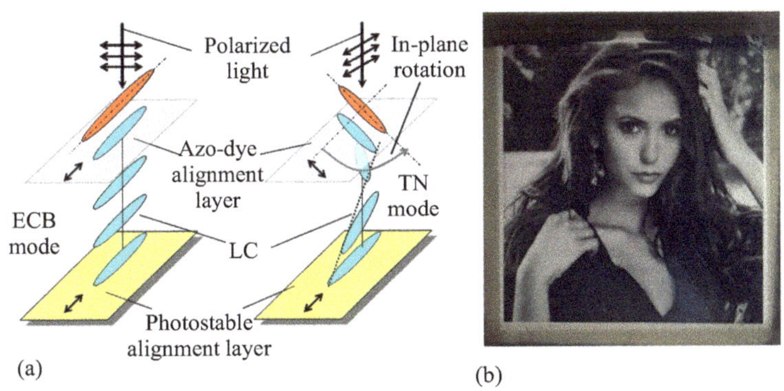

Figure 3.14 (a) Operation principle of LC ORW cell [5,7]. (b) 5 bit gray-level image on LC E-paper [3]

(ii) division of one pixel to several black or white parts to get gray levels, which reduces the spatial resolution of the image; and (iii) light polarization rotation for each pixel [23].

ORW is very tolerable to the cell gap variation as even 50 percent changing of the cell gap will not cause noticeable change in LC transmission value, while achromatic switching of all ORW gray levels can be obtained [5,7]. Every transmission level is stable and visualizes information with zero power consumption for a long time. Optical rewritable LC alignment is a good base for new types of LC displays, especially on plastic substrates, when no ITO films and electrodes are needed. 3D E-paper was also obtained by writing two images in one E-paper, providing 3D effect [24].

No backlight is required as reflective type polarizer is used as the bottom substrate. The image is truly stable, can be written to any gray level with the saturation and rewritten a large number of times with a high reproducibility of properties. The overall reflection is about 30 percent, while the aspect ratio is 1. The unique property of being cell-gap non sensitive allows to maintain proper performance even when the device is bent. The ORW E-paper potential applications are light printable rewritable paper, labels and plastic card displays.

3.3.2 Low-voltage driving-tunable LC lens using photoalignment method

Lenses with an electrically tunable focal length and/or are positionable focus are of broad interest for a number of applications. Because there are no moving parts in such a lens, they can be more reliable and able to withstand mechanical shock. Such tunable lenses can be applied in adaptive binocular glasses, autofocus cameras, LED light steering applications, 2D/3D switchable LCDs [5,7].

Many photonic devices required lenses with variable focal distance. A zoom lens system usually consists of a group of lenses, the separation distances between which are mechanically adjusted. Mechanical adjusting processes are, as a rule, complicated and bulky. That is the reason for developing a new type of tunable lenses that are compact, lightweight, low-cost and efficient. Such lenses are highly desirable and urgently needed for purposes of adaptive optics, optoelectronics, machine vision, stereo displays and eyeglasses applications. Due to the property to be controlled with an external electrical field, LC cells are especially attractive for tunable lenses.

Spatially varying the LC pretilt angle can be a good alternative to achieve low power-consuming tunable LC lenses [5,7]. The pretilt angle of the LC alignment is an important parameter to define the characteristic of LC devices. A tunable LC lens based on the approach of controlling the pretilt angle profile by means of the irradiance doses of a photoalignment (PA) layer was reported [25]. Such a PA layer provides precise control on the pretilt angle over a wide range of the pretilt angles, i.e., 1°–89°, which provides a great potential for applications not only in the tunable LC lens but also for other photonics and display devices. The PA material CPL024 (DIC, Japan) was used, which is a cross-linking polymer, providing a broad range of pretilt angles (1°–89°) depending on the irradiance doses. The pretilt angle profile for the CPL024 at different irradiance has been studied by making a 10-μm thick-antiparallel

LC cell with several distinct areas having different irradiance (Figure 3.15). The fabrication process flow is shown in the top insertion of Figure 3.15 [25].

The pretilt angle depends on the irradiance. Therefore, spatially variable UV irradiance patterns can be employed to achieve any desired pattern. The same approach has been used to fabricate the LC tunable lens (Figure 3.15, right) [25]. The schematics of the experimental setup and cell structure are shown in Figure 3.15(a) and (b), respectively. To generate the spatially variable UV irradiance, a mask with circular structure of radius R was placed at a distance D from the substrate (Figure 3.15, right) [25]. The light intensity exposed on the substrate was distributed in circular symmetry with the smallest intensity at the center and the largest at the border of the circle. All of these retardation parameter profiles show parabolic character that is also the best case for any lens design. Only 3 V was enough to tune f from 2.5 to 20 cm, which can be decreased further by optimizing the LC parameters and cell gap of the lens. Furthermore, given the easy fabrication procedure in association with low operating voltages, such devices could find applications in many modern optical and photonic devices. The proposed approach can also be employed to fabricate the bistable LC displays, photonic elements, e.g., beam steering devices, diffraction gratings, etc. In spite of having several advantages, further research is required to improve the lens profile, which could include use of certain grayscale masks or control of exposing environments, but these are simple technological issues that can be resolved by further improvement in the technology.

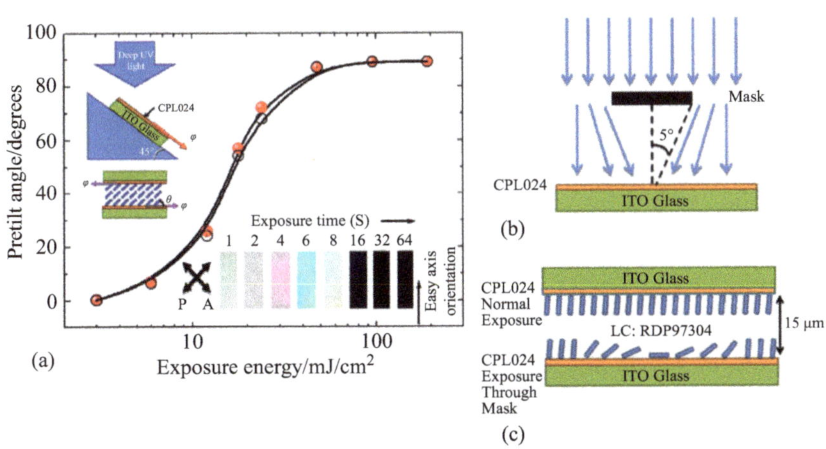

Figure 3.15 *(a) Pretilt angle dependence on the exposure energy (solid circles) [25]. The top insertion shows the simple fabrication procedure to make the LC cell to analyze different PA parameters, whereas the bottom insertion shows the photograph of the prepared LC cell under the crossed polarizers for distinct areas with different irradiance and pretilt angles. Right: (b) Exposure setup to achieve spatially variable light intensity on the exposing plane. (c) The fabricated LC lens structure [25]*

3.3.3 Nanoscopic optically engineered polymeric thin film polarizer

The concept to fabricate the polarizer with optical efficiency ~100 percent is based on the nanoscopic manipulations of the liquid crystal polymer (LCP) director to form a diffraction grating with complex 2D alignment patterns, in order to diffract incident non-polarized light into two distinguished circularly polarized beams and separate them spatially at an extremely short distance [26]. The proposed structure is an assembly of the two specially nanoscopically engineered films, first the 4-domain two-dimensional (D) polarization grating (PG) and second the 2 domain 2D quarter wave plate (QWP). The non-polarized light after passing through the spatially nanoscopically engineered patterned PG diffracts into left hand circular polarized (LHC) and right-hand circular polarized (RHC) beams corresponding to the −1st and +1st diffraction orders with diffraction efficiency ~100 percent (Figure 3.16). The two distinguished circular polarized light diffracted from the four domains film are separated spontaneously and collected separately at specific QWP domain (Figure 3.16) [26]. Thereafter the specially designed QWP converts the two distinguished circularly polarized lights in to the linearly polarized light. Unlike the simple PG, the proposed structure provides advantages for uniformity and compactness that could secure application in modern photonic devices.

The novelty of the concept, which lies in the 4-domain 2D patterned PG, offers a real experimental challenge to fabricate the nanoscopically defined alignment. Most common alignment technologies including the micro rubbing suffer from experimental limitations to design 2D patterned PG with small pitch ~2 μm without any serious defect and misalignment of different domains. Small PG pitch, domain size and the defect line in between the two domains are critical parameters that decide the optical quality of the system. Photoalignment (PA) technology provides a high-resolution nano scale alignment domains, which is the essential requirement in the fabrication of the proposed 4-domain 2D patterned PG (Figure 3.16).

The theoretical maximum efficiency, i.e., 100 percent can be achieved by precisely controlling the nano-scale alignment pattern and the film thickness with optical retardation equal to half of the light wavelength $\lambda/2$. The measured diffraction efficiency of the ±1st orders in the present case ~99.3 percent [26].

The novelty of the concept lies in the 4-domain 2D patterned PG which has been achieved by the nanoscopic PA technology. Experimentally measured extinction ratio of the converted impinging light, from the proposed thin film polarizer, is ~130:1 with efficiency >98 percent with incident acceptance angle 0°–13° from normal, which can be further improved (Figure 3.16). The thin film polarizer with ~100 percent polarization conversion efficiency, uniformity and compactness offer perfect solution to overcome the theoretical limits of the conventional thin film polarizers and therefore it could secure application in modern photonic and optoelectronic devices. One of the most important applications is LCDs, where the overall efficiency can be increased up to 2–3 times than the existing devices that will surely improve the energy consumption statistics of the existing LCDs. However, certain

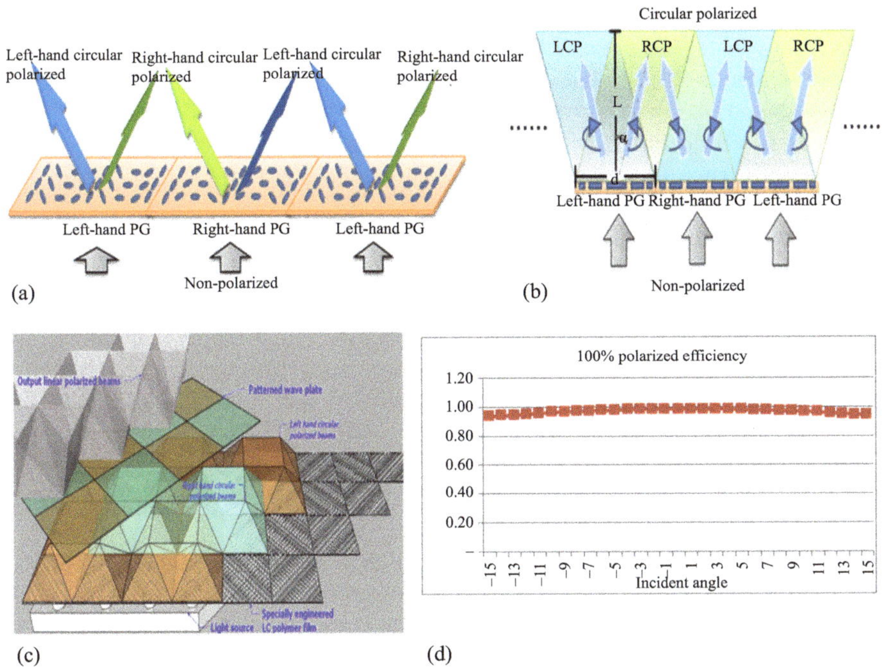

Figure 3.16 *Top: polarization grating (PG) separating the non-polarized light after passing through the spatially nanoscopically engineered patterned PG diffracts into left hand circular polarized (LHC) and right hand circular polarized (RHC) beams corresponding to the −1st and +1st diffraction orders with diffraction efficiency ~100%. The two diffraction orders are propagating in different directions corresponding to the diffraction angles defined by the equation* $\Lambda \sin \theta_m = m\lambda$*, where* Λ *represents the grating pitch,* λ*-light wavelength,* θ_m *represents the diffraction angle of the mth order, and* $m = \pm 1$ *for PG. Bottom from left to right: conversion of the circular polarized light to linear polarized light by a patterned QWP. The polarization efficiency as a function of the light incidence angle is also shown below*

parameters have to be optimized before it could be applied for the real applications. Moreover, it is critically important to improve the wavelength dependence of the system that can be done by multilayer structure. Although such films could immediately find application in LEDs, LED LCD backlight or other photonic devices where only narrow wavelength light band is needed.

3.3.4 Fast ferroelectric LCD

FLCD is the fastest LC mode, which can work with both fast responses time and low driving voltages and is highly suitable for field sequential (FSC) applications [27].

Fast switching ferroelectric liquid crystal (FLC) displays (FLCD) is a good candidate for the new generation of field sequential color (FSC) LCD, which proved to be better in response time than usually used nematic LC. The best FLC parameters can be obtained on the basis of electrically suppressed helix (ESH) mode [7,27].

Recently a reflective display cell which is suitable for projection displays based on electrically suppressed helix ferroelectric liquid crystal (ESHFLC) with fast response time, which is quite suitable for field sequential display was proposed [27]. The pulse width modulation (PWM) technique is used to control the residual light and thus provide several gray levels.

The electrooptical (EO) performance of reflective cell at the electric field of 3 V and the frequency of 5 kHz was set to achieve a high number of colors. The response time dependence of driving frequency under different driving voltage is shown in Figure 3.17. Even with small driven voltage, reflective FLC cell still maintains enough contrast ratio (CR) for display CR > 10,000:1. The response time of reflective FLC cell driven at 5 kHz and with the electric field of 5V was about 14μs. Such a fast response time enables us to drive the FLC cell even at very high driving frequency $f = 5$ kHz (Figure 3.17).

Novel photoaligned FLC devices may include FSC FLC with a high resolution, low power consumption and extended color gamut, which can be used in the screens of portable PCs, mobile phones, PDAs. The switchable goggles and lenses based on new FLC prototypes can be efficiently applied in the new generations of switchable 2D/3D LCD TV. FSC FLC microdisplays, which is now one of the most advanced technologies for pico-projectors can be also made on the basis of new materials and electrooptical modes in FLC [7]. The photoalignment technology enables to solve the key problems usually faced in FLCD applications, such as (i) quality of FLC alignment on sufficiently large surface area; (ii) appropriate adjustable anchoring energy and pretilt angle; and (iii) low losses in the alignment layers due to their small thickness, etc. [7,27].

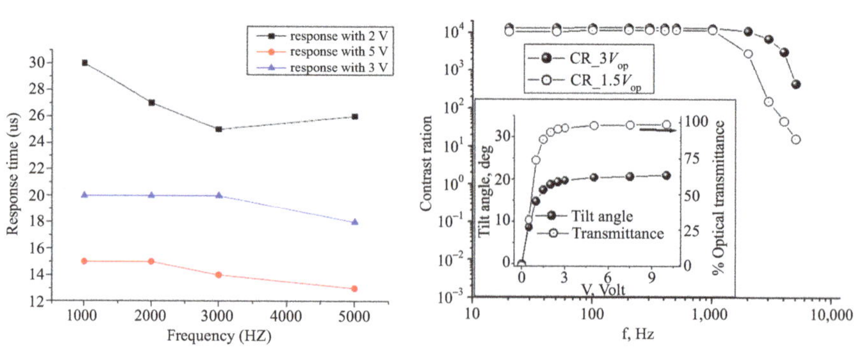

Figure 3.17 Left: Response time dependence of driving frequency under different driving voltage [27]. Right: Electrooptical response of photoaligned ESHFLC

The further the development of novel photoaligned fast FSC FLCD is aimed at: (i) further fundamental study of the new appropriate electrooptical modes used for switching; (ii) further understanding of physical mechanisms of FLC interaction with a photoaligned surface of different photosensitive nature to produce a stable alignment with a controllable anchoring energy and pretilt angle over sufficiently large surface area; (iii) development of new fast responded FLC materials with fast switching and a sufficient number of switchable gray levels (V-shape switching); (iv) implementation of the working prototypes of novel FSC FLC displays; and (v) investigations of regimes of operation to allow to use efficient addressing of FLCD.

3.3.5 *In plane switching (IPS) azo-dye photoalignment*

A photo-stable azo-dye photoalignment was also demonstrated for in-plane switching (IPS) LC mode using the stacked layer stabilization of sulfonic dye SD1 [28]. The alignment surface has a structure of stacking a thin photo-polymerizable LCP on top of an azo-dye photoalignment layer (Figure 3.13a). The LCP acts as a passivation layer for the azo-dye photoalignment layer. The alignment direction of the azo-dye cannot be changed once the LCP is polymerized. Hence, this synthetic alignment surface was robust, highly reproducible and has excellent stability, including photostability, thermal-stability and aging-stability. In addition, measurement results show the alignment properties of this synthetic alignment surface are comparable to conventional polyimide (see also Table 3.1). As a result, the proposed alignment layer is suitable for in plane switching LCD which requires high anchoring energy and very small residual direct current in order to have fast response time and avoid image sticking.

The ESHFLC display (see Section 3.3.4) mode also provides an alternative for IPS-LCD with extreme optical quality and low power consumption. The driving voltage ~1 V/μm is enough to drive the ESHFLC with optimal quality and fast response time ~200 μs. Thus ESHFLC display cell could be a perfect alternative for the high-resolution display with cheaper and relatively simpler fabrication process (Figure 3.18) [29].

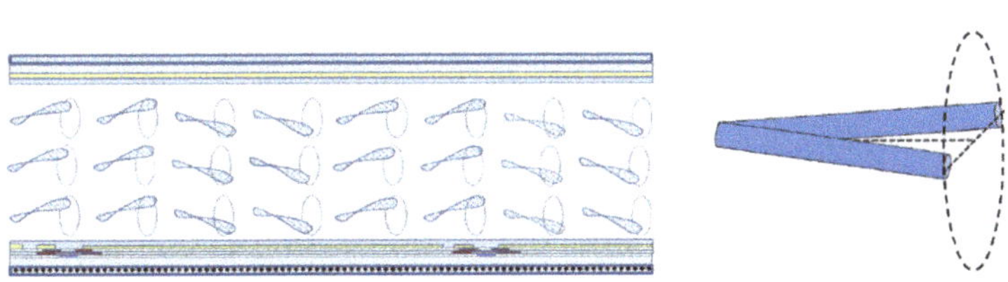

(a) (b)

Figure 3.18 The schematic of the (a) ESHFLC display cell and (b) represents the ESHFLC cone representing the two switching positions for different polarity of the display [29]

3.4 Summary

Photoalignment and photopatterning technology based on azo-dyes is a very promising direction in LC display and photonics devices due to the: (i) reliable and simple fabrication; (ii) high resolution; (iii) multi-functional applications, including highly efficient polarizers, optical lenses with voltage controllable focal distance, optically rewritable E-paper, fast ferroelectric LCD, etc.; and (iv) possibility to use glass and plastic substrates. We demonstrated several recent examples of the applications in this technology is LC display and photonics devices. New photovoltaic, optoelectronic and photonic devices based on highly ordered thin organic layers may also be possible. Examples of such applications are single molecules organic light emitted (OLED) films, solar cells, optical data storage and holographic memory devices. The novel and highly ordered layer structures of organic molecules may exhibit certain physical properties, which are similar to the aligned LC layers. As we have shown, all the characteristics of azo-dye photoaligning materials, such as VHR, RDC, anchoring energy, image sticking, photo and thermos stability, are comparable with the rubbed PI layers (see also Table 3.1). At the same time, azo-dyes provide the highest possible resolution among the other photoalignment materials, which makes them very useful for the applications in new display and photonics devices.

References

[1] Ichimura K., Suzuki Y., Seki T., Hosoki A. and Aoki K. 'Reversible change in alignment mode of nematic liquid crystals regulated photochemically by "command surfaces" modified with an azobenzene monolayer'. *Langmuir 4*, 1988, Vol. 4, p. 1214.

[2] Gibbons W.M., Shannon P.J., Sun S.T. and Swetlin B.J. 'Surface mediated alignment of nematic liquid crystals with polarized laser light'. *Nature*, 1991, Vol. 351, p. 49.

[3] Schadt M., Schmitt K., Kozenkov V. and Chigrinov V. 'Surface-induced parallel alignment of liquid crystals by linearly polymerized photopolymers'. *Jpn.J. Appl. Phys.*, 1992, Vol. 31, p. 2155.

[4] Chigrinov V., Pikin S., Verevochnikov A., *et al.* 'Diffusion model of photoaligning in azo-dye layers'. *Phys. Rev. E*, 2004, Vol. 69, pp. 061713-1-061713-10.

[5] Chigrinov V.G., Kozenkov V.M. and Kwok H.S. *Photoalignment of liquid crystalline materials: Physics and applications*. Wiley, 2008.

[6] Yaroshchuk O., Reznikov Yu. 'Photoalignment of liquid crystals: Basics and current trends'. *J. Mater. Chem.*, 2012, Vol. 22, pp. 286–300.

[7] Chigrinov V.G. *Liquid Crystal Photonics*. Nova Science Publishers, 2014.

[8] Chigrinov V., Prudnikova E., Kozenkov V., *et al.* 'Synthesis and properties of azo dye aligning layers for liquid crystal cells'. *Liq. Cryst.*, 2002, Vol. 29, pp. 1321–1327.

[9] Chang K.-H., Zhang C., Song Sh. and Chien L.-C. 'Command splay-bend switching in an optically-compensated bend cell with polymerized liquid crystal photoalignment layers'. *J. of Appl. Phys*, 2017, Vol. 121, pp. 114901-1-7.

[10] Konovalov V.A. , Chigrinov V.G. , Kwok H.S., Takada H. and Takatsu H. 'Photoaligned vertical aligned nematic mode in liquid crystals'. *Jpn. J. Appl. Phys.*, 2004, Vol. 43, pp. 261–66.

[11] Chigrinov V., Muravski A., Kwok H.S., Takada H., Akiyama H. and Takatsu H. 'Anchoring properties of photoaligned azo-dye materials'. *Phys. Rev. E*, 2003, Vol. 68, pp. 061702-1-5.

[12] Kiselev A.D., Chigrinov V.G. and Huang D. 'Photoinduced ordering and anchoring properties of azo-dye films'. *Physical Review E*, 2005, Vol. 71, pp.061703-1-061703-12.

[13] Li X., Kozenkov V.M., Yeung F., Xu P., Chigrinov V.G. and Kwok H.S. 'Liquid-crystal photoalignment by superthin azo dye layer'. *Jpn. J. Appl. Phys.*, 2006, Vol. 45, pp. 203–05.

[14] Kwok H.S., Chigrinov V.G., Takada H. and Takatsu H. 'New developments in liquid crystal photoaligning by azo-dyes'. *IEEE/OSA J. of Disp. Techn.*, 2005, Vol. 1, pp. 41–50.

[15] Hasebe H., Maruyama K., Takashima M., *et al.* 'Performance of novel LC-photo-aligning cinnamoyl side-chain polymers'. SID 2014 Digest, pp. 389–92.

[16] Shteyner E.A., Srivastava A.K., Chigrinov V.G. and Kwok H.S., Afanasyev A.D. 'Submicron-scale liquid crystal photo-alignment'. *Soft Matter*, 2013, Vol. 9, pp. 5160–65.

[17] Flexible Flat Panel Displays, edited by G.P. Crawford, (Wiley, 2005).

[18] Osterman J., Tong A., Skarp K., Chigrinov V. and Kwok H. S. 'Properties of azo-dye alignment on plastic substrates'. *J. of SID*, 2005,Vol. 13, pp. 1003–09.

[19] Maruyama K., Ono Y., Suzuki Y. and Ikariya T. 'A plastic color TN-LCD using a photo alignment method'. IDW'05 Digest, 2005, pp. 141–43.

[20] Shi Y., Zhao Ch., Ho J., *et al.* 'Exotic property of azobenzenesulfonic photoalignment material based on relative humidity'. *Langmuir*, 2017, Vol. 33, pp. 3968–74.

[21] Tseng M.-C., Yaroshchuk O., Bidna T., Srivastava A.K., Chigrinov V.G. and Kwok H.S. 'Strengthening of liquid crystal photoalignment on azo dye films: passivation by reactive mesogens'. *RSC Adv.*, 2016, Vol. 6, pp. 48181–88.

[22] Meng C.-L., Tseng M.-C., Lee C.-Y., Srivastava A.K., Chigrinov V.G. and Kwok H.S. 'Azo dye, liquid crystals polymer composite photo-alignment layer for modern liquid crystal displays'. SID'16 Digest, 2016, pp. 1566–69.

[23] Zhang W., Sun J., Fan F., Srivastava A.K., Chigrinov V.G. and Kwok H.S. 'Grayscale generation for optically rewritable E-paper based on polarization rotator'. 2014, IDW'14 Digest, pp. 48–49.

[24] Sun J., Srivastava A.K., Zhang W., Wang L., Chigrinov V.G. and Kwok H.S. 'Optically rewritable 3D liquid crystal displays'. 2014, *Opt. Letts.*, Vol. 39, pp. 6209–12.

[25] Fan F., Srivastava A.K., Du T., Tseng M.C., Chigrinov V. and Kwok H.S. 'Low voltage tunable liquid crystal lens'. *Opt. Letts.*, 2013, Vol. 38, pp. 4116–19.

[26] Zhao C.X., Fan F., Du T., Chigrinov V.G. and Kwok H.S. 'Multilayer photo-aligned thin-film structure for polarizing photonics'. 2015, *Opt. Letts.*, Vol. 40, pp. 2993–96.

[27] Shi L., Ma,Y., Srivastava A.K., Chigrinov V.G. and Kwok H.S. 'Field sequential color displays based on reflective electrically suppressed helix ferroelectric liquid crystal'. 2015, SID'15 Digest, pp. 1627–28.

[28] Tseng M.-C., Lee C.-Y., Yaroshchuk O., Srivastava A.K., Chigrinov V.G. and Kwok H.S. 'Photo-stable azo dye photoalignment polymer surface for in plane switching liquid crystal displays'. 2015, SID'15 Digest, pp. 997–1000.

[29] Srivastava A.K., Tseng M.-C., Chigrinov V.G. and Kwok H.S. 'Electrically suppressed helix ferroelectric liquid crystals a better alternative for the IPS displays'. SID'15 Digest, pp. 1001–3.

Chapter 4

Alignment materials and their surface deformation induced by manufacturing processes

Yukio Hirano[1]

Liquid crystal displays (LCDs) are used for displaying still and moving images in all types of large or small display media, including monitors for laptop computers or desktop personal computers, display panels for TVs, mobile phones, smartphones, on-board navigation, viewfinders for video cameras, preview screens for digital cameras, projection displays and information display monitors in public spaces.

In order to utilize liquid crystal (LC) material in LCDs for such applications, an electro-optical response is applied, in which the transmitted light intensity is controlled by the application of a set voltage to the driver devices. For this purpose, the alignment control of LC molecules is essential. In this chapter, alignment materials and their surface deformation anisotropy for alignment control of LC molecules are discussed.

4.1 Introduction

Alignment film material and manufacturing processes realize alignment control for enhancing the utility value of display devices. Till date, the response time has been improved by considering the molecular designs of the LC material. Screens have been enlarged owing to the progress in alignment control technology in uniform and large areas, and display quality indicators, such as the light-dark contrast and viewing angle characteristics, have been improved.

Alignment has been controlled mainly by controlling the alignment direction of the major axis (easy axis) of the LC molecules and the rising angle (pre-tilt angle) from the substrate surface, inducing anisotropic interaction between the alignment film surface and the LC molecules. Based on the alignment direction, alignment is generally classified into two types: alignment in which the major axis of the LC molecules is directed perpendicularly to the substrate surface of these

[1]Ichihara Research Center, JNC Petrochemical Corporation, Japan

molecules (homeotropic alignment) and alignment in which the major axis of the LC molecules is parallel to the substrate surface (homogeneous alignment). The alignment type is determined by the alignment film material and the manufacturing methods, in association with the application and driving modes.

The display mode of homeotropic alignment, which has excellent light-dark contrast relative to the front of the display, is mainly formed by controlling the surface energy of the alignment film; however, the display quality is poor in the oblique direction owing to considerable light leakage and the generation of colour registration error. On the other hand, in the homogeneous alignment display mode, especially the in-plane switching (IPS) [1,2] or fringe field switching (FFS) mode [3], the display quality indicators, such as the viewing angle characteristics, are excellent. In the homeotropic alignment display mode, a drop injection method has been adapted for the LC filling process in the manufacturing process, by which the size can be increased. This method has been adapted for several large TVs. On the other hand, in the homogeneous alignment display mode, it is essential to induce anisotropic interaction with the substrate surface for uniformly aligning the directions in which the major axes of the LC molecules are arranged within the substrate plane. Processes such as the rubbing process and photoalignment process are used in the manufacturing process; therefore, the number of manufacturing processes in the homogeneous alignment display mode tend to be greater than that in the homeotropic alignment display mode, in which the manufacturing load is high.

In recent years, in photoalignment technology, while technological innovations in the alignment film material and the manufacturing process have progressed and have been applied practically, rubbing technology has so far played a leading role in the alignment technology for LCDs. In the following sections, the author describes the rubbing technology, polyimide alignment film, anisotropy of the surface deformation induced by the rubbing process, and the alignability for arranging LC molecules.

For interaction between the LC alignment direction (easy axis) and the LC molecules, the alignment mechanism of the LCs on the polyimide alignment film to which the rubbing process or other alignment treatment has been applied has been summarized by Shoichi Ishihara [4].

4.2 Rubbing technology

The rubbing process involved in fabricating an LC display (LCD) device is a technology that dates back to 1911. It enables interaction with the anisotropy of the alignment film surface by mechanically scrubbing the surface with materials such as cloth in a particular direction such that the LC molecules are aligned in a direction, producing an LC medium with a uniform alignment direction [5]. In an LCD factory, most of the LCDs are fabricated by employing this method as the alignment treatment method for polyimide alignment films, whereby a large area of the film can be treated at a high speed. The rubbing process involves treatment by a metal roller wound with fabric implanted with fibres of nylon, rayon, polyester, etc., which contact the alignment film when the roller is rotated, scrubbing the surface in a particular direction. Thereby, the direction in which the LC molecules are aligned can be regulated to achieve uniform alignment. A large area of the film can be treated by increasing the length of the rubbing roller; consequently, the size of the rubbing apparatus increases for the alignment treatment of a large substrate.

4.2.1 Rubbing strength

As an indicator for quantitatively determining the rubbing process, the following was devised [6,7]. The rubbing strength, L, is represented by the following expression using various parameters, as shown in Figure 4.1.:

$$L = Nl\left(1 + \frac{2\pi rn}{60v}\right)$$

where N denotes the number of times a substrate is rubbed, l is the contact length [mm] of the rubbing roller with the substrate, r is the roller radius [mm], n is the roller rotational speed [rpm] and v is the stage movement speed [mm/s]. The rubbing cloth scrapes the polyimide alignment film, and if the rubbing operation is repeated, grooves are formed on the alignment film surface and the rubbing cloth surface acquires a slight yellow tinge.

4.2.2 Problems in the rubbing process

The problems in the rubbing process include the static electricity generated by scrubbing the alignment film, pollution of the manufacturing environment by the dust generated while scraping the alignment film surface, etc. Moreover, dust is attached onto the substrate surface by the generated static electricity, causing defects in the LCD device. The static electricity generated in the rubbing process reaches several tens to hundreds of volts, causing the breakage of the thin film transistor (TFT) device used as the driving circuit; irregular electrical faults are caused by the remaining static electricity. The generated dust necessitates an increase in the number of apparatuses in the production process for dust removal in the cleaning step and the treatment of the cleaning liquid.

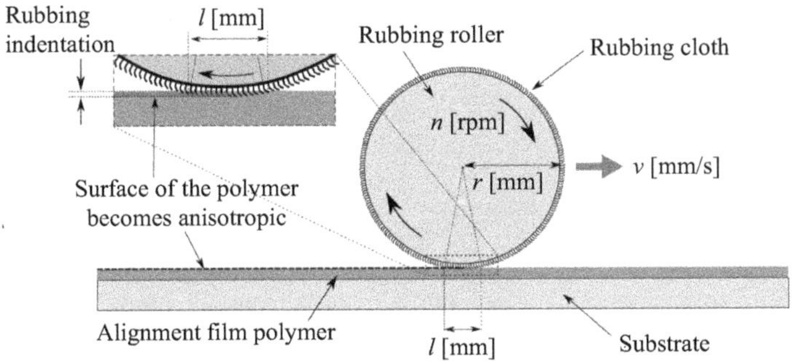

Figure 4.1 Schematic of the rubbing process

4.2.3 Polyimide alignment film

During the initial developmental stages of LCD devices, various materials other than polyimide alignment films, such as an oblique SiO [8] deposition film and a vertical alignment (VA) film of a silane coupling agent, were studied as LC alignment films. However, no other material is practically used because of the requirements of heat resistance, chemical resistance (resistance to LCs), printability, LC alignment stability, electrical properties, optical transparency, etc. Currently, display applications mainly use polyimide-based LC alignment films. Polyimide thin films can be generally fabricated by printing through the spinner method, flexography method, inkjet printing method, etc. A solution of polyamic acid is placed on a glass substrate on which a TFT device or colour filter is provided; the polyamic acid is then dehydrated by heat calcination to cause ring closing. Alternatively, the film can be fabricated by printing a solution of soluble polyimide and evaporating the solvent. The polyamic acid type has features such as a wide selection range of the polymer material and excellent margins in the manufacturing process, including in the printability, adhesion, and heat resistance, because polyamic acid with high solubility is used. On the other hand, the polyimide type has the capability of film formation at a temperature lower than that of the polyamic acid type owing to the absence of the dehydration process by heating as well as ring closure; however, it has problems in the printability and coating film strength, in many cases, because the structure of the polymer material is limited. In particular, compatibility with the nozzle of the inkjet head causes problems in inkjet printing, but the compatibility can be appropriately adjusted by the composition of the solvent, concentration of the alignment film material, etc.

Polyamic acid is generally synthesized through the reaction of tetracarboxylic dianhydride with diamine in a solvent, as shown in Figure 4.2 (see Table 4.1 for examples of 'X' and 'R' in the Figure 4.2). Soluble polyimide is synthesized by adding toluene, etc. to the obtained polyamic acid solution to azeotropically dehydrate the resulting material, or by using catalysts such as pyridine. The obtained polyimide has high heat resistance and excellent durability (an example is given in Table 4.1). Polyimide has a high glass transition temperature; therefore, the LC alignment capability to be added in the rubbing process also has excellent stability at ordinary operating temperature (approximately 10–50 °C). Moreover, with respect to the productivity, polyimide is easily handled because the film can be easily formed by dissolving polyimide in the solvent and then printing and applying the resulting material.

During the initial developmental stages of the LCD device, wholly aromatic polyimide, typified by No. 1 in Table 4.1, was studied. However, owing to the expansion of LCD device applications, improvements were demanded in the colouring, imidization temperature, pre-tilt angle control, voltage holding characteristics required for active matrix driving and display quality characteristics. Hence, polyimides with various structures were synthesized. Wholly aromatic polyimide is yellowish brown owing to the spread of a conjugated system of π electrons, which is not preferable, particularly in the case of coloured displays. In order to reduce

Table 4.1 Structure, glass transition temperature and colour phase of the polyimides

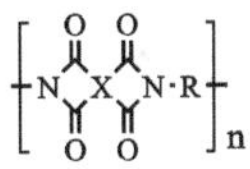

No.	X	R	Tg	colour phase
1	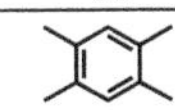	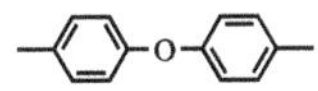	385	yellowish brown
2	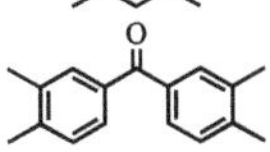	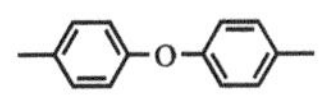	295	yellowish brown
3	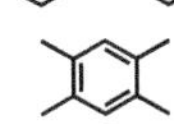	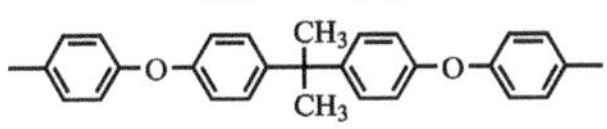	350	
4	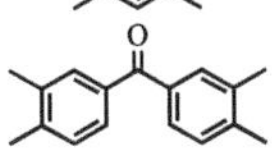	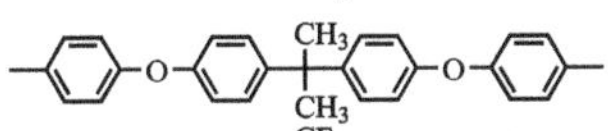	240	
5	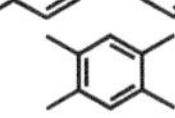	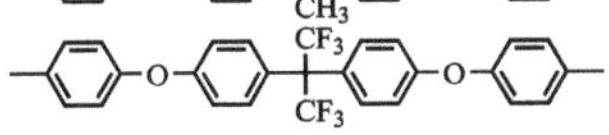	365	
6	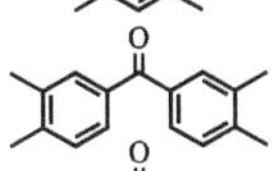	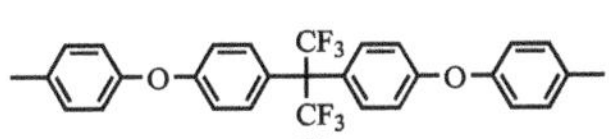	260	
7	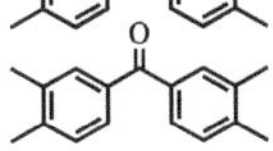	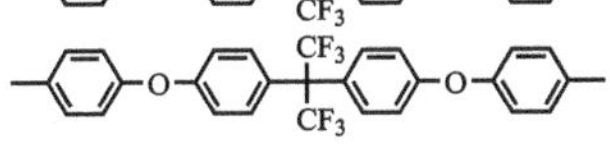	241	clear
8	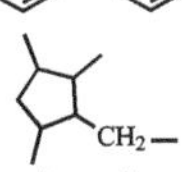	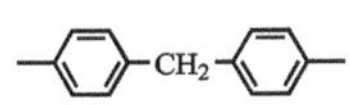	380	clear
9	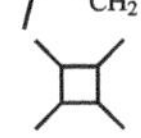	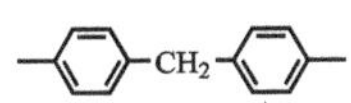	-	clear
10	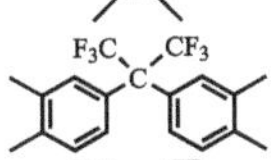	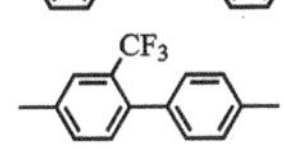	335	clear
11	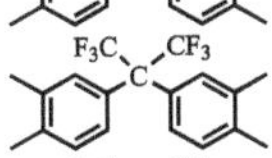	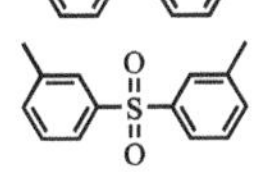	279	clear
12	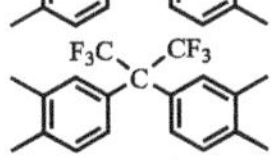	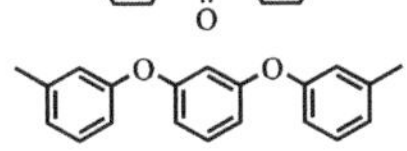	206	light yellow
13	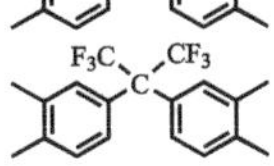	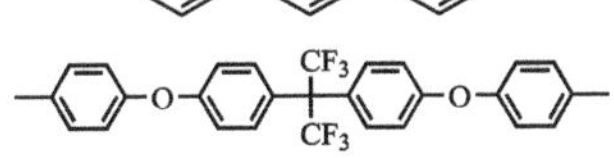	263	light yellow

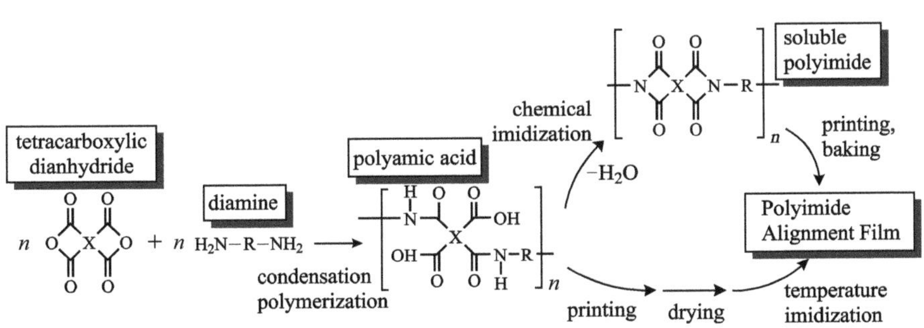

Figure 4.2 Synthesis pathway of the polyimide alignment film

this colouring, polyimides with the structures shown in Nos. 7–13 in Table 4.1 were developed. Moreover, a colour filter is installed for colour display; however, considering the heat resistance of the colour filter, the imidization temperature of polyimide, used as the alignment film, must be as low as possible. Further, if the calcination temperature is decreased in the polyimide thin film formed using ordinary polyamic acid, the imidization ratio is reduced, deteriorating the mechanical strength of the coating film, LC alignment capability, and electrical characteristics, in several cases. Polyamic acid with a low imidization temperature has been researched to solve this problem. Alternatively, soluble polyimide is used. The imidization temperature of polyamic acid is correlated to the glass transition temperature of polyimide [9], and can be decreased by controlling the glass transition temperature of the polyimide thin film by providing molecular flexibility to polyamide acid. Wholly aromatic polyimide is generally insoluble in a solvent, and the general techniques for improving the solubility without significantly sacrificing the heat resistance are introduced in Table 4.2 [10]. The rigidity of a polymer can be reduced by No. 1 in Table 4.2 introducing an easily rotatable linkage, such as $-O-$, $-SO_2-$, $-S-$, $-CO-$, $-C(CH_3)_2-$, and $-C(CF_3)_2-$, between the phenyl groups of the polyimide main chain. The solubility can be improved by No. 2 in Table 4.2 introducing a bulky group, phenyl group, $-CH_3$, or $-CF_3$ into the side chain of polyimide or by No. 3 in Table 4.2 reducing the symmetry of the molecules of an acid dianhydride or diamine component. In addition, the polyimide in No. 8 in Table 4.1 is a material corresponding to technique No. 3 in Table 4.2, in which an asymmetric acid dianhydride is used, and is soluble in γ-butyrolactone, N-methylpyrrolidone, N,N-dimethylformamide, etc. If the solvent is removed by heating, a completely imidized LC alignment film can be obtained because a polyimide film is formed on the substrate.

Table 4.2 General methods for improving the solubility of polyimide

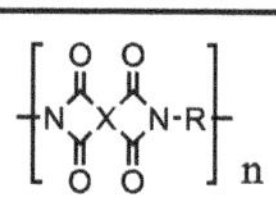

No.	Method	Substrate example
1	The structure which is easy to rotate between the phenyl groups of a principal chain is introduced -O-, $-SO_2-$, -S-, -CO-, $-C(CH_3)_2-$, etc.	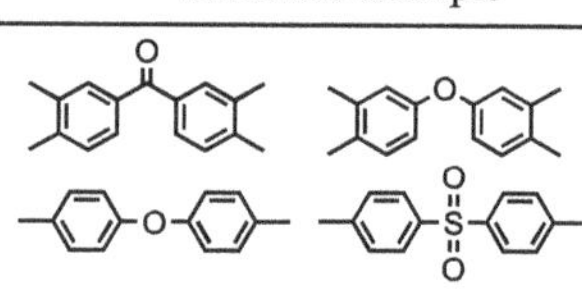
2	A bulky group, a phenyl group, $-CH_3$ and $-CF_3$ are introduced into a side chain.	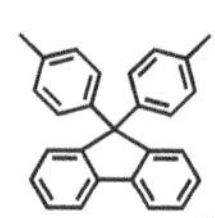
3	The symmetry of diamine and tetracarboxylic acid is lowered.	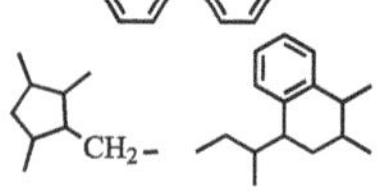

4.3 Surface deformation and optical birefringence of the polyimide material

The anisotropy of the surface deformation resulting from the rubbing process causes optical anisotropy to transmitted light in the form of optical birefringence. This surface deformation anisotropy can be measured in the visible light region as optical retardation by ellipsometry with transmission measurement, using devices such as spectroscopic ellipsometers. Although the grooves on the alignment film surface formed by the rubbing process are meaningful, considering the alignment mechanism of the LC molecules on the alignment film surface, they are not essential conditions for aligning the LC molecules. For example, even if the film does not have mechanical grooves, in the photo alignment process, it shows optical birefringence and the resulting anisotropic dispersion force primarily causes the LC alignment force (anchoring energy) to support the molecular alignment mechanism.

The optical birefringence of the alignment film is due to the occurrence of a type of anisotropy in which the polymer chains in the alignment film are plastically deformed by the shear stress on these chains due to the rubbing of the surface and are aligned in the scrubbed direction; the orientation directions of the atom group constituting the polymer are macroscopically arranged. That is, a specific atom group is macroscopically directed to a fixed direction in the rubbing process, causing polarization anisotropy.

4.3.1 Optical birefringence of the polyimide material structure

In the rubbing process, the optical retardation and anchoring energy in the alignment film exhibit a high correlation, when compared for the same type of alignment film material (Figure 4.3(a)). Figure 4.3(b) shows the influence of the length of a methylene group in a diamine compound used in a polyimide alignment film as an illustration of the dependence of the polymer material structure on optical retardation.

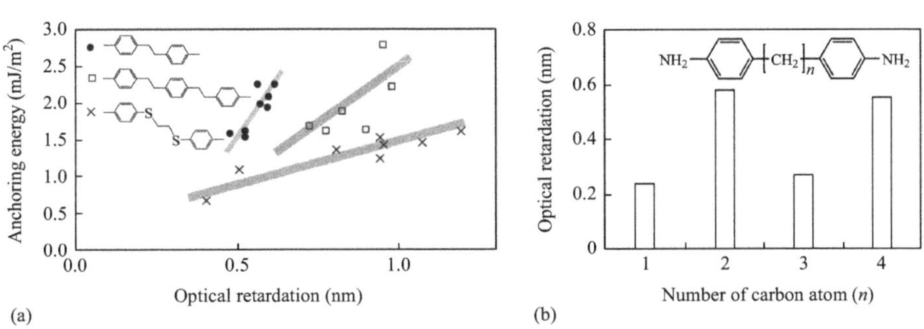

Figure 4.3 (a) Relationship between the anchoring energy degree and the optical retardation and (b) relationship between the number of carbon atoms (n) of diamine and the optical retardation. In the parity of the methylene length group, the optical retardation for odd numbers is lower, whereas that for even numbers is higher

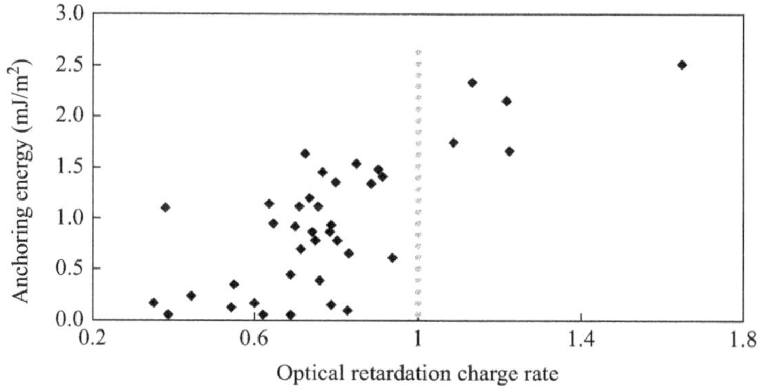

Figure 4.4 Relationship between the optical retardation change rate and anchoring energy. The optical retardation change rate can be used to estimate the rate of aging (170 °C, 2 h) from the initial optical retardation condition. The optical retardation was measured for the alignment film on the glass substrate with rubbing treatment by ellipsometry. For a material in which the optical retardation change rate is greater than unity, the optical retardation increases with thermal aging. Such a material exhibits a strong anchoring force

4.3.2 Thermal relaxation process of surface deformation anisotropy of the polyimide material

The resulting surface deformation anisotropy is reduced in the thermal relaxation process in several alignment film materials. In contrast, for an alignment film material that tends to exhibit crystallinity or liquid crystallinity, the polymer chains are aligned according to the ratio in the thermal relaxation process; thus, the surface deformation anisotropy increases. While surface deformation anisotropy is caused by the macroscopic external force exerted in the rubbing process, the thermal relaxation corresponding to the increase of surface deformation anisotropy originates from the microscopic dynamics of the molecules causing the thermodynamic phenomenon (self-organization).

For a material with optical retardation change rates less than unity, Figure 4.4 shows that the surface deformation anisotropy is reduced in the thermal relaxation process; in contrast, for materials with optical retardation change rates greater than unity, the surface deformation anisotropy is increased in the thermal relaxation process. Such an alignment film material that increases surface deformation anisotropy is known to exhibit a strong anchoring force to the orientation of LC molecules.

4.4 Surface deformation anisotropy and alignability

The alignability realized by the alignment materials and the manufacturing process is summarized into three types. The energy of the interaction between the LC and alignment film is expressed by the following expression:

$$\Delta F = \frac{1}{2} A_\phi \sin^2(\phi - \phi_0) = \frac{1}{2} A_\phi \sin^2 \Delta\phi$$

The three alignability types should be understood for each parameter that constitutes the following expression. The three factors are described as follows:

- The in-plane uniformity of the anisotropy of the alignment material surface deformation
  The anisotropy is required to be uniform within a place or target region in the substrate in-plane uniformity, ϕ_0
- Anchoring force
  A_ϕ denotes the intensity of the interaction between the anisotropy of the alignment material surface deformation and the LC. The energy, ΔF, increases depending on the magnitude of the coefficient, A_ϕ, and the displacement, $\Delta\phi$.
- Stability of the anisotropy of the alignment material surface deformation
  How stable is ϕ_0 against the stress?

If the uniformity of the alignment axis (easy axis) of the LC molecules within the substrate plane is high, the distribution of ϕ_0 will be narrow. Further, if the stability of surface deformation anisotropy is high, ϕ_0 will be stably maintained against various external mechanical loads. The voltages applied at the boundaries by the TFT device deform the directional orientation of the LC, which deformation induces an elastic restoring torque to an initial state of the surface deformation anisotropy. The ϕ_0 shift from an initial state caused by this elastic restoring torque would be small.

4.4.1 Relationship between the in-plane uniformity of the surface deformation anisotropy and the display contrast

For a display with excellent quality, surface deformation uniformity is required in a wider region in the rubbing process; therefore, the rubbing technique has been improved. Arrangement of the transmission axis of a polarizer (light in) and an analyser (light out) is a dominant factor of the display contrast. Therefore, in order to study the relationship between the in-plane uniformity of the surface deformation anisotropy and the display contrast, it is indispensable that the arrangement of the transmission axis of the polarizer and the analyser is optimized such that minimum light is transmitted. In the condition that the transmission axis of the polarizer and the analyser are optimized, the in-plane uniformity of the surface deformation anisotropy influences the leaking transmission light. If the anisotropy distribution of the surface deformation, from the micro to macro scale, is aligned, then the distribution of ϕ_0 is narrow (Figure 4.5); the alignment distribution of the LC molecules is also aligned, the light leakage is reduced, and the light-dark contrast of the LCD is improved.

Remark 1: *In general, the light-dark contrast of the photoalignment film is increased, because the uniformity is higher in photoalignment compared to that in rubbing.*

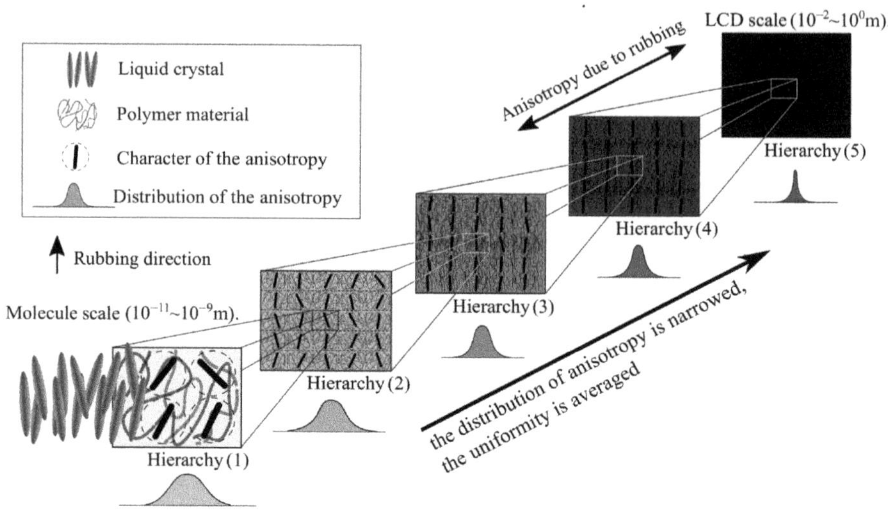

Figure 4.5 Anisotropy hierarchy scheme. The microscopic limit corresponds to the scale of the LC molecules ($10^{-11} - 10^{-9}$ m*), whereas the macroscopic limit corresponds to the display area of the LCD (*$10^{-2} - 10^{0}$ m*). The uniformity of the surface deformation anisotropy obtained in the rubbing process is not necessarily high, considering the scale of the LC or polymer molecules; as the hierarchy of the alignment scheme expands to a wider range, the uniformity is consequently improved ((1) → (2) → (3) → (4) → (5)). When the order parameter of the alignment is compared with respect to the surface deformation anisotropy between the bulk of the LC and the alignment film, it is lower in the surface deformation anisotropy of the alignment film; the order parameter of the alignment of the LCs in the vicinity of the interface of the LC and alignment film is reduced by this influence (Figure 4.6(a))*

LC molecules are aligned with each other into a stable state (Figure 4.6(a)). However, in the alignment film material, which originally has no anisotropy without the interaction by which the molecules are aligned with each other, the axis of the surface deformation anisotropy obtained in the rubbing process is not necessarily aligned uniformly. On comparing the order parameters of the molecular alignment, that of the LC is higher. Therefore, the order parameter of the surface deformation of the alignment film is lower than that of the LC in many cases of the rubbing process, and the LC molecules in the vicinity of the interface are influenced by this surface deformation; the order parameter is reduced in comparison with the LC layer of the bulk. In such a case, when the interaction between the LC and alignment film is strong, the alignment information of the surface deformation

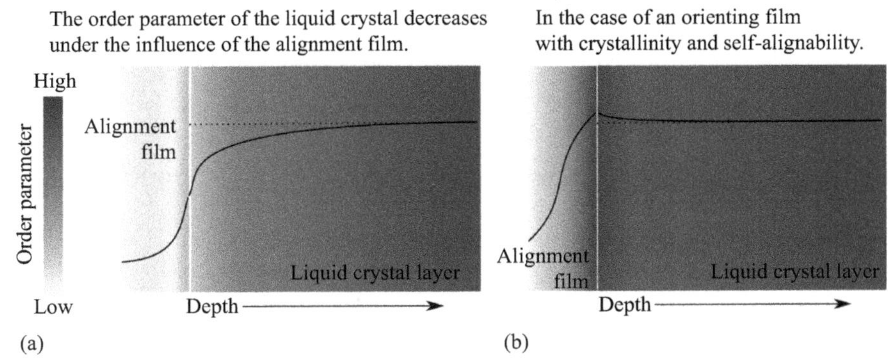

Figure 4.6 Order parameter of the alignment in the vicinity of the interface between the alignment film and LC layer

is clearly transferred from the surface to the bulk area of the LC. Conversely, if the interaction is weak, the influence of the surface deformation becomes weak, and the influence of the alignment of the LC layer of the bulk extends to the vicinity of the interface. Consequently, light leakage due to the influence of the surface deformation tends to be reduced.

In order to increase the order parameter of the alignment of the LC molecules in the vicinity of the interface, the order parameter of the surface deformation in the alignment film is required to be close to or higher than that of the LC, which is difficult to achieve using only the rubbing process.

The use of a material with crystallinity or liquid crystallinity as the alignment film material enhances the order parameter and the stability of the anisotropy of surface deformation on the alignment film surface (Figure 4.6(b)). In this case, the anchoring force and the order parameter of the alignment of the LC molecules in the vicinity of the interface increase (the region in which the optical retardation change rate is unity or greater, in Figure 4.3).

In the uniformity of the surface deformation anisotropy on the alignment film surface, in the anisotropy of a narrow region, the molecular structure of the alignment film material is a dominant factor (Figure 4.5: Hierarchy (1)). On the contrary, in the anisotropy of a wide region, a manufacturing process, such as the rubbing process, is the dominant factor (Figure 4.5: Hierarchy (5)). In order to reduce light leakage for improving the light-dark contrast in a high definition LCD, the surface deformation anisotropy is required to be uniform at any scale and any level in the anisotropy hierarchy.

Remark 2: *Moreover, the alignment film surface is damaged depending on the material or the rubbing process, generating alignment defects such as luminescent spots, and the display quality is adversely affected in several cases.*

4.4.2 Surface deformation anisotropy and anchoring force

For the interaction between the LC molecules and the alignment film caused by the interface effect, studies have been conducted on mainly two mechanisms, since the realization and practical use of the twisted nematic (TN)-LCD in 1971 [11]:

1. The surface shape effect caused by a micro-groove, which is generated in the rubbing process [12].
2. The intermolecular interaction between the uniaxially aligned alignment film in the rubbing process and the LC monomolecular layer in contact with it on the outermost surface [13].

It has been recently confirmed that the contribution of the surface shape effect '1.' is comparatively low, whereas that of the intermolecular interaction '2.' is dominant. The micro-groove theory '1.' originates from the smaller energy of the elastic deformation generated in the LC medium and the stabilization in the case where the LC molecules are aligned parallel to the micro-grooves formed in the rubbing process, compared to the case where the LC molecules are perpendicularly aligned.

If an increase in the twist energy, ΔF, generated by applying an external force, such as an electric field from the LC drive circuit, is represented by expression, ΔF is represented by the following expression [12]:

$$\Delta F = \frac{1}{2} A_\phi \sin^2 \phi = \frac{K}{4} a^2 q^3 \sin^2 \phi \tag{4.1}$$

where K is the elastic constant, and a and q are as shown in Figure 4.7. On a rubbed polyimide surface, as described in the rubbing process, cases with or without the formation of grooves are observed, depending on the polyimide material. The LCs can be sufficiently aligned, even if no grooves are observed. If the anchoring energy, A_ϕ, is determined by entering the values into (4.1) in the presence of grooves, $A_\phi = 10^{-6}$ J/m^2, which is significantly different from the measured value by $10^{-4} - 10^{-3}$ J/m^2. An increase in the energy caused by twist deformation, resulting from the anisotropic dispersion force '2', is represented by the following expression [14,15]:

$$\Delta F = \frac{h}{64\pi^2 d_0^2} \int_0^\infty \frac{\Delta\varepsilon^{(1)}(i\omega)\Delta\varepsilon^{(3)}(i\omega)}{\left\{\varepsilon_0^{(1)}(i\omega)+1\right\}^2 \left\{\varepsilon_0^{(3)}(i\omega)+1\right\}^2} d\omega \sin^2\phi \tag{4.2}$$

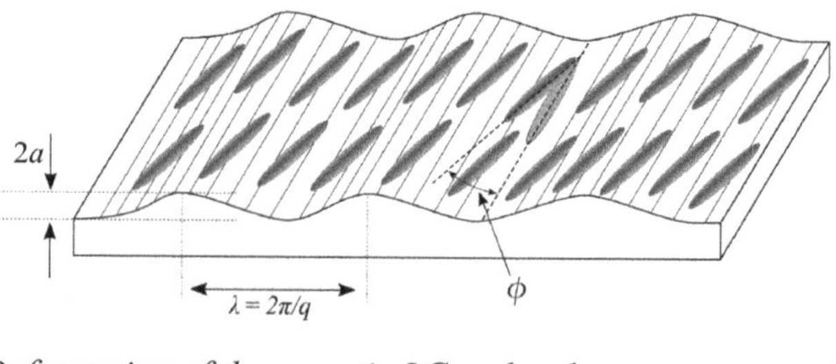

Figure 4.7 Deformation of the nematic LC molecule group arranged in alignment with the grooves

$$= \frac{1}{2} A_{\phi} \sin^2 \phi \tag{4.3}$$

where $\Delta\varepsilon^{(1)}(i\omega)$ and $\Delta\varepsilon^{(3)}(i\omega)$ are the anisotropies of the high-frequency complex permittivities of a base and LC medium, respectively; d_0 is the distance between media (1) and (3); and ε_0 is the mean value of ε.

Cases where the LC molecules are arranged in a direction perpendicular to the rubbing treatment direction have also been reported [16]. In such a case, the arrangement is different in the rubbing treatment direction or in the vertical direction, depending on the LC material constituting the LC molecules or the LC composition, and the alignment direction of the LCs is the main factor contributing to the anisotropic dispersion force.

4.4.3 Stability of the surface deformation anisotropy

It is important for high-quality LCDs to maintain ϕ_0 with respect to electrical stress from the drive voltage or thermal stress from the LED backlight. This concerns the stability of the surface deformation anisotropy on the surface of the polyimide alignment film. While the deformation related to the anchoring force is comprehended as elastic deformation, the stability of the easy axis is comprehended as the plastic deformation property.

The LC alignment is deformed by the electric field, and the alignment axis gradually causes plastic deformation with time. For example, when the electric field stress is turned off, the elastic deformation component in the LC rapidly returns to the original position. However, if the alignment axis of the alignment film is plastically deformed, the alignment axis of the LC does not return to the original position. This non-returning component causes deterioration in the LCD quality. In the alignment film material, it can be inferred that the easy axis stability is high if the anchoring force is high, specifically in the context of the relationship between the anchoring force and the easy axis stability.

4.5 Pre-tilt angle

As another important role, the LC alignment film tilts the LC relative to the substrate plane. The pre-tilt angle required for the LCD depends on the display mode. For example, the LC alignment film used for the VA-mode LCD has a desired pre-tilt angle of approximately 87°–90°, which is close to the vertical, relative to the substrate; the LC alignment film used for the optical compensated bend (OCB)-mode [17] LCD has a desired pre-tilt angle of approximately 7°–20°. Further, the desired pre-tilt angles of the LC alignment film used for the TN-mode or STN-mode LCD and that for the IPS-mode or FFS-mode LCD are approximately 3°–7° and 0°–3°, respectively. It is considerably difficult to stably maintain the pre-tilt angle in the range of 20°–80°.

Experiments demonstrate that the pre-tilt angle is correlated with the surface tension of the polyimide. For stably controlling the pre-tilt angle, the control of the imidization rate and glass transition temperature of the polyimide alignment film; the mixing of a long chain molecule, such as a metal complex or long chain alkyl amine, to polyimide-based alignment material; and the introduction of fluorine or a nonpolar group, such as an alkyl group, into the polyimide, by which the pre-tilt angle increases along with the reduction in surface energy in many cases, have been used. In a polyimide alignment film with long chain alkyl groups, the pre-tilt angle is considered to be determined by the influence of the surface energy as well as the interaction between the alkyl chain and the LC molecules. Meanwhile, it is reported that, in alignment films in which the cyanobiphenyl group has a skeleton similar to that of the LC molecules introduced into the polyimide side chain, the pre-tilt angle increases only for the cyanobiphenyl based LC, even though there is an increase in the surface tension, suggesting a dipole–dipole interaction between the LC molecules and the side chain of the polyimide alignment film The pre-tilt angle is generally adjusted by adjusting the structure of alignment film materials, mainly using diamine with a non-polar group or an acid dianhydride. The surface tension is varied by introducing a long chain alkyl group into the polyimide structure between a diamine long chain alkyl and an acid dianhydride. By varying the number of carbon atoms in the alkyl group, the pre-tilt angle is controlled.

4.6 Conclusion

The rubbing process of the polyimide alignment film causes surface deformation anisotropy and anisotropic interaction, which control the alignment of the LC molecules on the polyimide alignment film surface. There are cases with or without the formation of geometrical micro-grooves on the alignment film surface, depending on the alignment film material or the rubbing process. When the direction of the micro-grooves is matched with the slow axis of the refractive index, the LC molecules are arranged in the rubbing direction. However, when the micro-grooves and the slow axis of the refractive index are perpendicularly crossed, they compete with each other, and if the latter is dominant, the LC molecules are arranged in the slow axis direction of the refractive index. More specifically, the anisotropic dispersion force determines the alignment direction of the polymer in the alignment film surface. Then, the uniformity of the surface deformation and its stability play important roles in the light-dark contrast or display quality in high definition LCDs.

The surface deformation anisotropy, generated by the rubbing treatment, is thermally relaxed and reduced in several materials. On the other hand, the incorporation of a material with crystallinity or liquid crystallinity into the alignment film material causes realignment in certain materials, with the anisotropy generated by the rubbing treatment as an indicator, resulting in improved optical retardation and anisotropy depending on the content. Alignment film materials with such characteristics have excellent alignment stability, exhibiting a high anchoring force and are suitable for high-quality LCDs.

References

[1] Kiefer WBWF K and Bauer G. In-plane switching of nematic liquid crystals. Proceedings of the 12th International Display Research Conference (Japan Display '92); No. 2-30:547–550.

[2] Oh-e M and Kondo K. Electro-Optical characteristics and switching behavior of the in-plane switching mode. Applied Physics Letters. 1995;67(26): 3895–3897.

[3] Lee SH, Lee SL and Kim HY. Electro-optic characteristics and switching principle of a nematic liquid crystal cell controlled by fringe-field switching. Applied Physics Letters. 1998;73(20):2881–2883.

[4] Ishihara S. How far has the molecular alignment of liquid crystals been elucidated? Journal of Display Technology. 2005;1(1):30–40.

[5] Seo DS, Matsuda H, Oh-ide T, *et al.* Alignment of nematic liquid crystal (5CB) on the treated substrates: Characterization of orientation films, generation of pretilt angles, and surface anchoring strength. Molecular crystals and liquid crystals science and technology section A Molecular Crystals and Liquid Crystals. 1993;224(1):13–31.

[6] Uchida T, Hirano M and Sakai H. Director orientation of a ferroelectric liquid crystal on substrates with rubbing treatment: The effect of surface anchoring strength. Liquid Crystals. 1989;5(4):1127–1137.

[7] Sato Y, Sato K and Uchida T. Relationship between rubbing strength and surface anchoring of nematic liquid crystal. Japanese Journal of Applied Physics;31(5A):L579.

[8] Janning JL. Thin film surface orientation for liquid crystals. Applied Physics Letters. 1972;21(4):173–174.

[9] Numata S, Fujisaki K and Kinjo N. Studies on thermal cyclization of polyamic acids. In: Mittal KL, editor. Polyimides. 1st ed. New York: Plenum Press; 1984. p. 259.

[10] Cassidy PE. History of heat-resistant polymers. Journal of Macromolecular Science: Part A – Chemistry. 1981;15(7):1435–1460.

[11] Schadt M and Helfrich W. Voltage-dependent optical activity of a twisted nematic liquid crystal. Applied Physics Letters. 1971;18(4):127–128.

[12] Berreman DW. Alignment of liquid crystals by grooved surfaces. Molecular Crystals and Liquid Crystals. 1973;23(3-4):215–231.

[13] Geary JM, Goodby JW, Kmetz AR, *et al.* The mechanism of polymer alignment of liquid-crystal materials. Journal of Applied Physics. 1987;62 (10):4100–4108.

[14] Okano K, Matsuura N and Kobayashi S. Alignment of a liquid crystal on an anisotropic substrate. Japanese Journal of Applied Physics;21(2A):L109.

[15] Sugiyama T, Kuniyasu S, Seo D, *et al.* A simple model for pretilted nematic liquid crystal medium and its torsional surface coupling strength. Japanese Journal of Applied Physics. 29(10R):2045.

[16] Yamaguchi R, Goto Y and Sato S. Liquid crystal alignment surface with two easy axes induced by unidirectional rubbing. Applied Physics Letters. 2003;82(25):4450–4452.

[17] Miyashita T, Yamaguchi Y and Uchida T. Wide-viewing-angle display mode using bend-alignment liquid crystal cell. Japanese Journal of Applied Physics;34(2A):L177.

Chapter 5
Analysis and inspection of alignment membrane in LCDs

Michio Kitamura[1]

The role and importance of the alignment layer in liquid crystal displays (LCDs) are presented in theoretical point of view. Principle and validity of the reflection polarimetry for evaluating directly the property of the alignment layers are shown. The technique for evaluating the image sticking, which is considered to be brought from deterioration of the alignment membrane, is briefly presented with some examples.

5.1 Introduction

An LCD is a flat-panel display which utilises light-modulating properties of liquid crystal (LC). LCDs are non-light-emitting devices. They produce images by switching the light from the backlight units. An LCD consists of a pair of parallel glass plates and (nematic in general) LC filled in the narrow gap of several micro meters between them. Thin polymer film called alignment membrane is coated on each glass plate.

The alignment membrane restricts LC molecules in contact with its surface. LC molecules are elastically coupled and tend to be parallel to each other. When the electric field is applied, LC molecules tend to rotate, although the molecules at both boundaries (surface of alignment membranes) keep the original orientation. The LC molecules in the bulk align so as to minimise the free energy, which is the sum of the elastic coupling energy and the electrostatic energy. If the orientations of molecules at both boundaries are given, the molecular alignment is uniquely determined based on the principle of the free energy minimisation. In other words, the alignment of LC molecules is decided uniquely by the molecules at the surface of alignment membrane. The viewing characteristics, which depend exclusively on the molecular alignment, are determined by the molecules at the surface of alignment membrane.

Not only the static property mentioned above but also the dynamic characteristic is affected by the orientation of LC molecules at the surface of the alignment membrane. Figure 5.1 shows the dynamic responses of vertically aligned LC cells

[1]SHINTECH, Inc., Japan

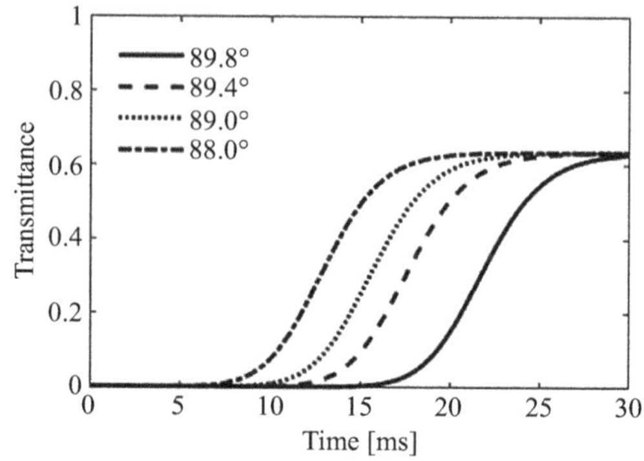

Figure 5.1 The dynamic response of VA cells with various pre-tilt angles. The voltage giving the maximum transmittance is applied to each cell

calculated by commercial modelling software package [1]. We find that a slight increase of the pre-tilt angle by 0.8° at 89.0° reduces the response by 5 ms or more.

Choosing the material of alignment membrane and controlling the alignment process are crucial in the manufacturing process of LCDs. The inspection of the alignment membrane in the early stage of the production line is required for improvement in the yield as well as for the quality control. The inspection will also give efficient information in developing the alignment materials. Non-destructive and non-contact inspection is strongly desired.

The evaluation of the alignment membrane by the ellipsometry was reported in detail by Hirosawa [2]. We review the optical characteristics of the alignment membrane and the polarisation transfer characteristics of it. The evaluation technique using a polarimeter is illustrated by practical examples. Additionally, we measured the parameters realised in the LC cells which were built up with glass substrates processed identically to those under the previous evaluation. The validity of the evaluation technique is presented with the results of the additional measurement. The image sticking has been regarded as a problem recently in high-quality TV monitors. Optical approach to evaluation of the image sticking is introduced briefly in the last section since the phenomenon could be a property of alignment membrane.

5.2 Optical properties of alignment membrane

5.2.1 Optical anisotropy derived by alignment process

Alignment membrane freshly coated on a substrate is optically isotropic. However, once the rubbing is processed on the surface of it, the optical anisotropy, or the birefringence, arises in the surface layer of it. The birefringence also appears in the surface irradiated with polarised UV light. It has been well known that there exists a correlation between the optical anisotropy induced by the alignment processing and alignment performances of LC molecules [3]. Accordingly, it is an effective approach for evaluating the aligning capability to observe the optical anisotropy which appears in a surface layer of an alignment membrane. Let the light beam with a defined polarisation state fall on the surface of the alignment membrane under test. The polarisation of the reflected light changes to other state if the surface has any anisotropy. Then observing the polarisation state of reflected light, we can determine the polarisation transfer characteristics of the alignment layer. The polarisation state in the transmitted light also contains information on the polarisation transfer characteristics. However, the transmitting light, for passing through the glass substrate, also contains unnecessary information which does not concern the alignment matter. Our goal is to evaluate the surface layer which interacts with LC molecules. The observation of reflected light is preferable.

5.2.2 *Optical model of alignment membrane*

We presume that all materials considered here are homogeneous, non-conducting and non-magnetic. As mentioned in the preceding section, the processed membrane has a double-layered structure: the surface layer in which the optical anisotropy is induced and the basic layer left in original isotropy. We assume that the anisotropy is uniaxial symmetry. The relative dielectric tensor of a uniaxial anisotropic medium is represented in the principal frame $\xi\eta\zeta$ by a diagonal matrix

$$\boldsymbol{\varepsilon}_{\mathrm{p}} = \begin{bmatrix} \varepsilon_{\parallel} & 0 & 0 \\ 0 & \varepsilon_{\perp} & 0 \\ 0 & 0 & \varepsilon_{\perp} \end{bmatrix} \tag{5.1}$$

where $\varepsilon_{\parallel}$ and $\varepsilon_{\perp}$are the relative dielectric constants corresponding to the electric fields parallel to and perpendicular to the principal axis ξ, respectively. The principal frame $\xi\eta\zeta$ is transformed to the reference frame XYZ that is fixed to the alignment membrane by the rotation $\mathbf{R}(\phi, \theta)$ as

$$\begin{bmatrix} X \\ Y \\ Z \end{bmatrix} = \mathbf{R}(\phi, \theta) \begin{bmatrix} \xi \\ \eta \\ \zeta \end{bmatrix} \tag{5.2}$$

where ϕ is the angle between the $\xi\zeta$ plane and the XZ plane, and θ is the angle formed by the principal axis ξ against the surface of the membrane. See Figure 5.2.

The dielectric tensor of the anisotropic layer in the xyz frame $\boldsymbol{\varepsilon} = [\varepsilon_{ij}]$ is given by the transform $\mathbf{R}\boldsymbol{\varepsilon}_{\mathrm{p}}\mathbf{R}^{-1}$:

$$[\varepsilon_{ij}] = \begin{bmatrix} \varepsilon_{\perp} + \Delta\varepsilon \cos^2\phi \cos^2\theta & \Delta\varepsilon \cos\phi \sin\phi \cos^2\theta & \Delta\varepsilon \cos\phi \cos\theta \sin\theta \\ \Delta\varepsilon \cos\phi \sin\phi \cos^2\theta & \varepsilon_{\perp} + \Delta\varepsilon \sin^2\phi \cos^2\theta & \Delta\varepsilon \sin\phi \cos\theta \sin\theta \\ \Delta\varepsilon \cos\phi \cos\theta \sin\theta & \Delta\varepsilon \sin\phi \sin\theta & \varepsilon_{\perp} + \Delta\varepsilon \sin^2\theta \end{bmatrix} \tag{5.3}$$

where $\Delta\varepsilon = \varepsilon_{\parallel} - \varepsilon_{\perp}$. The relative dielectric constant of non-magnetic material equals to the square of the refractive index. Every element ε_{ij} is referred in calculating the reflection transfer matrix of the alignment layer in the Section 5.2.3.

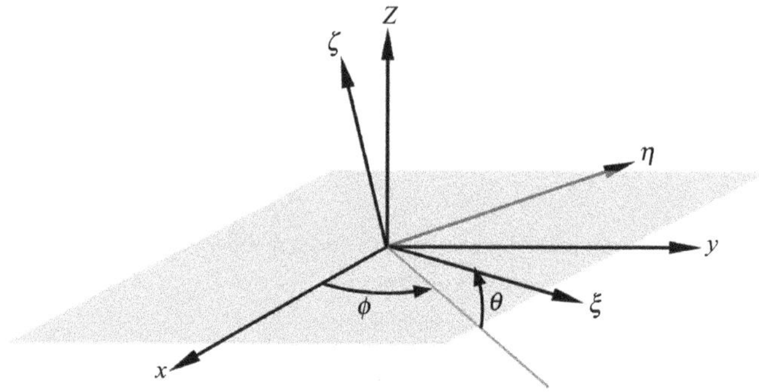

Figure 5.2 The principal frame ξηζ and the reference frame xyz fixed to the alignment layer

An isotropic medium may be considered as a special model of an anisotropic medium of which dielectric constants are identical, $\varepsilon_{\parallel} = \varepsilon_{\perp}$. The propagation of the polarisation in an isotropic material can be treated in the same way as described above.

5.2.3 Light propagation in alignment layer

The electromagnetic field of light is described by Maxwell's equations. Assuming a time harmonic wave, we write the electric and the magnetic fields generally as $\exp[-i\omega t]\boldsymbol{E}(x, y, z)$ and $\exp[-i\omega t]\boldsymbol{H}(x, y, z)$, respectively. We assume that neither charge nor current exists in the membrane. Maxwell's equations are reduced to the partial differential equations with respect to *x*, *y*, *z*:

$$i\omega\mu_0\boldsymbol{H} = \nabla \times \boldsymbol{E} \tag{5.4}$$

$$-i\omega\varepsilon_0\varepsilon\boldsymbol{E} = \nabla \times \boldsymbol{H} \tag{5.5}$$

The supplementing scalar equations are

$$\nabla \cdot (\varepsilon\boldsymbol{E}) = 0 \quad \nabla \cdot \boldsymbol{H} = 0 \tag{5.6}$$

Since all substances concerned here are non-magnetic, the magnetic permeability is equal to the constant μ_0, the permeability in vacuo. ε is the relative permittivity, which is represented by a tensor of rank two for anisotropic media as presented in Section 5.2.2, while by a scalar constant for isotropic ones.

The model through which the probe light is passing is shown in Figure 5.3. The refractive indices of the air, the isotropic layer in the alignment membrane and the glass substrate are n_0, n and n_g, respectively. The thickness of the whole membrane is d. The thickness of the anisotropic surface layer is ρd, where $\rho(\leq 1Z)$ denotes a fraction of depth of the anisotropic part to the whole membrane. We consider the electromagnetic field in four regions: (1) the open air space ($z < 0$), (2) the anisotropic surface layer of the alignment membrane ($0 \leq z \leq \rho d$), (3) the isotropic layer of it ($\rho d \leq z \leq d$) and (4) the glass substrate ($z > d$).

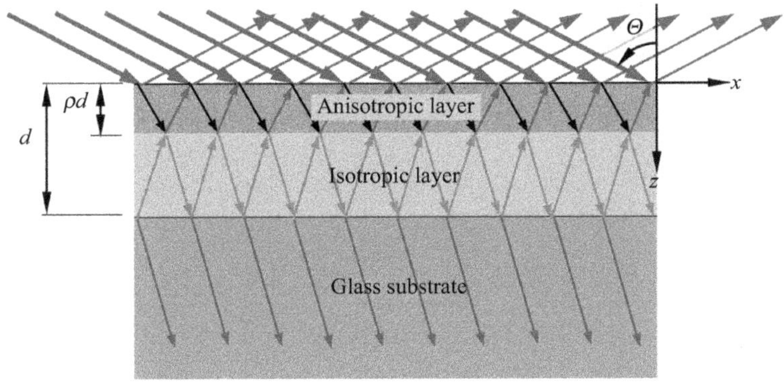

Figure 5.3 Stratified structure of alignment membrane on glass substrate

Suppose that a monochromatic plane wave with the wavelength of λ_0 in vacuo is propagating along the xz plane and incident upon the surface with the incident angle Θ. The electric field of the incident wave is written in the form

$$\boldsymbol{E}^{(i)}(x,z) = \boldsymbol{A}\exp[ik_0 n_0(\sin\Theta x + \cos\Theta z)] \tag{5.7}$$

where $k_0 = 2\pi/\lambda_0$, and n_0 is the refractive index of this region. Since $\boldsymbol{E}^{(i)}$should satisfy (5.6) or $\nabla\cdot\boldsymbol{E}^{(i)} = 0$, then we have $A_z = -\tan\Theta A_x$. According to (5.4), the magnetic field of the incident light is given by

$$\boldsymbol{H}^{(i)}(x,z) = \sqrt{\frac{\varepsilon_0}{\mu_0}}n_0\begin{bmatrix}\cos\Theta A_y\\ A_x/\cos\Theta\\ \sin\Theta A_y\end{bmatrix}\exp[ik_0 n_0(\sin\Theta x + \cos\Theta z)] \tag{5.8}$$

The electric field of the reflected wave is given by

$$\boldsymbol{E}^{(r)}(x,z) = \boldsymbol{R}\exp[ik_0 n_0(\sin\Theta x - \cos\Theta z)] \tag{5.9}$$

We also have the relation $R_z = \tan\Theta R_x$ from (5.6). Thus the magnetic field of the reflected light is given by

$$\boldsymbol{H}^{(r)}(x,z) = \sqrt{\frac{\varepsilon_0}{\mu_0}}n_0\begin{bmatrix}\cos\Theta R_y\\ -R_x/\cos\Theta\\ \sin\Theta R_y\end{bmatrix}\exp[ik_0 n_0(\sin\Theta x - \cos\Theta z)] \tag{5.10}$$

The electric and the magnetic field in the region are $\boldsymbol{E}^{(i)} + \boldsymbol{E}^{(i)}$ and $\boldsymbol{H}^{(i)} + \boldsymbol{H}^{(i)}$, respectively.

The transmitted wave from the isotropic layer to the glass substrate should be a plane wave. The electric field in the glass substrate is given by

$[\boldsymbol{T}\exp[i\{k_x x + k_y y + k_z(z-d)\}]$. We set $k_x = k_0\alpha$ and $k_y = 0$ from the phase continuity at the boundary plane $z = d$. Then, we have

$$\boldsymbol{E}^{(t)}(x,y) = \boldsymbol{T}\exp\left[ik_0\left\{\alpha x + \gamma_g(z-d)\right\}\right] \tag{5.11}$$

where $\gamma_g = k_0\sqrt{{n_g}^2 - \alpha^2}$. In the same manner as in the incident field, we have the magnetic field in the glass substrate

$$\boldsymbol{H}^{(t)}(x,z) = \sqrt{\frac{\varepsilon_0}{\mu_0}}\begin{bmatrix} -\gamma_g T_y \\ T_x {n_g}^2/\gamma_g \\ \alpha A_y \end{bmatrix}\exp\left[ik_0\left\{\alpha x + \gamma_g(z-d)\right\}\right] \tag{5.12}$$

We next consider the propagation of the electromagnetic wave in a thin anisotropic layer. Multiple reflection arises between two parallel boundary planes, $z = 0$ and $z = \rho d$. See Figure 5.3. The multiple reflection causes interference since the gap ρd between both boundary planes is small (usually <100 nm) compared with the wavelength. The algorithm suitable for this calculation is well-known 4×4 matrix formulation that was first presented by Dwight Berreman [4].

The electric field in the anisotropic layer of the alignment membrane is written as in the form

$$\boldsymbol{E}(x,z) = \boldsymbol{E}(z)\exp[in_0 \sin\Theta x] \tag{5.13}$$

$k_y = 0$ and $k_x = k_0 n_0 \sin\Theta$ should be held since the phase at the boundary surface $z = 0$ is coincident with that of the incident light. We define anew the electromagnetic field vector $\boldsymbol{\psi}$ by

$$\boldsymbol{\Psi}(z) = \begin{bmatrix} E_x(z) \\ \eta_0 H_y(z) \\ E_y(z) \\ -\eta_0 H_x(z) \end{bmatrix} \tag{5.14}$$

where $\eta_0 = \sqrt{\mu_0/\varepsilon_0}$ is put prior to the magnetic field to match the dimension. Maxwell's equations are reduced to the ordinary differential equation with $\boldsymbol{\Psi}(z)$

$$\frac{d\boldsymbol{\Psi}}{dz} = ik_0\boldsymbol{\Delta}_U\boldsymbol{\Psi}, \tag{5.15}$$

where $\boldsymbol{\Delta}_U$ is a 4×4 matrix of which elements are written explicitly as

$$\boldsymbol{\Delta}_U = \begin{bmatrix} -\alpha\varepsilon_{31}/\varepsilon_{33} & 1-\alpha^2/\varepsilon_{33} & -\alpha\varepsilon_{32}/\varepsilon_{33} & 0 \\ \varepsilon_{11}-\varepsilon_{13}\varepsilon_{31}/\varepsilon_{33} & -\alpha\varepsilon_{13}/\varepsilon_{33} & \varepsilon_{12}-\varepsilon_{13}\varepsilon_{32}/\varepsilon_{33} & 0 \\ 0 & 0 & 0 & 1 \\ \varepsilon_{21}-\varepsilon_{23}\varepsilon_{31}/\varepsilon_{33} & -\alpha\varepsilon_{23}/\varepsilon_{33} & -\alpha^2+\varepsilon_{22}-\varepsilon_{23}\varepsilon_{32}/\varepsilon_{33} & 0 \end{bmatrix} \tag{5.16}$$

α denotes $n_0 \sin\Theta$. ε_{ij} is given explicitly by (5.3). Since all ε_{ij} are constants, $\mathbf{\Delta}_U$ is a constant matrix. A general solution of (5.16) is written as

$$\boldsymbol{\Psi}_U(z) = \exp[ik_0 z\mathbf{\Delta}_U]\boldsymbol{\Psi}_U(0), \quad 0 \leq z \leq \rho d \tag{5.17}$$

The electromagnetic field vector in the isotropic layer ($\rho d \leq z \leq d$) is given as a solution of similar equation to (5.15):

$$\frac{d\boldsymbol{\Psi}}{dz} = ik_0\mathbf{\Delta}_I\boldsymbol{\Psi}, \quad \rho d \leq z \leq d \tag{5.18}$$

$\mathbf{\Delta}_I$ is given by replacing ε_{ij} in $\mathbf{\Delta}_U$ with n^2 if $i = j$, otherwise 0.

$$\mathbf{\Delta}_I = \begin{bmatrix} 0 & 1-\alpha^2/n^2 & 0 & 0 \\ n^2 & 0 & 0 & 0 \\ 0 & 0 & 0 & 1 \\ 0 & 0 & n^2-\alpha^2 & 0 \end{bmatrix}, \tag{5.19}$$

where n is the refractive index of the isotropic layer. We have a general electromagnetic field in the isotropic layer $\psi_i(z)$

$$\boldsymbol{\Psi}_I(z) = \exp[ik_0(z-\rho d)\mathbf{\Delta}_I]\boldsymbol{\Psi}_I(d), \quad \rho d \leq z \leq d \tag{5.20}$$

In general, the tangential components of the electric field and the magnetic field will be continuous across a boundary surface. This boundary condition gives the equations at the respective boundary surface; $\boldsymbol{\Psi}_U(\rho d) = \boldsymbol{\Psi}_I(\rho d)$, $\boldsymbol{\Psi}^{(i)} + \boldsymbol{\Psi}^{(r)} = \boldsymbol{\Psi}_U(0)$ and $\boldsymbol{\Psi}_i(d) = \boldsymbol{\Psi}^{(t)}$. $\boldsymbol{\Psi}^{(i)}$ denotes the electromagnetic field at the boundary surface $z = 0$, while $\boldsymbol{\Psi}^{(t)}$ at the boundary surface $z = d$. Then, we have a linear equation

$$\boldsymbol{\Psi}^{(t)} = \mathbf{P}(\boldsymbol{\Psi}^{(i)} + \boldsymbol{\Psi}^{(r)}) \tag{5.21}$$

We define the matrix $\mathbf{P}$ by

$$\mathbf{P} = \exp[ik_0(1-\rho)d\mathbf{\Delta}_I]\exp[ik_0\rho d\mathbf{\Delta}_U] \tag{5.22}$$

$\boldsymbol{\Psi}^{(i)}$, $\boldsymbol{\Psi}^{(r)}$ and $\boldsymbol{\Psi}^{(t)}$are given from (5.7), (5.8), (5.9) and (5.10).

$$\boldsymbol{\Psi}^{(i)} = \begin{bmatrix} A_x \\ n_0A_x/\cos\Theta \\ A_y \\ \alpha A_y \end{bmatrix}, \quad \boldsymbol{\Psi}^{(r)} = \begin{bmatrix} R_x \\ n_0R_x/\cos\Theta \\ R_y \\ \alpha R_y \end{bmatrix} \tag{5.23}$$

or transposing unknown vectors to be solved to the left-hand side,

$$-\boldsymbol{\Psi}^{(r)} + \exp[-ik_0(1-\rho)d\mathbf{\Delta}_I]\exp[-ik_0\rho d\mathbf{\Delta}_U]\boldsymbol{\Psi}^{(t)} = \boldsymbol{\Psi}^{(i)} \tag{5.24}$$

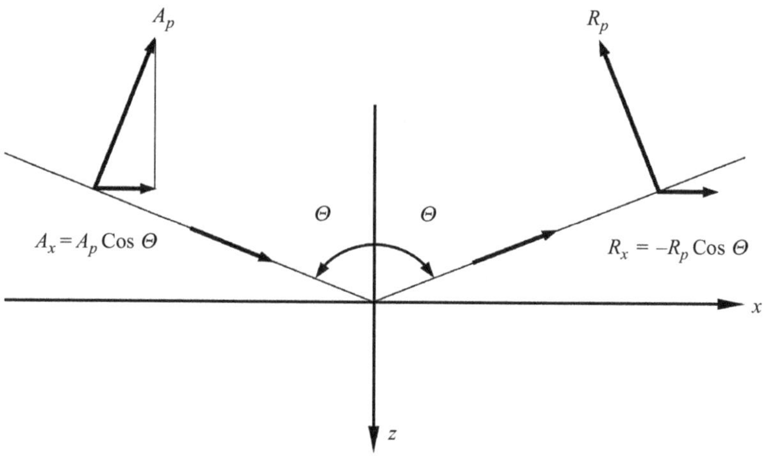

Figure 5.4 Definition of A_p, A_s, R_p and R_s

A_x, A_y are given. Equation (5.24) is a four simultaneous linear equation system with four unknown variables R_x, R_y, T_x and T_y. Thus, we obtain four elements r_{ij} solving the linear equation (5.24) for R_x and R_y.

$$\begin{bmatrix} R_x \\ R_y \end{bmatrix} = \begin{bmatrix} r_{11} & r_{12} \\ r_{21} & r_{22} \end{bmatrix} \begin{bmatrix} A_x \\ A_y \end{bmatrix} \tag{5.25}$$

It may be convenient to represent the electric field by the components parallel to and perpendicular to the plane of incidence instead of x and y components: $A_x = A_p \cos\Theta$, $A_y = A_x$, $R_x = R_p \cos\Theta$, and $R_y = R_s$. The definitions of A_p, A_s, R_p and R_s are illustrated in Figure 5.4. We have the new relation:

$$\begin{bmatrix} R_p \\ R_s \end{bmatrix} = \begin{bmatrix} 1/\cos\Theta & 0 \\ 0 & 1 \end{bmatrix} \begin{bmatrix} r_{11} & r_{12} \\ r_{21} & r_{22} \end{bmatrix} \begin{bmatrix} \cos\Theta & 0 \\ 0 & 1 \end{bmatrix} \begin{bmatrix} A_p \\ A_s \end{bmatrix} \tag{5.26}$$

Stokes parameters of the reflected light are given by

$$\begin{aligned} s_0 &= |R_p|^2 + |R_s|^2 \\ s_0 &= |R_p|^2 - |R_s|^2 \\ s_2 &= R_p R_s^* + R_s R_p^* \\ s_2 &= R_p R_s^* - R_s R_p^* \end{aligned} \tag{5.27}$$

5.3 Evaluation of alignment layer by reflection polarimetry

5.3.1 Principle of reflection polarimetry

The evaluation is carried out using a commercial general purpose polarimeter [5]. The optical system of the apparatus is shown in Figure 5.5. Incident monochromatic or white light with definite polarisation state is transmitted from the polarisation state generator unit indicated with PSG in Figure 5.5.

The light source supplies white or monochromatic light to PSG through the optical fibre. PSG, consisting of a quarter wave plate and a polariser, transmits light beam in an arbitrary state of polarisation, from linear to circular polarisation. The polarisation state of reflected light is detected by the polarisation state analyser unit indicated with PSA in the figure. PSA consists of a polariser mounted on the motorised revolver and a photodetector. PSG and PSA are attached individually to the coaxially rotatable respective arms so as that the incident angle is able to be chosen arbitrarily. The rotatable sample stage is provided in order to change the azimuth of the X-axis of the frame of a sample.

The polarisation state of the reflected light is detected by rotating analyser. The intensity of detected light is written as a sinusoidal function of the angle of the transmission axis $A = \omega t$, where ω is the angular velocity of the rotating analyser.

$$I(A) = I_0(1 + \alpha \cos 2A + \beta \sin 2A).$$

I_0, α and β can be obtained by the following integrations.

$$I_0 = \frac{1}{\pi}\int_0^{\pi} I dA, \quad \alpha = \frac{2}{\pi I_0}\int_0^{\pi} I \cos 2A dA, \quad \beta = \frac{2}{\pi I_0}\int_0^{\pi} I \sin 2A dA$$

These α and β are equivalent to (normalised) Stokes parameters:

$$\alpha = \frac{s_1}{s_0}, \quad \beta = \frac{s_2}{s_0} \tag{5.28}$$

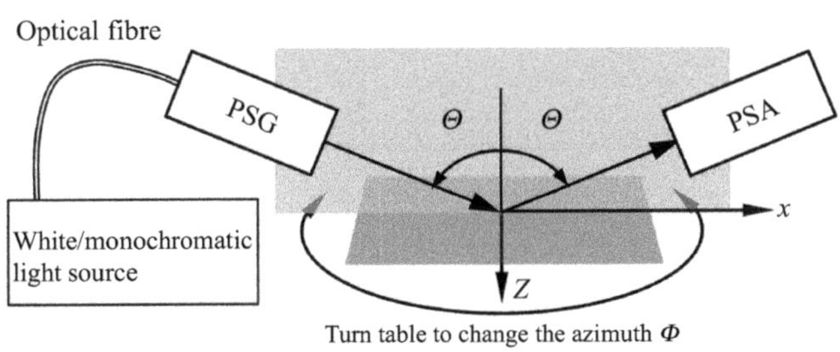

Figure 5.5 Schematic drawing of the polarimeter

5.3.2 Evaluation of alignment membrane using reflection polarimeter

An evaluation technique by observing the reflected light has been reported [2]. Another approach carried out in this work is presented in detail here.

We carried out the evaluation via following two-step procedures:

Step 1: Determining the azimuth angle ϕ and the tilt angle θ of the principal axis ξ.
Step 2: Determining the birefringence of the anisotropic layer.

Additionally, we measured cell parameters of the actual LC cells constructed with same substrates as those evaluated in Step 1 and Step 2.

Six samples were prepared: three samples provided by the rubbing process and the rest three provided by the photo-alignment process (irradiating linearly polarised UV light). The nominal thickness of the alignment membranes is 100 nm. The conditions of the rubbing treatment are summarised in Table 5.1. The translation direction was controlled strictly so as to be parallel to the base side of every glass substrate. In the photo-alignment process, the polarisation orientation was set strictly parallel to the base side of every glass substrate. LC molecules shall be aligned perpendicular to, or depending on the material parallel to this direction. Our material belongs to the former category. Table 5.2 shows an amount of energy irradiated on each sample.

5.3.2.1 Determination of azimuth and tilt: evaluation step 1

Set the polariser in PSG so as to transmit p-polarised light, or the light of which electric field is parallel to *xz* plane (the plane of incidence), and the polariser in PSA to detect s-polarisation as shown in Figure 5.6.

Table 5.1 The conditions of rubbing treated on the samples. Three samples are distinguished by ID. The rubbing direction is parallel to the base side of the substrate glass

ID	Revolution speed of rubbing roller (rpm)	Moving velocity (mm/s)	Pushing depth (mm)
RW	300	20	0.2
RM	500	30	0.3
RS	1,000	20	0.4

Table 5.2 The amount of exposed energy of UV light

ID	Energy irradiated [J/cm^2]
PW	300
PM	600
PS	900

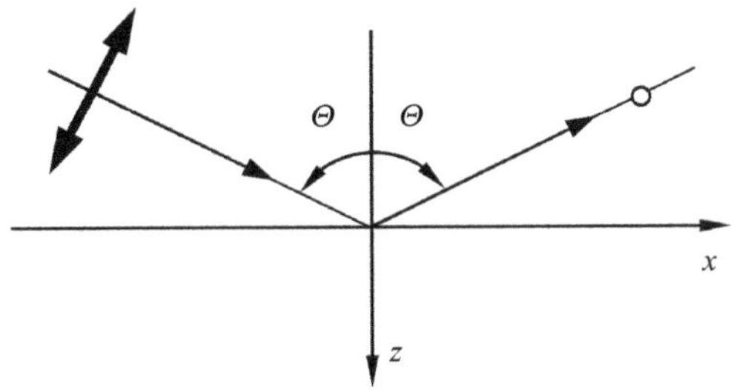

Figure 5.6 Alignment of polariser and analyser. The transmission axes of the polariser and the analyser are parallel to and perpendicular to xz plane

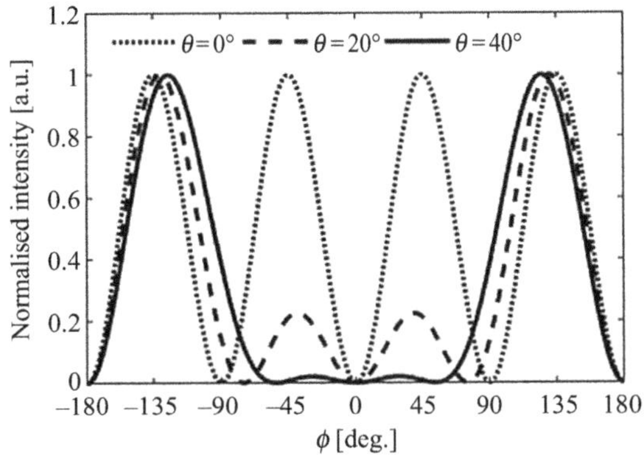

Figure 5.7 The intensity of s-polarisation of the reflected light from the model

Prior to the experiment, we carried out a numerical simulation based on the optical model illustrated in Section 5.3.2. The intensity of reflected light is calculated numerically using the algorithm presented in Section 5.3.3. The physical parameters defining the model and calculation conditions are as follows: $\varepsilon_{\parallel} = n_{\parallel}{}^2 = 2.89$ $\varepsilon_{\perp} = n_{\perp}{}^2 = 2.403$, $d = 100$nm and $\rho = 0.2$, $n_0 = 1.0$ and $n_g = 1.5$, and Θ=60°. We calculated the polarisation transfer matrix r_{21} and the intensity of s-polarisation $|R_s(\phi, \theta)|^2 = |r_{21}|^2 \cos^2 \Theta$ with three different tilt angles: $\theta = 0°$, 20° and 40°. The intensity, which is normalised by the maximum value, is plotted against the azimuth in Figure 5.7.

We find in Figure 5.7 that the intensity of s-polarisation $|R_s|^2$ takes the minimum value at $\phi = 0$, at which the principal axis ξ is parallel to the plane of incidence, irrespective of θ. Other azimuths that give local minimum depend on the tilt angles. $|R_p(\phi, \theta)|$ is symmetric with respect to the extinction azimuth $\phi = 0$. If the principal axis is parallel to the surface ($\theta = 0$), $|R_p(\phi, 0)|^2$ indicates four-fold rotational symmetry.

Therefore, in the experiment, if the intensity profile of detected s-polarisation shows periodicity with a period of 90°, it is estimated that the principal axis is parallel to the surface. Otherwise, the principal axis is estimated to be tilted.

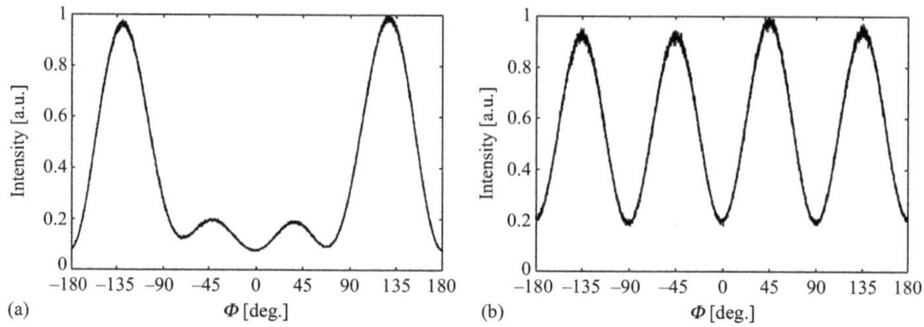

Figure 5.8 The intensity of s-polarisation of reflected light against to the azimuth of a sample: (a) is result from rubbed sample RS and (b) is from photo-aligned sample PS

We can find also that the larger the difference between larger local maximum intensity and smaller one, the steeper the principal axis slants. We estimated the tilt angle using curve-fitting technique.

We examined six samples which are treated by the process indicated in Tables 5.1 and 5.2. The results from the rubbed sample RS and the photo-aligned sample PS are shown in Figure 5.8(a) and (b), respectively. The intensity profile plotted in Figure 5.8(a) is similar to the curves with $\theta > 0$ in Figure 5.7. Other two samples treated by the rubbing, RM and RW, showed similar feature. We can conclude that the azimuth at which the intensity comes to minimum indicates the azimuth of the principal axis and that the rubbing process gives the tilt. On the other hand, the intensity profile in Figure 5.8(b) shows periodicity with a period of 90°. Other two photo-aligned samples PM and PW also showed similar feature. We recognise that photo-alignment process brings no or little tilt of the principal axis. There exists equally spaced four extinction azimuths: $\phi_m = \phi_0 + 90° \times m,\ m = 0, 1, 2, 3$. The azimuth of the principal axis is ϕ_0(or ϕ_2) or ϕ_2(or ϕ_4). It may be difficult to determine which is the azimuth of the principal axis ξ only by the intensity plot. However, we can choose easily the one since we know polarisation direction of the UV light approximately. It is obvious that ϕ_0(or ϕ_2) should be chosen in our case.

5.3.2.2 Determination of birefringence: evaluation step 2

The birefringence and other concerning parameters of the anisotropic layer are determined by minimising the error, or the distance between experimental result and result of modelling with postulated parameters. The procedure we carried out is illustrated below.

Let unknown parameters be the refractive indices $n_{\|}(= \sqrt{\varepsilon_{\|}})$ and $n_{\perp}(= \sqrt{\varepsilon_{\perp}})$, the thickness of the whole membrane d and a fraction ρ of anisotropic layer to the whole membrane.

1. Rotating a sample on the turntable in the apparatus, we measure the normalised Stokes parameters s_1/s_0 and s_2/s_0 at every azimuth of the sample: $\Phi_0, \Phi_1, \ldots, \Phi_N$. Two series of the normalised Stokes parameters $\{S_1^{\mathrm{exp}}[i]\}$ and $\{S_2^{\mathrm{exp}}[i]\}$, in which ith member corresponds to Φ_i, are obtained.

2. Substituting ($n_{\|}$, $n_{\perp}$, d, ρ) with estimated proper values, we define the optical model. We calculate the normalised Stokes parameters of the reflected light at every azimuth $\Phi_0, \Phi_1, \ldots, \Phi_N$, which are coincident with the experiment. Two series of the normalised Stokes parameters $\{S_1^{\text{calc}}(n_{\|}, n_{\perp}, \bar{n}, d, \rho)[i]\}$ and $\{S_2^{calc}(n_{\|}, n_{\perp}, \bar{n}, d, \rho)[i]\}$ are obtained.
3. The calculated values are compared with actually measured values. The distance between the actual values and calculated values is defined by

$$D = \left[\frac{1}{N}\sum_{i=1}^{N}\left|S_1^{\text{exp}}[i] - S_1^{\text{cakc}}(n_{\|}, n_{\perp}, \bar{n}, d, \rho)[i]\right|^2 + \frac{1}{N}\sum_{i=1}^{N}\left|S_2^{\text{exp}}[i] - S_2^{cakc}(n_{\|}, n_{\perp}, \bar{n}, d, \rho)[i]\right|^2\right]^{1/2} \tag{5.29}$$

If D is smaller than the predetermined value D_0, the postulated set of values $(n_{\|}, n_{\perp}, \bar{n}, d, \rho)$ is regarded as a set of the actual ones. Otherwise, correcting $(n_{\|}, n_{\perp}, \bar{n}, d, \rho)$ and carry out step 2 and 3. The procedure shall be repeated until convergence.

We carried out the procedure illustrated above on every sample. The measured Stokes parameters s_1/s_0 and s_2/s_0 of sample RS are plotted with open circles in Figure 5.9(a) and (b), respectively. The calculated curves at the convergence are imposed on the respective figure.

The obtained parameters of RS and PS are shown in Table 5.3 and Table 5.4. The retardation is given by $(n_{\|}^{\text{effect}} - n_{\perp})\rho d$, where $n_{\|}^{\text{effect}}$ is the effective refractive index defined by

$$n_{\|}^{\text{effect}} = \frac{n_{\|} n_{\perp}}{\sqrt{n_{\|}^2 \sin^2\theta + n_{\perp}^2 \cos^2\theta}} \tag{5.30}$$

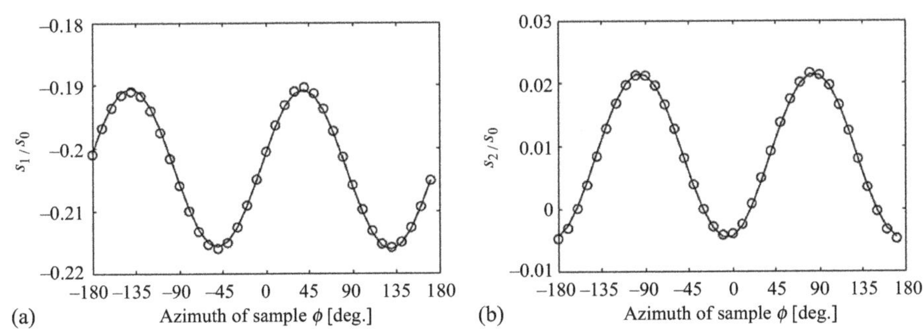

Figure 5.9 Measured data and calculated curves. The measured Stokes parameters s_1/s_0 and s_2/s_0 of sample RS are plotted with open circles in (a) and (b), respectively. The calculated curves at the convergence are imposed on the respective figure

Table 5.3 The result of Step 1

Item	RS	PS
Tilt θ of ξ axis [deg.]	31.20	0
$n_{\parallel}$	1.736	1.777
$n_{\perp}$	1.691	1.768
ρ	0.156	0.829
d [nm]	98.49	96.57
Retardation [nm]	0.496	0.671

Table 5.4 Summary of Step 1 and Step 2

	RW	RM	RS	PW	PM	PS
Tilt θ of ξ axis [deg.]	16.9	27.5	31.2	0*	0*	0*
Retardation [nm]	0.184	0.257	0.496	0.287	0.518	0.671

*The tilts of photo-aligned samples PW, PM, PS are negligibly small.

Table 5.5 Cell parameters determined by the rotating analyser polarimeter

Cell ID	RWC	RMC	RSC	PWC	PMC	PSC
Corresponding alignment process	RW	RM	RS	PW	PM	PS
Pre-tilt [deg.]	1.84	1.52	0.59	0.06	0.06	0.05
Twist [deg.]	−0.32	−0.92	−1.24	0.76	0.59	0.31
Gap [μm]	3.68	3.67	3.61	3.60	3.66	3.75

5.3.3 Complementary measurement

We prepared six homogenously aligned LC cells of which glass substrates were treated by the processes identical to those shown in Tables 5.1 and 5.2. All six cells are filled with LC MLC3019 (Merck Ltd., Darmstadt, Germany). One substrate of each cell is provided with a plane and a herringbone patterned electrodes for driving in the fringe field switching (FFS for short) mode. The pre-tilt angles and the twist angles of those LC cells were determined by the transmission polarimeter with rotating analyser [6,7]. The results are shown in Table 5.5. The six cells are distinguished by the ID name after the corresponding alignment process in Tables 5.1 and 5.2. The twist angles are designed so as to be 0°. The cause of non-zero twist angles in the table is the error while laminating the substrates.

5.4 Evaluation of image sticking by *V–T* property

The image sticking on LCD is a phenomenon where a faint outline of a previously displayed image remains visible on the screen even though the image has been changed. Image sticking is observed after a static image has been displayed for an exceedingly long period of time. We will not mention the cause or mechanism of the image sticking here. Instead, we suggest an approach to examine if alignment materials and conditions shall cause the image sticking on an LCD from a phenomenological viewpoint. The approach is based on comparing *V–T* properties before and after the stress that is the continued drive for a long period of time.

We examined six LC cells listed in Table 5.5. We set each sample between a pair of polarisers in crossed-Nicol alignment and measured the transmissivity with changing the voltage from 0 to 5 V. The transmissivity data were acquired before imposing the stress, immediately after and one hour after the stress removed. The stress was imposed by driving the cells with alternating square wave voltage continued for 24 h. After the stress period, the stress is removed and the electrodes are short-circuited. The results of the examination on the rubbing cell RSC and the photo-aligned cell PSC are shown in Figures 5.10 and 5.11, respectively.

The *V–T* property measured on the fresh LC cell is plotted with a solid line in Figure 5.10(a). The alternating voltage (the amplitude and the repetition rate are 4.5 V and 30 Hz, respectively) has been applied on the cell and continued for 24 hours. Then the voltage source is removed and the electrodes are short-circuited. Immediately, the *V–T* property is measured. The result is plotted out with a broken line on the same coordinates. Kept short-circuited for 1 h, *V–T* is measured again. The result is plotted out with a dotted line. Although those three curves seem to be overlapped, two curves obtained after the stress are shifted slightly from the base

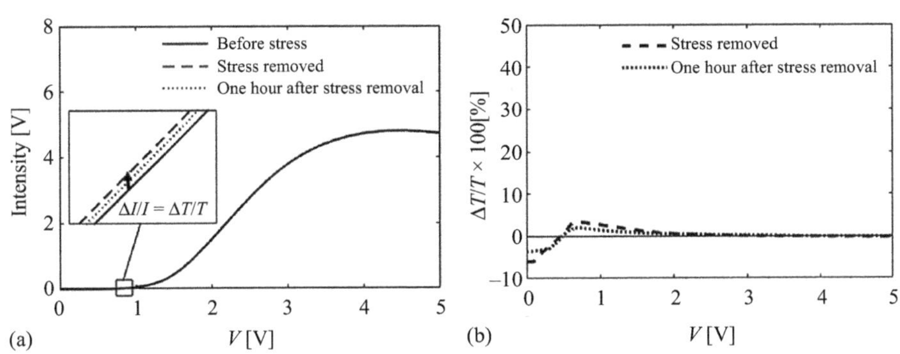

Figure 5.10 Transmissivity versus applied voltage of the sample cell RSC: (a) shows general view of V–T properties before the stress, immediately after the stress removed and after 1 h passed and (b) shows corresponding V–τ properties

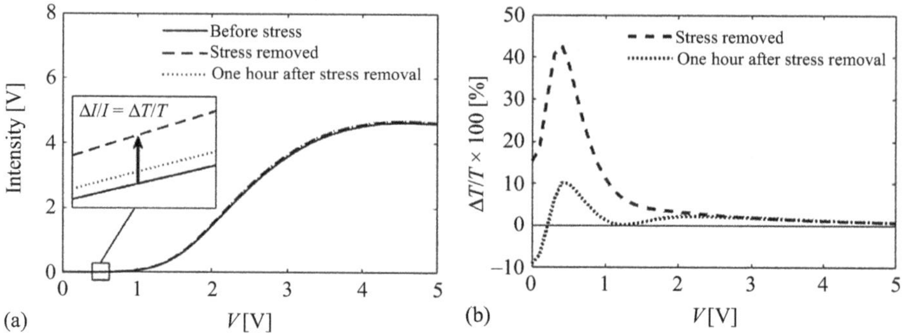

Figure 5.11 Transmissivity versus applied voltage of the sample cell PSC: (a) shows V–T properties before the stress, immediately after and one hour after the stress removed and (b) shows corresponding V–τ properties

curve (the solid line). In order to make those shift remarkable, we introduce a new parameter τ, which is defined by

$$\tau(V) = \frac{T(V) - T_0(V)}{T_0(V)} \tag{5.31}$$

where $T_0(V)$ is the transmissivity of the initial state. τ(multiplied with 100) is plotted against V in Figure 5.10(b).

The results of the examination on the photo-aligned LC cell PSC are shown in Figure 5.11.

If the stress leaves no residue in the cell, those three curves must overlap completely since the LC molecular alignment would return perfectly to the original state. Figures 5.10(b) and 5.11(b) show that the stress leaves some deformation in the cells. The residual effect appears especially in the rising phase of the transmissivity. The deformation derived from the stress relaxes very slowly.

A major cause of the image sticking is said widely impurity ions contaminated in the LC material. It arises from ionic charges accumulated at the surface of the alignment membrane after continuous driving. However, those distinctive features in $V-\tau$ curves suggest another cause. The analysis by $V-\tau$ profile presented here is thought to be an effective approach in developing the material as well as in LCD manufacturing process

5.5 Summary

The evaluation technique which consists of two steps of procedures using the reflection polarimeter is presented. We have carried out those procedures and come to the following conclusion. Both the rubbing and the photo-alignment give the optical anisotropy in the alignment membrane. The azimuth of the principal axis (ξ axis) agrees with the direction of translation in the rubbing process, while parallel or perpendicular to the direction of the polarisation plane of the UV light. This is because no tilt occurs by the photo-alignment. The azimuth of LC molecules on the surface (so-called easy axis) of LC cell constructed with the rubbing processed substrates agrees with the azimuth of the principal axis. The easy axis realised in the photo-aligned LC cell is, depending on the chemical property of the alignment membrane, either parallel or perpendicular to the principal axis. Our sample shows the perpendicular to the polarisation plane of the UV light.

The photo-alignment process does not tilt principal axis irrespectively of the exposure energy of UV light. The membrane with zero tilt of the principal axis realises zero pre-tilt alignment in the LC cell. On the other hand, the rubbing process tilts the principal axis. The tilt of the principal axis in the alignment membrane induces the pre-tilt in the LC cell. Referring to Tables 5.4 and 5.5, the steeper the tilt in the membrane, the slower the pre-tilt angle realised. However, we should not draw from this result the conclusion that this trend is the actual trend.

Both alignment processes of the rubbing and the photo-alignment give small retardation of one thousandth of the wavelength of the visible light. The deeper the pushing depth and the larger the exposed energy, the steeper the principal axis slants in the anisotropic layer. However, the quantitative relation between the retardation and the pre-tilt angle to be realised is not clear. The thickness of the anisotropic layer brought by the rubbing process is thinner ($\rho < 0.2$), while that by the photo-alignment is thicker $\rho > 0.8$. The result is reasonable since the rubbing creates the anisotropy by a mechanical contact, while UV light penetrates deep into the alignment membrane.

It may be difficult to estimate the pre-tilt angle that will be realised in the actual LC cell from the tilt angle of the principal axis that appears in the alignment membrane. The anisotropy that arises in the alignment membrane does not give straightforward the final cell parameters. However, the relation between a measured value of the tilt angle in the membrane and the pre-tilt angle that will be realised in the actual LC cell can be obtained in good reproducibility. For example, the alignment membrane of which principle axis has a tilt of 31° is estimated to realise the pre-tilt angle of 0.6° in the LC cell as referred to Tables 5.4 and 5.5. In this regard, the evaluation due to the procedures illustrated above is meaningful. Monitoring the anisotropy that appears in the alignment membrane using the polarimeter is an effective approach for quality control in the manufacturing process of LCDs. This evaluation is useful and effective also for manufacturer supplying the materials of alignment membrane.

We present an evaluation technique of the image sticking phenomenon by examining the V–τ characteristics. The technique is based on the examination of the profile of V–T curve. The residual effect left by the stress, or a long-time application of the electric field, brings slight but finite change in the V–T property. The change is too slight to find out by a usual V–T measurement. However, the V–τ property, presented in the Section 5.5 emphasises the slight change appears in the V–T profile. The V–τ characteristic is thought to be a valid and helpful index for evaluation of the image sticking.

Acknowledgement

The author is very grateful to Dr. Kimiaki Tsutsui, Nissan Chemical Corporation, for his contribution to prepare the sample substrates and cells. The author is also grateful to Mr. Atsushi Ikemura, SHINTECH, Inc., for his contribution in acquisition and analysis of the experimental data.

References

[1] LCD Master, modelling software package for LCD designers produced by SHINTECH, Inc., http://www.shintech.jp
[2] Hirosawal I. 'Relation between Molecular Orientation and Rubbing Strength Observed by Reflection Ellipsometry', *Japanese J. Appl. Physics, Part 1.* 1997, vol. 36(8), pp. 5192–96.
[3] Sawa K., Sumiyoshi K, Hirai Y., Tateishi K. and Kamejima T. 'Molecular Orientation of Polyimide by Infrared Dichroism', *Japanese J. Appl. Physics, Part 1.* 1994, vol. 33(8), pp. 6273–76.
[4] Berreman D. 'Optics in Stratified and Anisotropic Media: 4 × 4-Matrix Formulation'. *J. Opt. Soc. Am.* 1972, vol. 62(4), pp. 502–10.
[5] OPTIPRO, general purpose polarimeter produced by SHINTECH, Inc., http://www.shintech.jp
[6] Kitamura M., Kobayashi S. and Mori K. 'New Measuring Method for determining Alignment Parameters in Twisted-Nematic, and Super-Twisted-Nematic Cells', *J. SID*, 2005, vol. 13(3), pp. 261–66.
[7] Kitamura M., Kobayashi S. and Mori, K. 'Accurate Determination of Pretilt Angle in Twisted Nematic Liquid Crystal Cells by using a Rotating Analyzer', *J.SID*, 2006, vol. 14(5), pp. 509–14.

Chapter 6

Electrical evaluation of alignment layers and LC panels

Masaru Inoue[1]

This chapter describes a basic overview of electrical measurements such as voltage holding ratio (VHR), ion impurity, residual DC (RDC) voltage and capacitance–voltage (*C–V*) for elastic constant measurement of liquid crystals. Furthermore, ion impurity measurement in a thin film transistor liquid crystal display (TFT-LCD) panel is also described to find a root cause of image-sticking phenomena.

6.1 Introduction

It is well known that alignment layers play an important role for aligning liquid crystal molecules. Alignment layers are also applied voltage with liquid crystal layer, so it is required to know the electrical characteristics of these layers to avoid any display defects. Since 1998, there are many electrical measurement approaches to characterise an impact of alignment layers for TFT-LCD application such as VHR measurement [1], transient current measurement which applies square wave voltage to liquid crystal (LC) cell [2] and *C–V* hysteresis measurement [3]. However, these approaches had been proposed to measure materials at an early stage, so measurement of commercialised latest materials with high sensitivity was required. This chapter describes about standardised electrical measurement methods for alignment layers in detail.

6.2 Basic overview of electrical measurements

6.2.1 Voltage holding ratio

VHR measurement is one of the effective measurements to judge liquid LC cell's reliability [1,4]. This is a quick and easy measurement in which the measured voltage is related to the discharge phenomena in an LC cell, and the circuit is equivalent to the actual one pixel of TFT panel. If VHR becomes smaller, the mobile ions exist in the LC layer that comes from the LC itself, alignment layer, sealant and some other materials, or the resistivity of materials, which exist

[1]TOYOTech LLC, USA

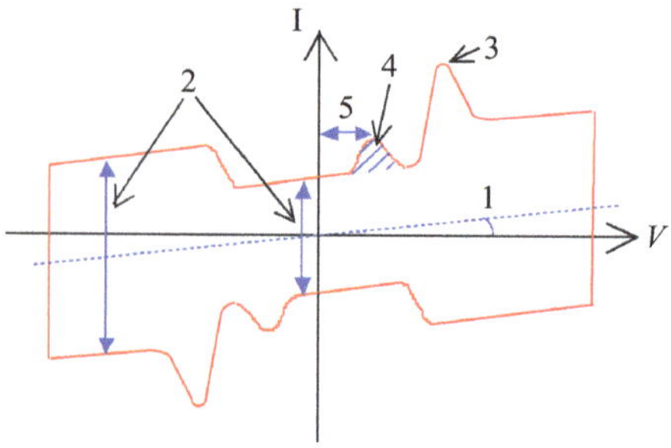

Figure 6.1 Typical ion impurity measurement result

between each electrode, is quite small [5]. VHR value is a total result so that it is not easy to find a root cause from the value.

6.2.2 Ion impurity

Ion impurity measurement has been performed by applying a triangle voltage waveform to an LC cell and by measuring the current through the cell [5]. The result of this measurement is displayed by a voltage vs. current curve to see the LC switching voltage, ion mobility and so on. The number of mobile ions is calculated from the area of current peak that appears at the appropriate position. This measurement has higher sensitivity than VHR.

Figure 6.1 shows a typical result of ion impurity measurement. The horizontal and vertical axes correspond to the applied voltage and measurement current, respectively. Ion impurity in LC cell is calculated from the area of current peak (4) that appears before LC switching peak (3). The other parameters such as panel resistance (1), capacitance (2) and ion mobility (5) are also calculated from this result.

6.2.3 Residual DC (RDC) voltage

RDC voltage measurement has been performed to investigate the cause of flicker phenomena. There are several kinds of measurement methods such as '*C–V* hysteresis' [6], 'flicker free' and 'dielectric absorption' [7]. In this section, dielectric absorption method is introduced and discussed.

Dielectric absorption method applies DC voltage to an LC cell as a defined stress, and an LC cell is shorted electrically, and the residual DC voltage is measured. This method also makes it possible to estimate the relaxation time of the LC and alignment layer from the residual DC curve.

Figure 6.2 shows the time chart of residual DC measurement. T_1, T_2 and T_3 mean stress, short circuit and measurement time, respectively. The change of voltage at the measurement time is the relaxation phenomena of the LC and alignment layers.

Figure 6.3 shows the voltage potential profile of a test cell at the three steps of residual DC voltage measurements. Once the stress voltage is applied to an LC cell, the voltage potential of the LC and alignment layer is changed. The potential is relaxed to the resistive ratio of each layer. When the LC cell is shorted electrically,

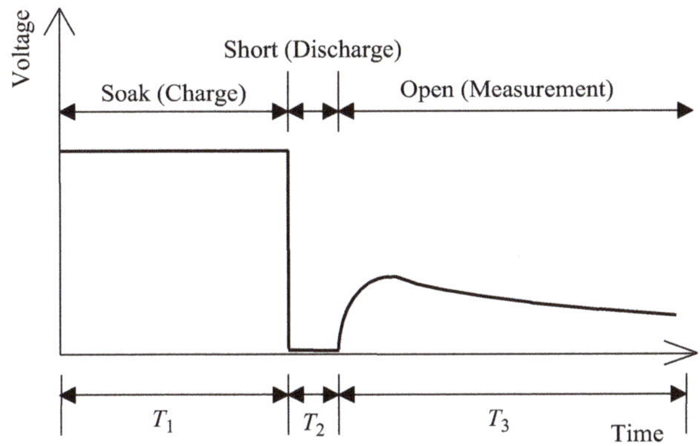

Figure 6.2 Time chart for residual DC voltage measurement

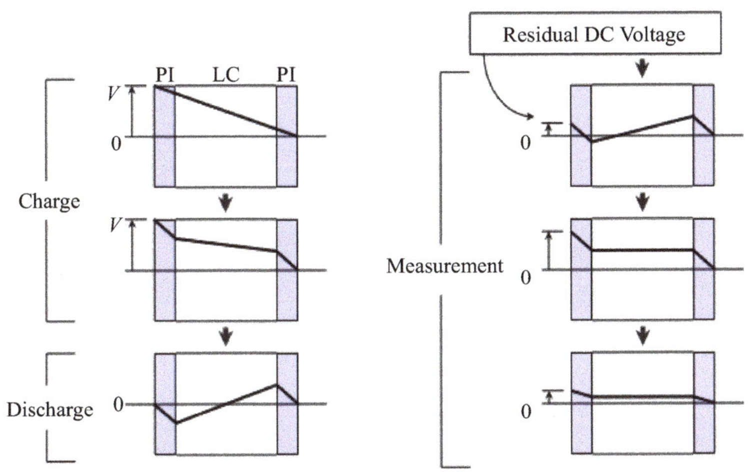

Figure 6.3 Voltage potential profile of residual DC voltage

then the voltage potential of the LC layer becomes an opposite slope direction due to the zero potential of each electrode. After the LC cell is electrically shorted, the residual DC voltage appears due to the time constant difference of the LC and alignment layer. The relaxation of the LC layers having a short time constant appears as an increasing voltage, and then the relaxation of the alignment layer having a long time constant appears as a decreasing voltage.

6.2.4 Capacitance-voltage (C–V) for elastic constant

Elastic constant is one of the important parameters of liquid crystal material for designing LCD panels and is obtained from *C–V* measurement. However, if a conventional LCR meter is used to measure *C–V* curve, there is an influence of DC charge to an LC cell because of one polarity bias voltage application.

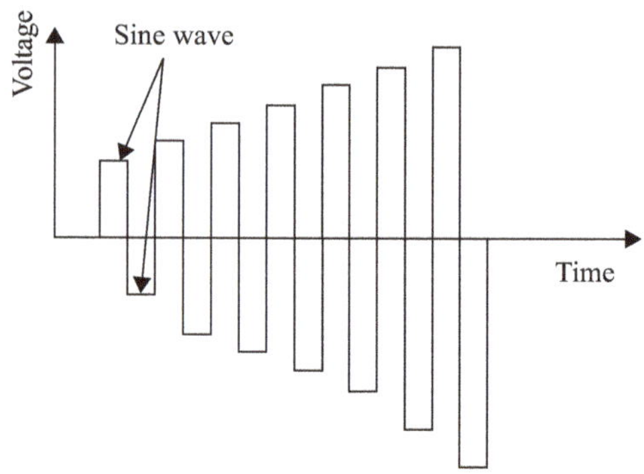

Figure 6.4 Alternative bias voltage for C–V measurement

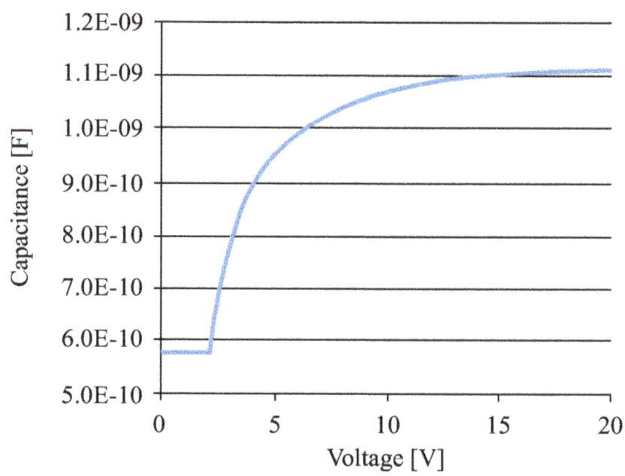

Figure 6.5 C–V Characteristics of LC mixture

In this case, the alternative bias voltage method was proposed to obtain precise elastic constant by avoiding the influence of DC charge [8]. Figure 6.4 shows a bias square waveform; simultaneously, a small amplitude sinusoidal waveform is applied on the square waveform to measure the capacitance.

The value of capacitance for each bias voltage step is calculated by using the sinusoidal wave signal. Normally, the frequencies of the sine wave that is a probe signal and the square wave for a bias signal are 1 kHz and 10 Hz, respectively. If LC cell has contaminations such as mobile ions, *C–V* measurement is done by increasing the frequency of the square wave because mobile ions are difficult to reach to the surface of alignment layers at a high frequency.

Figure 6.5 shows a typical C–V characteristic of a liquid crystal mixture. In case of a negative liquid crystal mixture, the bend elastic constant is obtained from the threshold voltage by using (6.1) [9].

$$V_{\mathrm{th}} = \pi\sqrt{\frac{K_3}{\varepsilon_0|\Delta\varepsilon|}} \tag{6.1}$$

where,

V_{th} Threshold voltage
d Cell gap
K_3 Bend elastic constant
ε_0 Vacuum dielectric constant
$\Delta\varepsilon$ Dielectric anisotropy

6.3 Experiments for alignment layers

6.3.1 Correlation between VHR and ion impurity

VHR and ion impurity were measured for test cells that have three kinds of alignment layers and the same liquid crystal [9]. The sample cells and the conditions of each measurement are as follows:

Test cells: alignment layer A, B and C, liquid crystal ZLI-4792 (Merck), cell gap 5 μm, area of electrode 1 cm^2

VHR: 60 μs pulse width, 16.67 ms frame frequency, ±5 V voltage, room temperature

Ion impurity: 0.1 Hz triangle waveform, ±10 V voltage, room temperature

Figure 6.6 shows the correlation between ion impurity and VHR. The VHR value starts to decrease more than 200 pC/cm^2 ion impurity value. Most of the PI-A samples have more than 99 per cent VHR value, but ion impurity values are distributed from 2 pC/cm^2 to 1,000 pC/cm^2. The PI-B sample has more than 1,000 pC/cm^2 ion impurity and less than 98 per cent VHR value. The PI-C sample has more than 99 per cent VHR values but the ion impurity values are from 30 to 80 pC/cm^2. The VHR measurement is a common evaluation method to see the material and cell process reliability, but ion impurity measurement is also needed to investigate reliability in detail. If an LC cell has high VHR but also has high ion impurity, then this might have some problem in the future.

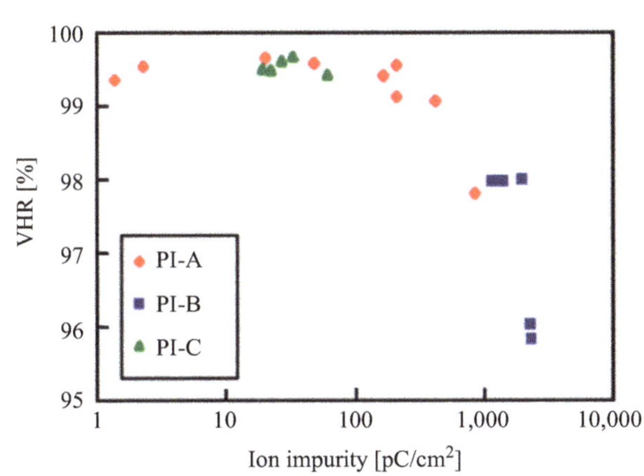

Figure 6.6 Ion impurity vs. VHR

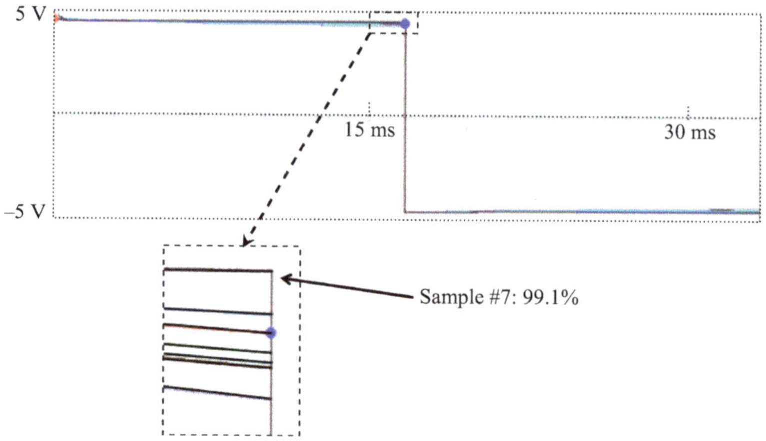

Figure 6.7 VHR comparison

6.3.2 Correlation between VHR, ion impurity and RDC

Figures 6.7, 6.8 and 6.9 show the results of the measurements of VHR that are the decay of positive area, ion peak position of ion impurity measurement and the measurement part of residual DC voltage, respectively [9]. The sample no.7 has the largest VHR (99.1 per cent) and the smallest ion impurity. However, this sample has the largest residual DC voltage because of the highest resistivity of the alignment layer.

6.3.3 Cell evaluations having a single-sided alignment layer

Figure 6.10 shows a profile of an LC cell that consists of liquid crystal and a single-sided alignment layer [10]. The liquid crystal and alignment layer were ZLI-4792 (Merck) and the alignment film for vertical alignment, respectively. In this condition, the deformation of the LC layer becomes a hybrid alignment.

The cells having different doping nanoparticle materials and conditions were prepared. SiO_2 and MgO were doped into LC or the alignment layer. The combinations of alignment layer with nanoparticles are as follows:

- Alignment layer + SiO_2 (50 weight per cent)
- Alignment layer + MgO (26 weight per cent)

As a reference, non-doping nanoparticle LC cells were also prepared. Figure 6.11 shows the voltage vs. current data of the reference LC test cell that consists of liquid crystal and a single-sided alignment layer (PI + LC). There are two data results: one is the measurement result just after fabricating the cell, and the other is the result after 150 days since cell fabrication. These waveforms show that ion peaks increased with time, probably because materials in the LC test cell generated mobile ions into the liquid crystal layer. The difference between positive

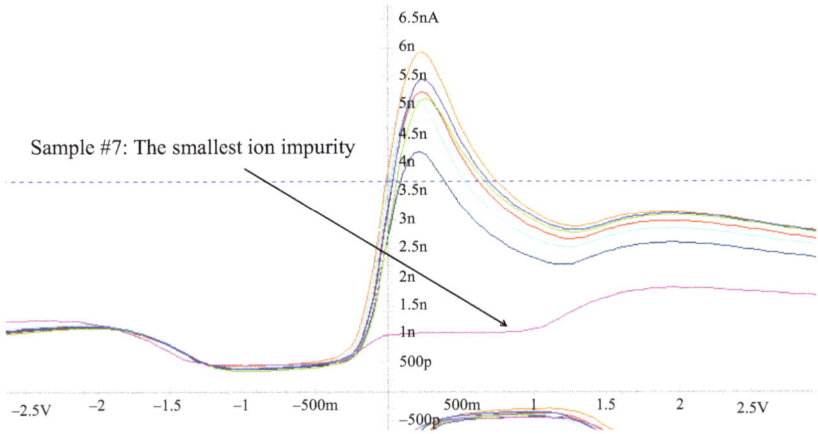

Figure 6.8 Ion impurity comparison

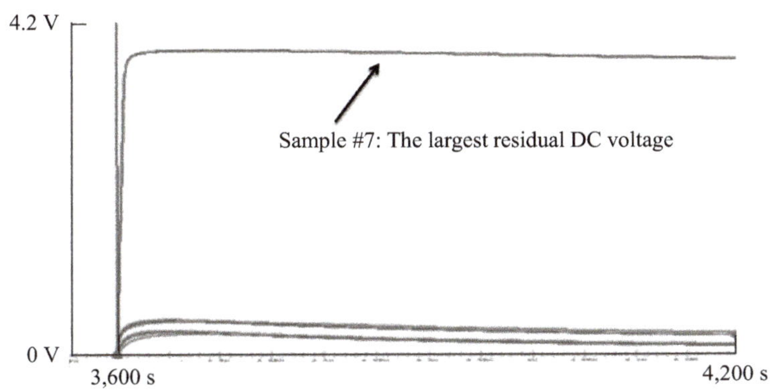

Figure 6.9 Residual DC comparison

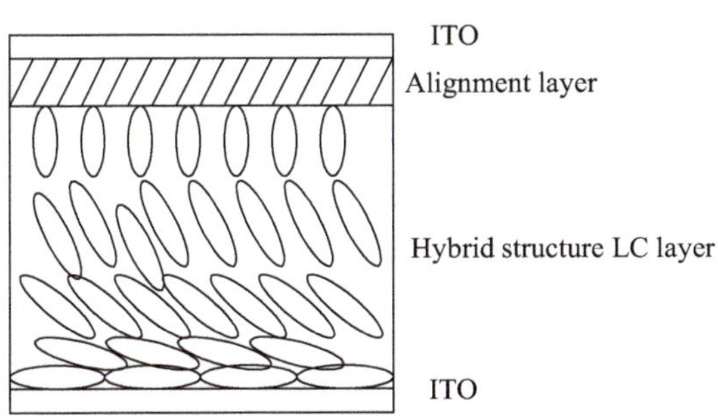

Figure 6.10 Structure of an LC cell

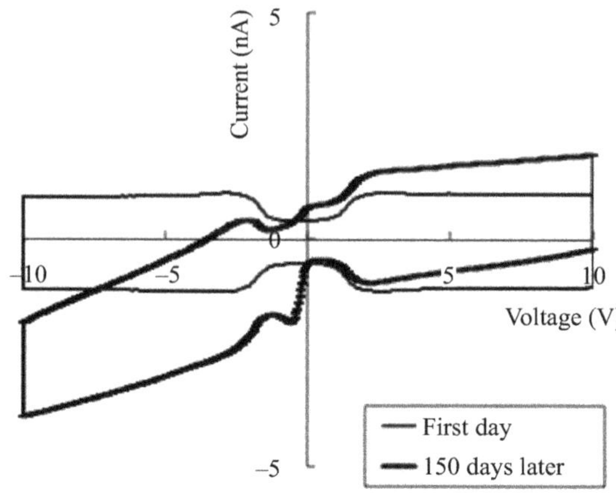

Figure 6.11 Voltage vs. current characteristics of reference cell

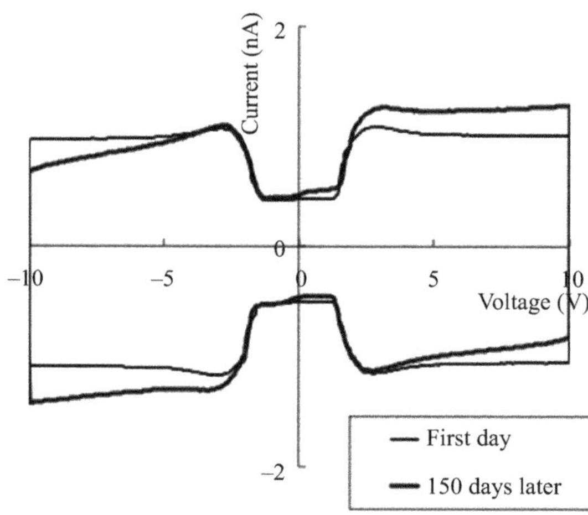

Figure 6.12 LC + Alignment layer with SiO_2

and negative ion peaks corresponds to the polarity of the ion. Especially, the number of positive ions that appear in the negative voltage application is greater than that of the negative ions in the reference LC test cell.

Figure 6.12 shows the measurement results of the LC test cell that consists of liquid crystal and a single-sided alignment layer with 50 weight per cent of SiO_2. Compared with the measurement result of the reference LC test cell, the ion peak that appears at the negative voltage application decreased. By doping SiO_2 nanoparticles into the alignment layer, positive ions are reduced in the liquid crystal layer.

Figure 6.13 shows the measurement result of the LC test cell that consists of liquid crystal and a single-sided alignment layer with 26 weight per cent of MgO.

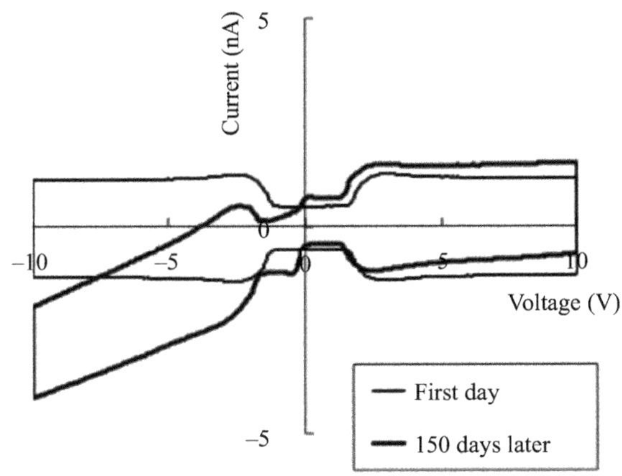

Figure 6.13 LC + Alignment layer with MgO

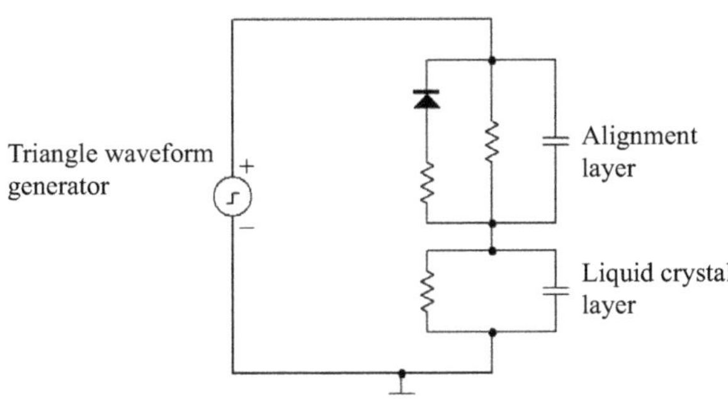

Figure 6.14 SPICE simulation circuit

Even if MgO nanoparticles were doped into the alignment layer in the LC cell, there was no change of measurement result compared with the reference LC cell. This result also explains that MgO has no function to catch positive ions.

In this experiment, the asymmetric voltage vs. current curve of the LC cell having a single-sided alignment layer was measured as shown in Figure 6.12. In the case of measurement of a conventional LC cell having alignment layers on both ITO glass substrates, the measurement results of the voltage vs. current curve are symmetric and of the same slope level. However, there are different slope levels at both positive and negative voltage applications. These results explain that an alignment layer has some diode characteristics, so the impact of an alignment layer as a diode model was investigated on an analogue circuit simulation program,

Figure 6.14 shows an equivalent circuit for SPICE simulation of liquid crystal and a single-side alignment layer [11]. It is assumed that an alignment layer

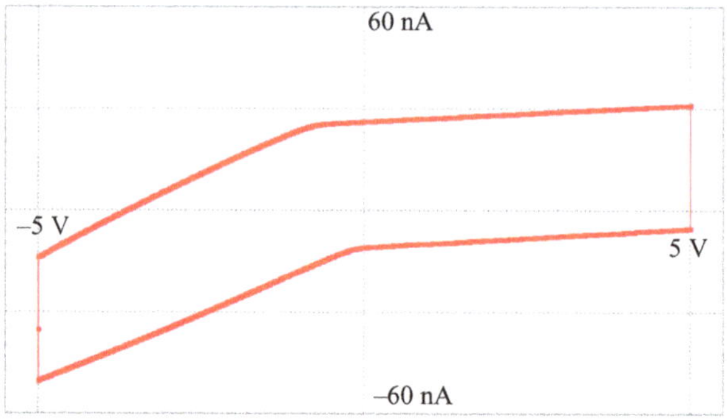

Figure 6.15 Simulation result

consists of resistors, a capacitor and a diode. This circuit applies triangle voltage waveform and calculates the current that flows to the LC cell.

Figure 6.15 shows a simulation result voltage vs. current curve of a liquid crystal and a single-sided alignment layer. The horizontal and vertical axes correspond to the applied voltage and calculated current, respectively. This graph shows that there is a good correlation in the slope of the waveform between an experiment and a simulation result. It is confirmed that an alignment layer in thin film condition has some diode characteristics.

6.3.4 Threshold voltage (V_{th}) impact by alignment layers

Figure 6.16 shows the cell gap dependency of threshold voltages for three kinds of alignment layers with same vertical alignment (VA) liquid crystal mixture [9]. The dots and lines are the measurement results of LC cells having a different cell gap and calculations from (6.2), respectively. This result suggests that the influence of surface on alignment layers cause the change of the threshold voltage. So, (6.1) should be modified to (6.2) that takes into account the negative extrapolation length (d_e).

$$V_{\text{th}} = \frac{d\pi}{d - 2d_{\text{e}}}\sqrt{\frac{K_3}{\varepsilon_0|\Delta\varepsilon|}} \tag{6.2}$$

where,

d_e Negative extrapolation length

6.4 Evaluations of TFT-LCD panels

Ion impurity and VHR measurements are effective methods to investigate the cause of panel defects by fabricating a test panel that has no TFT pixels. It has already explained that ion impurity and VHR values become larger and smaller, respectively, when panels to be measured have contaminations that come from the materials or fabrication process.

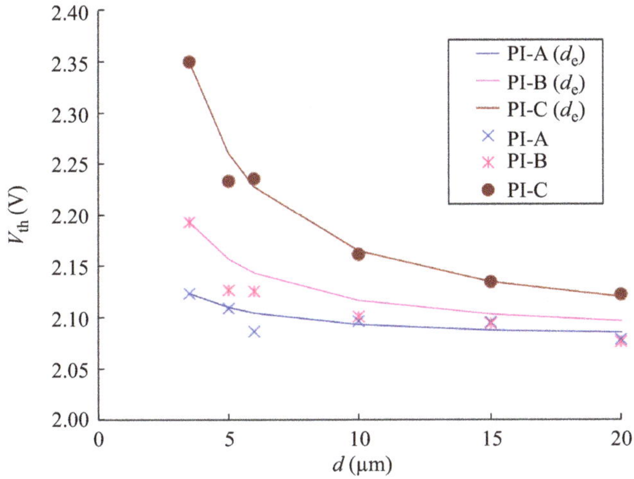

Figure 6.16 Cell gap dependency of V_{th}

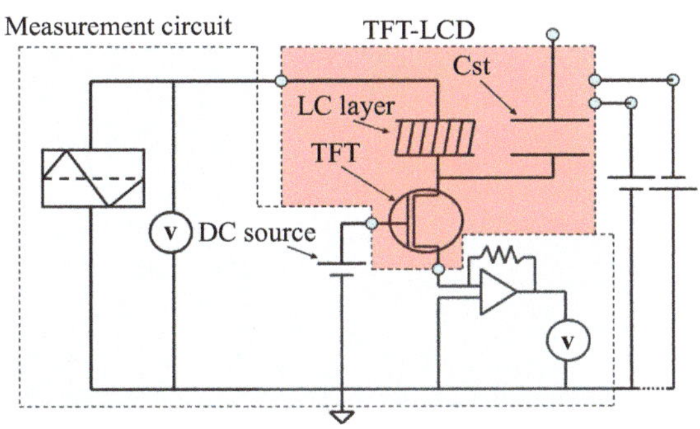

Figure 6.17 Equivalent measurement circuit for ion impurity in TFT-LCD

However, test panel measurements are not enough to investigate the cause of panel defects such as area and line image-sticking phenomena in a real TFT-LCD panel due to the difference of substrates, panel design, materials, fabrication process and so on. So, it is required to measure an ion impurity that has unexpected contaminations in real TFT-LCD panels.

6.4.1 Measurement principle of ion impurity for TFT-LCD

Figure 6.17 shows an equivalent measurement circuit for a TFT-LCD panel [12]. The measurement circuit consists of a triangle waveform generator, a DC power supply for gate input signal, a current measurement circuit (operational amplifier and voltmeter) and a voltmeter to observe an applied voltage. A triangle waveform is applied to the common electrode (V_{com}) of the TFT-LCD panel and the current

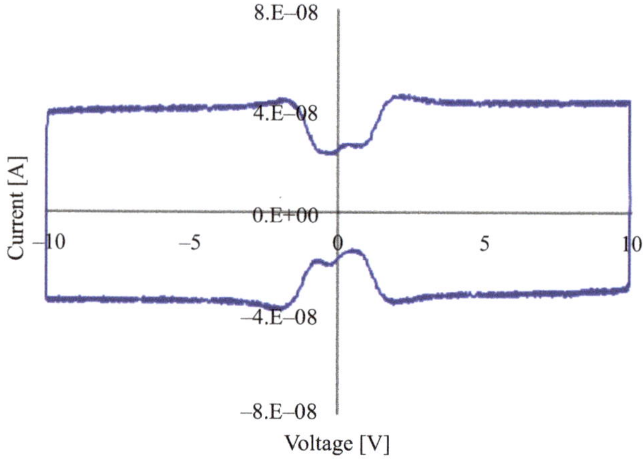

Figure 6.18 Small ion impurities

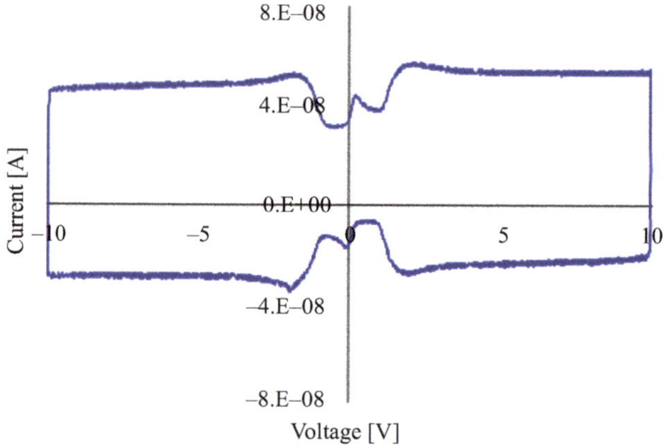

Figure 6.19 Medium ion impurities

that flows into the LC layer is measured through the drain and source in TFT pixels while applying appropriate DC voltage to the gate electrodes.

Figures 6.18 and 6.19 show measurement results of the LC layer in real TFT-LCD panels for mobile phone application that are different from the amount of ion impurities. The measurement conditions were 0.01 Hz, 10 V triangle waveform at 70 °C. The amount of ions in Figure 6.19 is larger than that of Figure 6.18, so that this measurement method could be used for the evaluation of real TFT-LCD panels.

6.4.2 *Ion impurity vs. image sticking*

There are five kinds of TFT-LCD panels (A, B, C, D and E) that differ between alignment materials and process [13]. Two kinds of alignment materials are prepared,

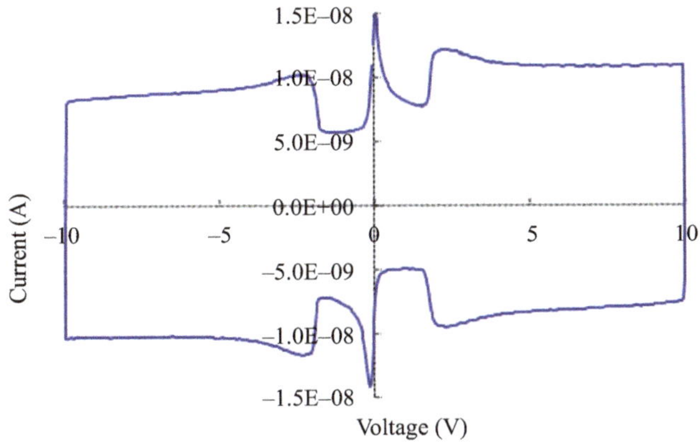

Figure 6.20 Large ion impurities

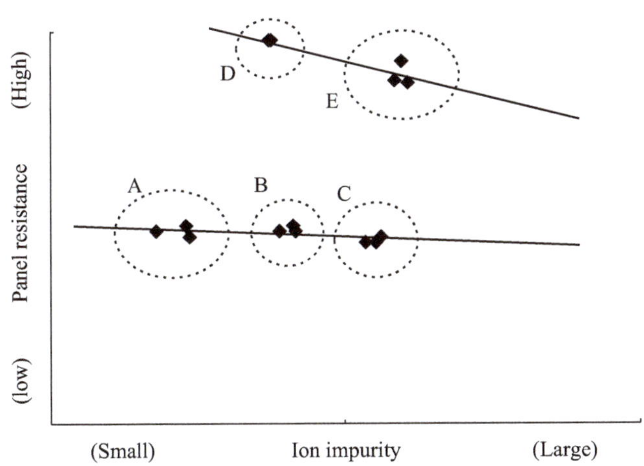

Figure 6.21 Ion impurity vs. panel resistance

one material is for panels A, B and C, and the other material is for panels D and E. The panel resistance and ion impurities were measured at 60 °C.

Figure 6.20 shows a typical measurement result of a TFT-LCD panel having large ion impurities. Ion impurity and panel resistance is calculated from an area of peak that appears at around the 0 V region and a slope level of voltage vs. current characteristics, respectively.

Figure 6.21 shows a correlation between ion impurity and panel resistance. By decreasing the amount of ion impurity, the panel resistance is increased. The result also shows that the difference of alignment material causes a different panel resistance. As a result, panel D and E, having higher resistance than panel A, B and C, may cause an area of image sticking due to long discharge time. Panel E and C, having many ions, may also cause an area of image sticking due to the internal DC

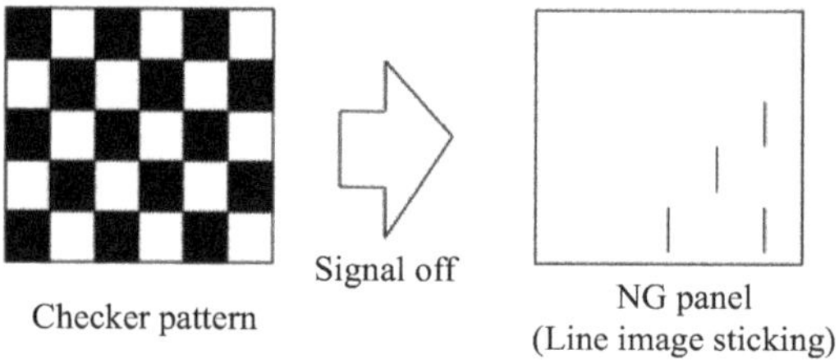

Figure 6.22 Line image sticking

voltage that is made by the electric double layer on the surface of the alignment layer. Moreover, a panel having many ions has a possibility to have line image sticking by means of lateral voltage difference between drive and non-driving pixels as shown in Figure 6.22. These border areas have a higher ion concentration than other areas and make a voltage potential even if all pixels are non-driving. These results suggest that small ion impurity and low resistivity of the alignment layer are required to avoid the image-sticking problem for TFT-LCD panels.

6.5 Conclusions

1. An alignment layer is a key material to align liquid crystal molecules in LCD panel, and is required to have small ion impurity and a similar order of resistivity as liquid crystal to avoid flicker, image sticking and Mura failures.
2. By doping SiO_2 nanoparticles into an alignment layer, ion impurity can be reduced in LC cells.
3. An alignment layer causes a voltage drop in the LC layer due to the existence of negative extrapolation length, so it is required to know the characteristics of the alignment layer for the measurement of elastic constant value of the liquid crystal.
4. It is confirmed that ion impurity measurement in a real TFT-LCD panel can be used for a process optimisation of the alignment layer.

References

[1] Mizushima S., Shimasaki T., Minezaki S., Yano K. and Hijikigawa M. 'Storage Characteristics of TFT-LCD (II)'. Proceedings of Japanese Liquid Crystal Society annual meeting; Sendai, Japan, 1988 pp. 80–81.

[2] Wakemoto H., Ishihara S. and Matsuo Y. 'Transient Electric Current in LCD Cells'. Proceedings of Japanese Liquid Crystal Society annual meeting; Sendai, Japan, 1988 pp. 212–13.

[3] Miyake S., Adachi K., Takasago H. and Yamazaki T. 'Effects of Alignment Layer on Electro-optical Properties of TFT-LCD'. Proceedings of Japanese Liquid Crystal Society annual meeting; Sapporo, Japan, 1991 pp. 288–89.

[4] Shimasaki T., Mizushima S., Minezaki S., Yano K. and Hijikigawa M. 'Storage Characteristics of TFT-LCD (I)'. Proceedings of Japanese Liquid Crystal Society annual meeting; Sendai, Japan, 1988 pp. 78–79.

[5] Inoue M. and Nakanowatari J. 'Mobile Ion Frequency Dependency in LC cells and Correlation Between Mobile Ion and Voltage Holding Ratio'. Proceedings of Japanese Liquid Crystal Society annual meeting; Sendai, Japan, 1995 pp. 362–63.

[6] Nishikawa M., Suganuma T., Tsuda Y. and Bessyo N. 'Electrical Properties of LC cells with Soluble Polyimide Alignment Layers'. Proceedings of Japanese Liquid Crystal Society annual meeting; Tokyo, Japan, 1993 pp. 108–09.

[7] Inoue M., Manabe N. and Nakanowatari J. 'Correlation Between Flicker Minimizing Method and Dielectric Absorption Method for Residual DC Measurement'. Proceedings of Japanese Liquid Crystal Society annual meeting; Tokyo, Japan, 1997 pp. 216–17.

[8] Inoue M., Nomura T., Yamaguchi T., *et al.* 'Novel Measurement Method of Elastic Constant for Negative Dielectric Anisotropy LC Mixture' Presented at 20th International Liquid Crystal Conference; Ljubljana, Slovenia, 2004.

[9] Inoue M., Takatoh K. and Kobayashi S. 'Recent Measurement of Liquid Crystal Characteristics'. Proceedings of the 13th International Display Workshops; Otsu, Japan, Dec. 2006, Tokyo: IDW; 2006 pp. 647–50.

[10] Inoue M., Tasaka S., Imaoka H., Sano S., Takatoh K. and Kobayashi S. 'Nanoparticle Behaviors in LC Cells Having a Single-Sided Alignment Layer Structure'. Proceedings of the 14th International Display Workshops; Sapporo, Japan, Dec. 2007, Tokyo: IDW: 2007 pp. 1613–16.

[11] www.spctrum-soft.com [Accessed 16 May 2018].

[12] Inoue M. 'A Novel Measurement Method of Ion Density in TFT-LCD Panels'. Proceedings of the 15th International Display Workshops; Niigata, Japan, Dec. 2008, Tokyo: IDW: 2008 pp. 47–50.

[13] Inoue M. 'Analysis of Image Sticking Phenomena in TFT-LCD Panels'. Proceedings of Japanese Liquid Crystal Society annual meeting; Fukuoka, Japan, 2010.

Chapter 7
Analysis method of image sticking

Yonghwan Shin[1]

In this chapter, we present a method to analyze the image-sticking phenomenon of the liquid crystal display (LCD). Image-sticking phenomenon is one of the severe problems that deteriorate image quality of not only LCD but also other display systems such as OLED and CRT.

First, we have categorized the main factors of the image sticking and then simulated the proportion of these factors by using the experimental data. We expect that this method of analysis can be utilized to understand the basic mechanism of the image sticking and to solve the chronic problem of LCDs.

7.1 Introduction to image sticking

Image sticking or image-retention phenomenon can be observed after aging (display fixed image for a long time). Table 7.1 shows how image sticking can be seen on the display. Initially intended image (as shown in the first line) is identical with an actually displayed image. During a certain amount of time, display image like the second line (called an aging from now on). Then intended image and actual image on display are no longer identical to the third line. Because this phenomenon looks like the previous image to be stuck on display, this is called an image sticking. Image sticking usually can be shown as (1) area-type image sticking and (2) line-type image sticking. The actual image on the third line of Table 7.1, can be found in both area-type and line-type image sticking. Area-type image sticking can be induced from the difference of brightness between (A) and (B) areas. On the other hand, line-type image sticking particularly prominent at the boundary area of (A) and (B) [1,2].

In some cases, we could observe severe line-type image sticking than area-type image sticking [3]. But the causes of line-type image sticking are different from the causes of area-type image sticking. Therefore, a different analysis method for them should be applied. In this section, we would like to focus on the analysis method of area-type image sticking.

There are many different factors to affect this phenomenon, also it is difficult to understand how they effect on. Therefore, there are many papers published about

[1]OLED A Development Team, Samsung Display Co., Ltd., Republic of Korea

Table 7.1 Comparison of intended and actual display images before/during/after the aging period

	Intended image	Actual image on the display
Initial		
During aging		
After aging		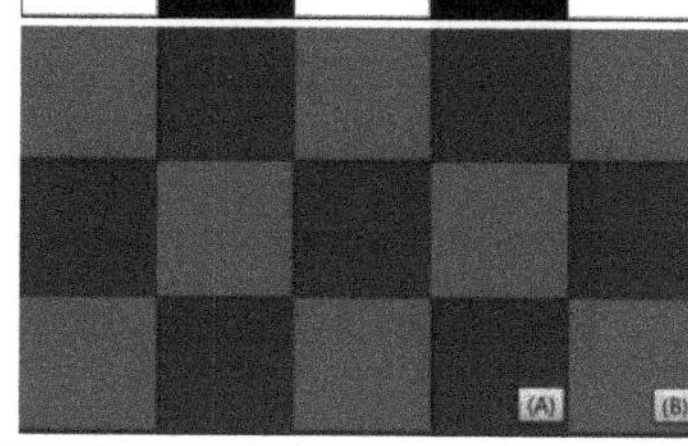

improving image-sticking problem of LCD. The transition of pretilt angle [4], VHR [5] (voltage holding ratio), and R-DC [6] (residual DC) can be a cause of image sticking. But, there is no research on how much each cause is affecting on image sticking. In this section, we would like to explain how to calculate the portions of three causes of image sticking especially on the vertical alignment (VA) mode LCD.

It is not sure that this method can be applied to analyze the causes of image sticking for other LC mode. But it is expected that this analysis method will serve as a guideline of analysis for image sticking.

7.2 Driving scheme of TFT-LCD

To understand the causes of image sticking, it is required to comprehend the driving scheme of TFT-LCD. Figure 7.1 shows the equivalent circuit of typical TFT-LCD.

The voltage of the storage and the pixel capacitors are charged to the level of source electrode which is supplied by source driver IC when gate switch is ON. On the other hand, when the gate switch is OFF, the charge stored at the storage and pixel capacitors is kept while the gate is turned on again at the next frame.

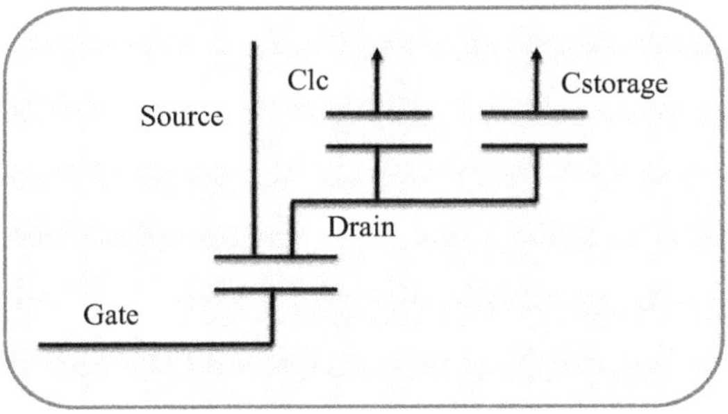

Figure 7.1 Equivalent circuit of a pixel of typical TFT-LCD

After gate signal of the odd frame is ON, the channel of TFT is open. Then the storage and pixel capacitor can be charged to the certain voltage level near that of the source driver IC.

When a gate signal is OFF, the voltage level of a pixel was changed. It is called a kickback and it induces to decrease the voltage level of pixel and storage capacitors.

Ideally, during the rest gate OFF state, the charge stored in the storage and pixel capacitors is kept. But in fact, there is some dissipation due to the leakage current through the (1) LC/PI layers, (2) storage capacitor and (3) TFT. This dissipation is related with VHR. If VHR is 100 percent, the charge which is stored in the pixel and storage capacitors should be maintained during a frame, and not be dissipated. Generally, VHR has been used the characteristics of LC and PI layer of TFT-LCD. But in this section, we also consider as leakage current through the storage capacitor and TFT.

When data signal polarity is inverted, kickback occurs due to drain signal polarity inversion.

After next even frame starts, and the gate signal is turned ON again, the storage and pixel capacitors are charged. Then the voltage level of storage and pixel capacitors is decreased because the polarity of the source driver IC signal is inverted at the even frame.

When the gate signal is turned OFF, the kickback occurs again. Also, this kickback decreases to the voltage level of storage and pixel capacitors.

And voltage dissipation occurs again during the rest of the frame. This is how circuits in the TFT-LCD work in two frames.

If we set the V_{com} voltage level somewhere like Figure 7.2, the effective voltage of the pixel electrode at the odd frame (positive polarity) is proportional to the ▨ filled area. And the effective voltage which is applied at the even frame (negative polarity) is proportional to the ▧ filled area. Therefore the ratio between odd and even frame's effective voltage will be changed when V_{com} voltage or R-DC of PI/LC layers are changed.

The kickback by the change of gate voltage always causes to the voltage drop. Therefore the alteration of R-DC or the kickback voltage will cause the DC effect. On the other hand, the alteration of the VHR or pretilt angle will cause the AC effect.

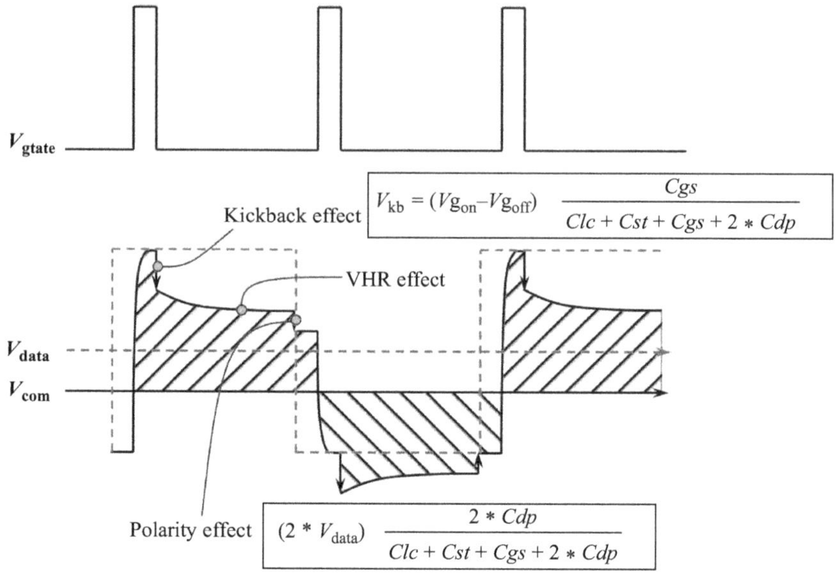

Figure 7.2 Driving scheme of TFT-LCD

7.3 Cause of image sticking

Causes of image sticking can be categorized according to the change of (1) pretilt angle (Δpretilt), (2) residual-DC (ΔR-DC) and (3) VHR (ΔVHR). First of all, we will explain how each factor induces an image sticking.

7.3.1 Pretilt angle

LC in Patterned VA [7] mode LCD aligns vertically at the alignment layer (We assume the pretilt angle of this status to be 90°). But LC in optically aligned VA mode [8] or polymer-stabilized VA (PSVA) [9] mode has some pretilt angle. The voltage–transmittance curve (we call *V–T* curve from now on) of this kind of VA mode LCD depends on the pretilt angle like following Figure 7.3. If the pretilt angle is close to 90°, the V-T curve is shifted in the right direction. It means that lower pretilt angle makes higher transmittance with the same voltage. This tendency is definite near 2~3 V.

Therefore if there is some change on the pretilt angle after aging, it could be a cause of image sticking.

7.3.2 Residual DC

In general, materials of LC and alignment layer for TFT-LCD are electrically neutral. But either LC or alignment layer or both have local charges for some reasons, these local charges make an internal electric field. Then effective electric field in the LC layer will be increased or decreased to depend on whether the

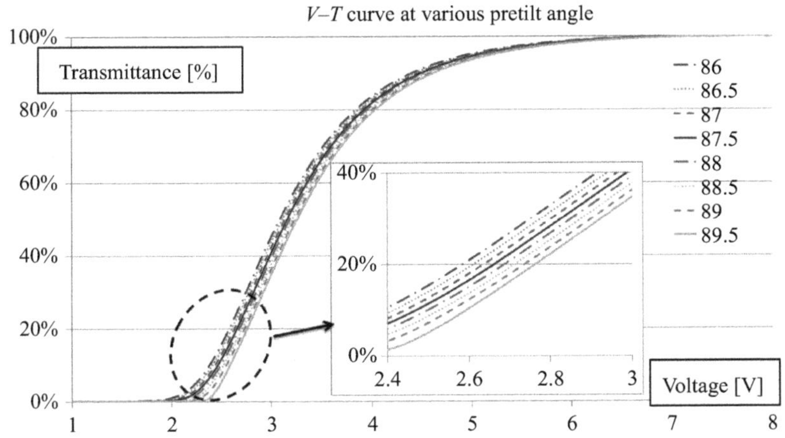

Figure 7.3 The V–T curve at various pretilt angle simulated by Twist Cell Optics Ver. 4.1 developed by former Prof. Jack Kelly at Kent State University

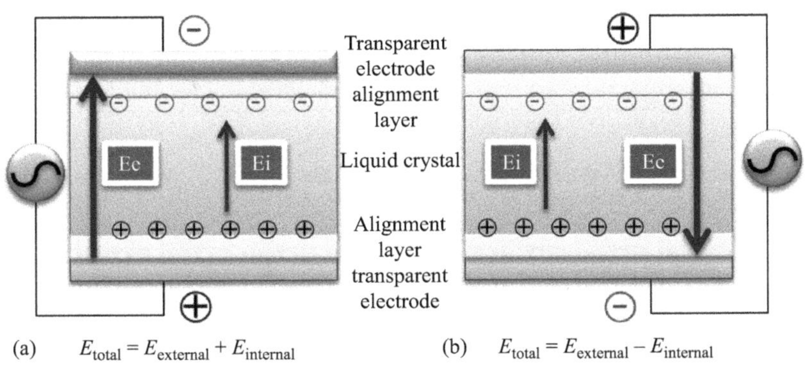

Figure 7.4 Internal (induced by R-DC in LC or PI layer) and external electric field (induced by source driver IC)

direction of the applied external electric field is matched or not as shown in the Figure 7.4.

Although R-DC values of (A) and (B) in Figure 7.4 are identical, the total strength of the electric field (E_{total}) are different at the positive and negative frame. Because, if the direction of external electric field matches with the internal electric field by R-DC (case (A)), the strength of the total electric field is a summation of internal and external electric field. On the contrary, if the direction of the external electric field is opposite direction of the internal electric field, then the total electric field is calculated to subtract the internal electric field from the external electric field.

Therefore, if an R-DC value is changed for some reason [10] after aging, it could be a cause of image sticking.

7.3.3 Voltage holding ratio (VHR)

When the gate is OFF, TFT should isolate LC and storage capacitor's charge from leaking. VHR is related to the index to indicate how much to keep the charge during the frame time. And it is very important in TFT-LCD. If the VHR is low, effective electric field applied in LC will be decreased.

When the VHR of some pixels drops, their transmittance should be decreased than other pixels. This phenomenon cause to (1) decrease transmittance, (2) decrease contrast ratio, (3) increase driving voltage and power consumption and (4) cause various reliability issues [11].

In general, the definition of VHR is restricted only for the characteristics of LC and alignment layer. But in this section, we would like to apply this terminology on how much charge is kept from leaking not only the LC and alignment layer but also TFT and storage capacitor in the pixel.

The threshold voltage of a-Si TFT can be shifted due to induced voltage and illumination [12]. This threshold voltage shift can change the transmittance. When the gate is OFF, the leaking current that leaks through TFT is determined by off current (current between drain and source electrode when the gate is off voltage). If this leaking current is changed, the voltage induced in LC can be changed. Therefore, the transmittance can be changed. In the respect of current leakage, we regarded as the same characteristic for both VHR of LC and TFT leakage.

7.4 Transmittance change after aging

Figure 7.5 shows the measurement result of γ–curve before and after aging, at each (A) and (B) area of Table 7.1. The difference between the two area's γ-curve is recognized as image sticking. If this difference is bigger, the image sticking would be observed clearly. Although there are many quantifying methods of image sticking [13,14], we call this image sticking figure of merit (ISFoM) in this chapter.

7.5 AC and DC image sticking

AC causes of image sticking have been related with frame frequency. However, DC causes have not been related to (just like the internal field in Figure 7.4). Therefore if the transmittance is decreased (or increased) by AC causes, both positive and negative polarity have the same tendency. But if increase the transmittance at positive polarity (the size of ▨ filled area) due to DC causes then the transmittance of negative polarity (the size of ▧ filled area) will be decreased as described in Figure 7.2.

If the ratio between these two voltages (ratio between the size of ▨ and ▧ filled area) is changed, the transmittance at each frame could be different, although the average value of the induced voltage in LC at the positive and negative frame (or the total size of ▨ and ▧ filled area) is maintained.

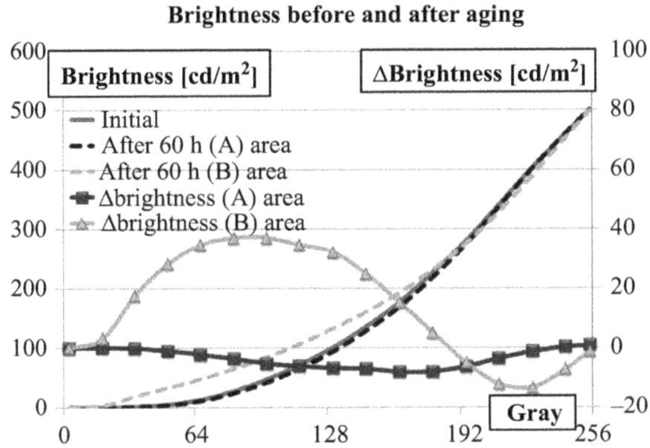

Figure 7.5 Gamma curve before and after aging at (A) and (B) area in Table 7.1. Aging time is 60 h

If V_{com} was set at the optimal V_{com} voltage and the additional voltage α due to DC cause was induced in the LC/PI layer, then the effective voltages at the positive and negative frame are changed to $V + \alpha$ and $V - \alpha$, respectively.

1. If the response time of LCD is fast enough compared to the frame time, the temporal average of transmittance can be a just average value of transmittance at the positive and negative frame. If V is classified as the steep/linear/gentle region of the V–T curve, then the transmittance at $V(T_V)$ is lower/identical/higher than the average value of $T_{V+\alpha}$ and $T_{V-\alpha}$, as shown in Figure 7.6.
2. Otherwise, transmittance can be strongly affected by DC alternation and decided as RMS value of $T_{V+\alpha}$ and $T_{V-\alpha}$.

7.5.1 Transmittance dependence on V$_{com}$ *voltage*

As mentioned before, if there are changes of R-DC in LC or alignment layer or storage capacitor, these ΔR-DC can induce the effective voltage change in the pixel. This effective voltage change can be compensated by modifying the V_{com} voltage. Also, if DC image sticking occurs V_{com}-Brightness curve will be moved horizontally. While AC image sticking will be moved vertically in the V_{com}-brightness curve.

7.5.2 Separation of AC and DC image sticking

Figure 7.7 shows V_{com}-brightness dependency of at initial and after 60 h of aging at (A) and (B) area. From this relationship, both (A) and (B) area has the different contribution of AC and DC image sticking, V_{com}-brightness curve is shifted to the x-direction if DC causes are affecting on. While the curve is shifted to the y-direction if AC causes are affecting on.

From Figure 7.7, if we change the V_{com} value of the panel, the image sticking due to DC cause can be changed or even be vanished at the specific V_{com} value. V_{com}-brightness curve at area (A) is shifted about −300 mV and −2.3 cd/m^2 to the x- and y-directions, respectively. Also at (B) area, this curve is shifted about

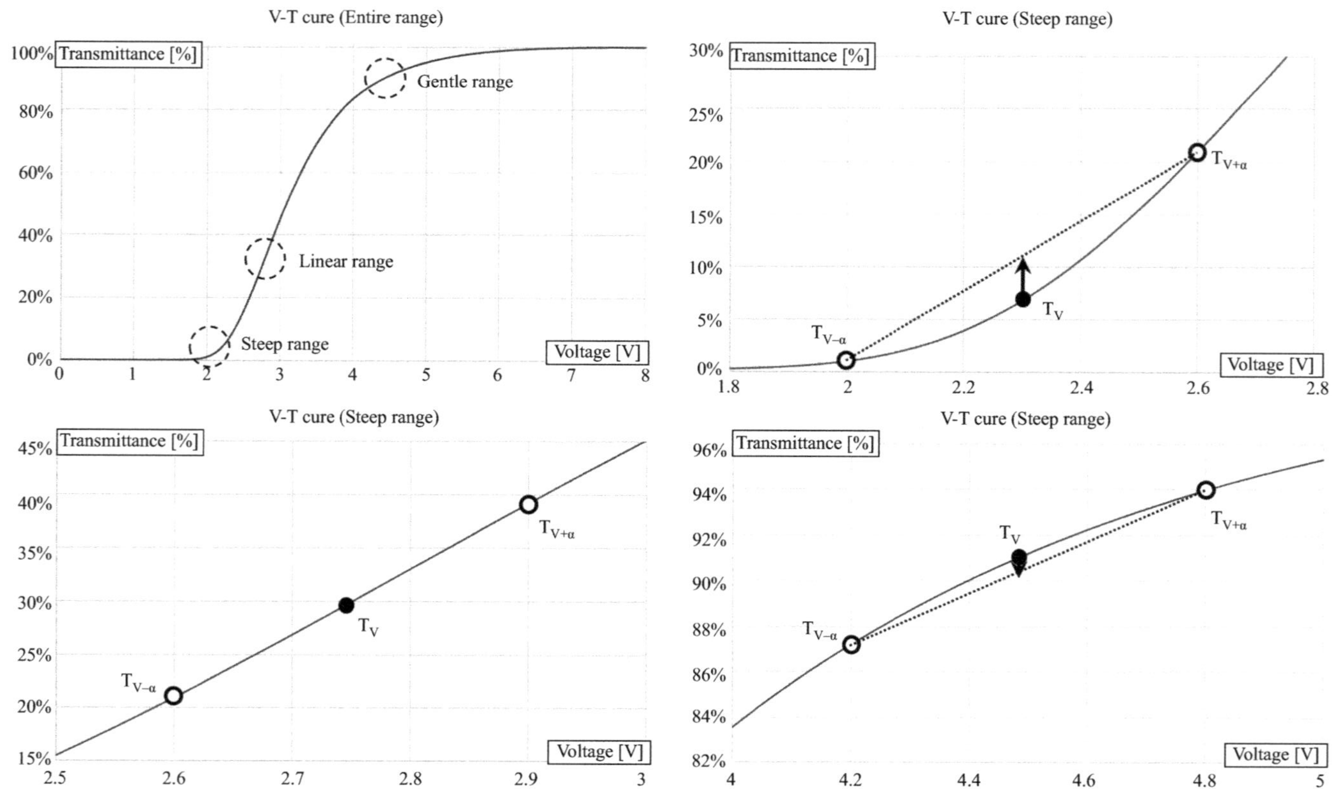

Figure 7.6 Transmittance change due to DC causes when the response time is fast enough. At steep/liner/gentle region of the V–T curve

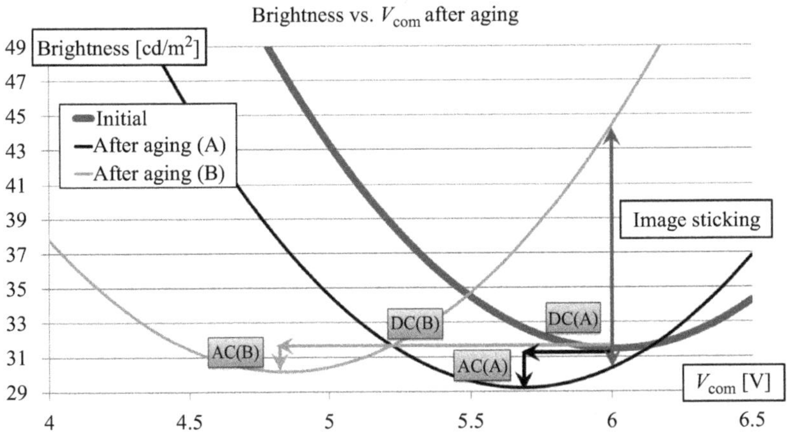

Figure 7.7 Brightness dependency of the V_{com} voltage

−1,100mV and −1.3 cd/m². This analysis was suggested to calculate how much each DC and AC cause is contributed to image sticking.

7.6 Image sticking due to ΔR-DC

To obtain the transmittance differences due to ΔR-DC, it is necessary to know how much R-DC is changed. To know this, we should measure the optimal V_{com} voltage before and after aging[1]. At the optimal V_{com} voltage, the size of the ▨ filled area and the ▧ filled area are should be equal. When we set the V_{com} voltage at the optimal voltage, the difference of the effective voltage between the positive and negative frame is minimized and the flicker can be minimized.

The optimal V_{com} voltage could be measured by flicker or brightness minimization method. In this section, the optimal V_{com} voltage was measured by flicker minimization method as shown in the Figure 7.8.

The optimal V_{com} voltage is the V_{com} voltage where the flicker is minimized. We should measure this at each gray.

Figure 7.9 shows how much ΔR-DC can be changed by the different aging condition. Each (A) and (B) area has the same aging time, while each display pattern is 0 and 255 gray, respectively.

After 60 h of aging, R-DC in the (A) area is changed about 300 mV while R-DC in the (B) area is changed about 1,110 mV.

V_{com}-brightness curve is shifted −300/−1,100mV due to DC causes at (A) and (B) area, respectively, as shown in Figures 7.7 and 7.9. Therefore if we shift V_{com}-brightness curve into the reverse direction, we can compensate ΔR-DC effect in the V_{com}-brightness curve and remove the effect of ΔR-DC. V_{com}-brightness dependency before and after the ΔR-DC compensation applied is shown in Figure 7.10.

[1]We can also calculate the optimal V_{com} voltage from Figure 7.7 that how much V_{com}-brightness curve is shifted.

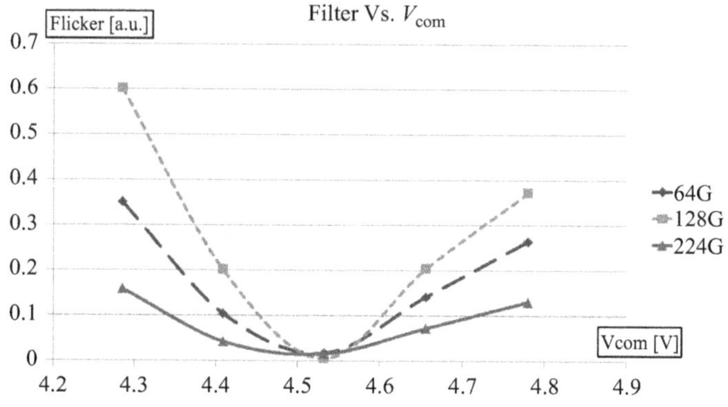

Figure 7.8 Optimum V_{com}: measured by flicker minimization method

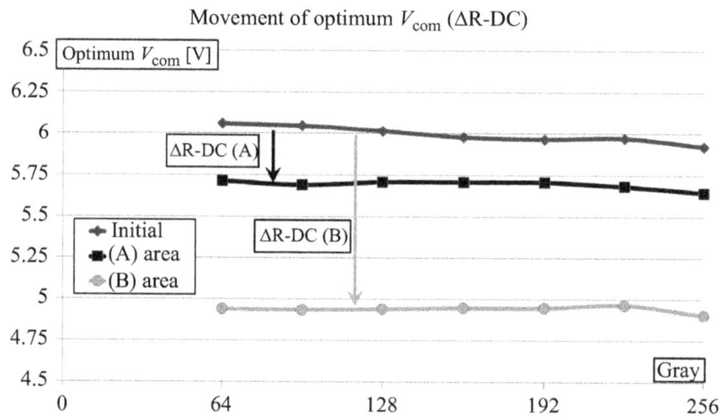

Figure 7.9 Movement of optimum V_{com} voltage depending on aging condition. Aging time 60 h

DC image-sticking contribution can completely be removed like the above method. And we can calculate the brightness value when image sticking due to DC cause is eliminated from the V_{com}-brightness curve.

This DC image sticking could be generated by (1) internal field change due to ion impurities in LC and PI layer, (2) temporal polarity relaxation mismatching of a multi-layer stack of dielectric layers (LC/PI/LC) [15], (3) V_{com} or V_{st} voltage (normally induced from the outside of panel) change or (4) optimal V_{com} voltage change, etc.

7.7 AC Image sticking due to Δpretilt

Figure 7.10 shows V_{com} dependency of brightness after aging and compensation of DC cause. The major causes of AC image sticking are Δpretilt and ΔVHR. First, we

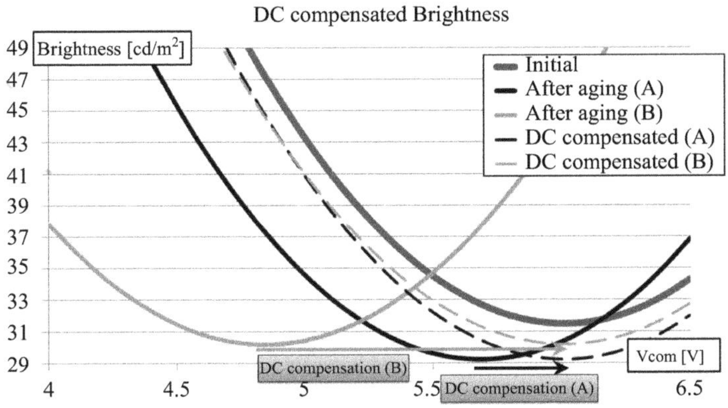

Figure 7.10 Movement of the V_{com}-Brightness curve depending on aging condition

Table 7.2 The brightness of initial and after aging, when image sticking due to DC cause is eliminated

Initial brightness	**DC compensated brightness after aging**	
Both (A) and (B) area	**(A) area**	**(B) area**
31.5	29.2	30.2

would like to separate out the effect of image sticking by the change of the pretilt angle.

Test cells, as well as real panels, were used to measure the pretilt angle of LC at the alignment layer. The test cell was designed to measure electrical and optical characteristics easier because it has no TFT. In addition to, we had already verified that the pretilt angle of the test-cell matched with that of real-panel.

From the test-cell measurement results, the pretilt angle of each area is changed (A) 89.5→89.5° and (B) 89.5→87.9°, respectively. It means that only pretilt angle at (B) area is lower after aging than before aging.

From Figure 7.11 when the pretilt angle changes from 89.5° to 87.9°, then the brightness will be 23.5 percent brighter at 128 gray. If we decrease the (B) area's brightness at the V_{com}-brightness curve 23.5 percent, we can eliminate or compensate the Δpretilt cause of image sticking as shown in Figure 7.12.

7.8 AC Image sticking due to ΔVHR

When VHR is low, charges stored in the storage and LC capacitor would be dissipated rapidly. If so, the effective voltage applied to the pixel will be decreased. Then the transmittance of the pixel is also changed. VHR of a panel may be

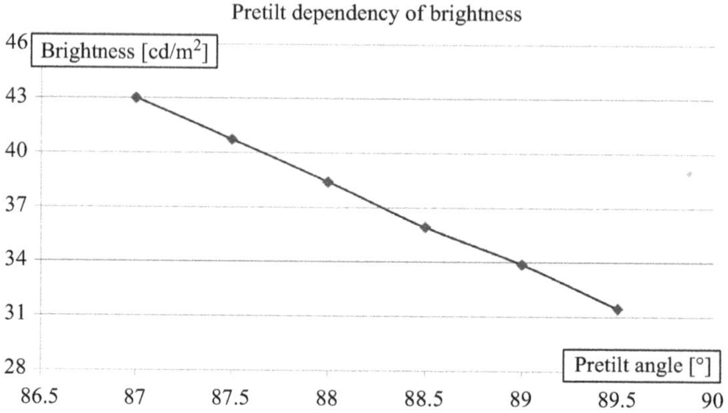

Figure 7.11 Pretilt angle dependency of brightness at 128 gray. Simulated by TCO (Test Cell Optics) S/W

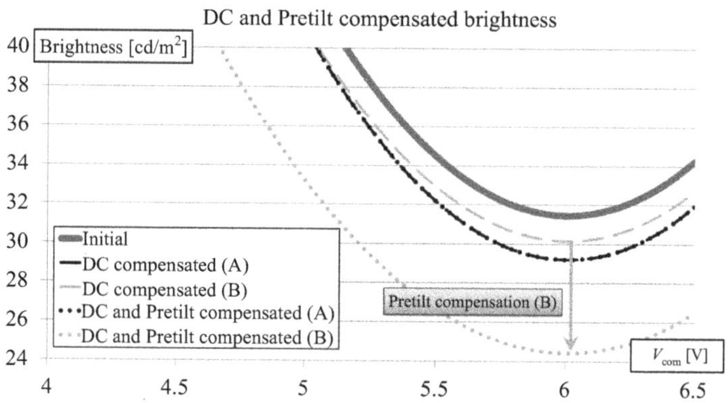

Figure 7.12 Brightness dependency of V_{com} voltage before and after aging. ΔR-DC and Δpretilt angle effects are compensated

decreased for various reasons, just like (1) increasing the number of ionic impurities in LC or (2) increasing the leakage current through TFT, and so on.

To know whether image sticking due to ΔVHR happens or not, the real-time brightness measurement could be a solution. Because direct measurement of the voltage level of the pixel electrode of TFT-LCD is very difficult. Real-time optical measurement of brightness gives us information about VHR of TFT-LCD [16].

These measurement results show that there are different leakage characteristics between (A) and (B) areas. Figure 7.13 describes that the decreasing rates of brightness between two areas are different. In Figure 7.13, (A) area's transmittance decreasing speed is rather fast. Therefore we could know that (1) the two area's VHR characteristics are different and (2) (A) area's VHR characteristics are lower. These characteristics are more easily observed when TFT-LCD is driven at the lower frame frequency.

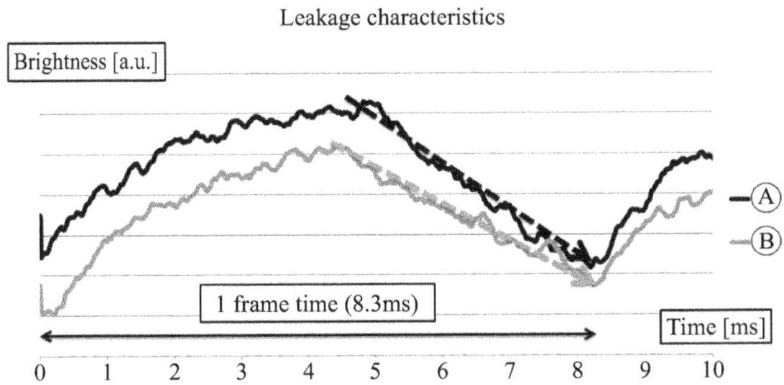

Figure 7.13 Comparison of leakage characteristics by optically measured at (A) and (B) area

Table 7.3 The brightness change due to each image sticking causes at 128 gray

	(A) area	(B) area
Initial	31.5	31.5
After aging	30.3	44.4
DC compensated	29.2	30.2
DC and pretilt compensated	29.2	24.4
DC and pretilt and VHR compensated	31.5	31.5

If causes of image sticking due to R-DC and pretilt angle change are eliminated, there should exist image sticking only due to VHR. The difference between the initial brightness versus DC and pretilt compensated brightness is image sticking only due to VHR. Then we can complete Table 7.3.

7.9 Improvement of image-sticking characteristics

When all the ΔR-DC, Δpretilt, and ΔVHR are zero, image sticking cannot be observed anymore. But actually, it is difficult to improve all of the causes of image sticking. If the image-sticking characteristics of some devices are worse than others and have to improve that, we recommend calculating the portion of brightness change before and after aging and calculating each portion of causes as shown in Table 7.3. Then, we can find which materials and process conditions are effective for improvement.

We would like to explain two cases for image-sticking improvement. From these cases, this analysis method could prove useful.

7.9.1 ΔVHR improvement

Assumption:Current status of ISFoM is identical with Figure 7.5 in Section 7.4. (2) ΔVHR can be improved 50 percent compared to the result of Section 7.4 at each

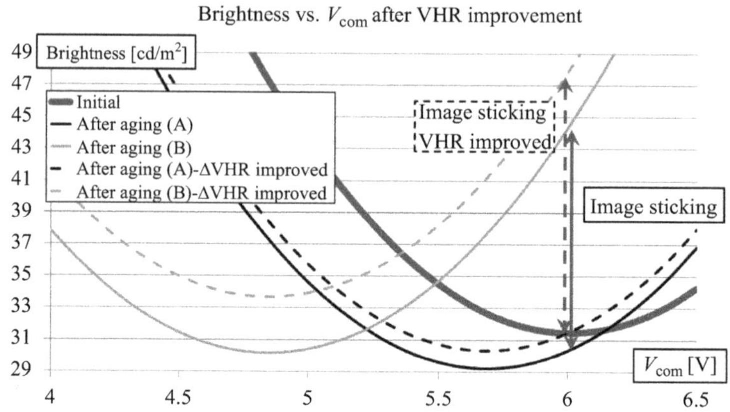

Figure 7.14 Transmittance dependency of V_{com} voltage after improving ΔVHR

(A) and (B) areas. (3) VHR can be improved by optimizing alignment material and process condition.

In Table 7.3, the brightness difference before and after aging due to ΔVHR/Δpretilt are −2.3/0 and −7.1/5.8 cd/m^2 at each (A) and (B) area. If ΔVHR can be improved by 50 percent compared to the previous one, this brightness values will be changed to −1.2/0 and −3.6/5.8 cd/m^2.

Before improvement of ΔVHR, the brightness difference due to AC cause (sum of brightness change due to ΔVHR and Δpretilt) is −2.3 cd/m^2 and −1.3 cd/m^2 at (A) and (B) area, respectively. After improving ΔVHR 50 percent, the differences are changed to −1.2 and 2.2 cd/m^2. Therefore, DC and pretilt compensated brightness curve is changed from Figure 7.12 to Figure 7.14. Normally improvement of image sticking is expected when ΔVHR is improved. But in this case, image sticking (brightness difference between (A) and (B) area) without DC cause is worse than before (1.0→3.4 cd/m^2). And image sticking with AC and DC causes is also worse than before as shown in Figure 7.14.

ΔVHR decrease brightness at (B) area and Δpretilt increase that at (B) area. Therefore, brightness difference due to Δpretilt is compensated due to ΔVHR and image sticking is improved. But with ΔVHR improvement, brightness difference due to Δpretilt less compensated compared to earlier. Therefore the image sticking goes worse than before.

7.9.2 ΔR-DC improvement

Assumption: (1) Current status of ISFoM is identical with Figure 7.5 in Section 7.4. (2) ΔR-DC was improved equivalent value (800 mV) in both (A) and (B) area. (3) This improvement was done by improving the alignment layer material or optimizing the alignment process condition or driving method.

In this case, ΔR-DC improvement is effective for improving image sticking. The larger ΔR-DC difference is, the bigger the brightness difference between (A) and (B) area is. Moreover, if ΔR-DC of each (A) and (B) area change from 300 and

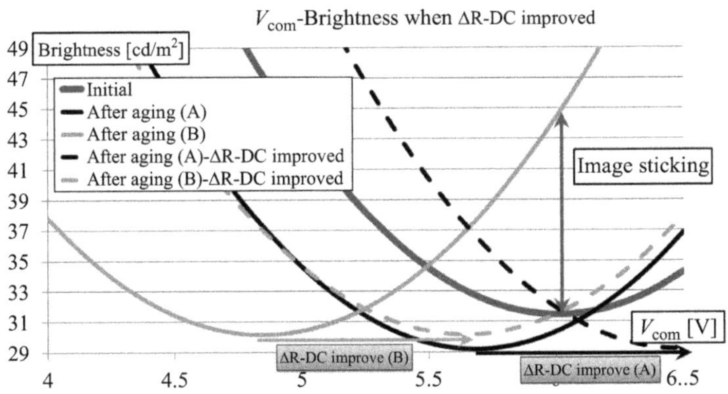

Figure 7.15 Transmittance dependency of the V_{com} voltage after improving ΔR DC

Table 7.4 Brightness contribution due to each image sticking causes at 128 gray

	DC	Pretilt	VHR	Total
(A) area	1.1	0.0	−2.3	1.2
(B) area	14.2	5.8	−7.1	12.9

1,100 mV to −500 and 300 mV, respectively, then image sticking cannot be observed anymore, although Δpretilt and ΔVHR characteristics are the same as shown in Figure 7.15.

From these two examples, it is important to calculate the relationship and contribution of each cause of image sticking. We can calculate how much image sticking could be improved if we improve one of the causes of image sticking. And we can choose the effective cause to improve image sticking in the current situation.

7.10 Conclusion

Using this analysis method, we could calculate how much ΔR-DC, Δpretilt, and ΔVHR contribute to image sticking as shown in Table 7.4.

It is important to calculate the contribution of each cause of image sticking to improve this issue.

Acknowledgments

1. This work presented at SID 2017 in Los Angeles, USA.
2. Data in this section is measured by test products for test purpose only.
3. Thanks to Mr. Hiroyuki Kamiya (FPD Solutions Co. Ltd), who gave valuable advice for this research.

References

[1] Stojmenovik G. 'Ion Transport and Boundary Image Retention'. The thesis of Ph.D., University of Gent, 2005.
[2] Park G.H, Lee J.S, Kwon N.O and Shin S.T. 'Measurement Method for Image Sticking using CSF'. IDW 2011 proceedings, pp. 333–36, 2011.
[3] Liu L, Liao Y, Song J, *et al.* 'Research on Line Image Sticking for Large Size High Resolution TFT-LCD Panel'. IMID 2015 Digest, pp. 341–45, 2015.
[4] Tsutsui K, Sakai T, Goto K, Sawahata K, Ishikawa M and Fukuro H. 'An Image Sticking-Free Novel Alignment Material for IPS-LCD'. Digest of SID, Vol. 34, Issue 1, pp. 1166–69, 2003.
[5] Park Y.G, Kim S.K and Lee E.S. 'A Study on Reducing Image-Sticking Artifacts in Wide-Screen TFT-LCD Monitors'. *Journal of SID*, Vol. 15, Issue 11, pp. 969–73, 2007.
[6] Mizusaki M, Miyashita T, Uchida T, Yamada Y and Ishii Y. 'The Mechanism of Image Sticking on LCD and Its Evaluation Parameters Related to LC and Alignment Materials'. Digest of SID, Vol. 37, Issue 1, pp. 673–76, 2006.
[7] Kim K.H and Souk J.H, 'Patterned Vertical Alignment (PVA) Mode for High Performance LCD Monitors'. Proceedings of Euro Display 99, pp. 115–19, 1999.
[8] Miyachi K, Kobayashi K, Yamada Y and Mizushima S. 'The World's First Photo Alignment LCD Technology Applied to Generation Ten Factory'. Digest of SID, Vol. 41, Issue 1, pp. 579–82, 2010.
[9] Kuo Y.P, Jiang S.C.F, Liu L.P, Lin C.H, Lin C.C and Hu C.J. 'Novel Advanced MVA Technology for Mobile Application'. IDW 08 Digest, 1573, 2008.
[10] Mizusaki M, Miyashita T and Uchida T. 'Generation Mechanism of Residual Direct Current Voltage in a Liquid Crystal Display and Its Evaluation Parameters Related to Liquid Crystal and Alignment Layer Materials'. *Journal of Applied Physics*, Vol. 102, Issue 1, pp. 014904.1–014904.6, 2007.
[11] Ishihara S. 'Factors Influencing the Voltage Holding Ratio'. *Technical Report of Sharp*, pp. 11–16, 2004.
[12] Powell M.J. 'The Physics of Amorphous-Silicon Thin-Film Transistors'. *IEEE Transactions on Electron Devices* , Vol. 36, Issue 12, pp. 2753–63, 1989.
[13] Chen P.L, Chen S.H and Su F.C. 'An Effective Method for Evaluating the Image-Sticking Effect of TFT-LCDs by Interpretative Modeling of Optical Measurements'. *Liquid Crystals*, Vol. 27, Issue 7, pp. 965–75, 2000.
[14] Su J.C. 'Quantifying the Image-Sticking Phenomenon for the Checkerboard Stimuli: Contrast, Spatial Frequency, Edge Effect, and Noise Interference'.

Proceedings of SPIE 7240, Human Vision and Electronic Imaging XIV, Vol. 7240, pp. 72401X.1–72401X.13, 2009.

[15] Nakazono Y, Takagi T, Ichinose H, Sawada A and Naemura S. 'Residual DC Evaluation of LC Materials with Negative Dielectric Anisotropy'. *Technical Report of ITE*, 22.16, pp. 19–24, 1998.

[16] Palmer S. 'The Optical Response of Liquid Crystal Cells to a Low Frequency Driving Voltage'. *Liquid Crystal Journal*, Vol. 36, Issue 10, pp. 587–98, 1998.

Chapter 8
New alignment process and its application

Munehiro Kimura[1]

Until recently, it has been thought that for LCD fabrication it was necessary to coat an LC alignment film, such as a polyimide (PI) film, on the substrate before filling the LC material. Generally, plastic substrates for flexible displays are incompatible with the thermal processes for curing a PI layer after the coating process on the substrate. Recently, LC alignment techniques that do not require a PI layer have been proposed. In this chapter, the slit coater method as a PI-free LC alignment technique is introduced.

8.1 Introduction

As described in Chapters 1 and 2, controlling the LC molecular alignment is critical for LCDs because the performance of the LCD depends on the regularity of the LC molecular alignment. The ordering of the LC molecules can be controlled by an external field such as electric, magnetic, elastic force or flow force fields. Most LCDs obtain optical contrast by a deviation from the state of initial LC alignment. The surface LC alignment on the substrate is governed by an alignment film. Generally, the alignment film should be prefabricated on the substrate before filling the LC material. The most common alignment film material is polyimide (PI), because of its versatility [1]. However, the prefabrication process for PI contributes significantly to the cost of manufacturing. If this process can be eliminated, the cost of materials, instruments and energy use can be reduced.

Various LC alignment techniques that do not require an LC alignment film have been demonstrated. Polymer-dispersed LCs (PDLCs) and polymer-stabilised blue phase (PSBP) LCs do not require an LC alignment film. In addition, various orientation techniques to replace the PI-rubbing method have been proposed. Among them, photo-alignment technology is the most promising, which will be described in Chapters 1 and 2. Photo-alignment films called 'photo-cross-linkable polymer LC films' composed using mesogen are the subject of active study [2]. Recently however, employing the intrinsic self-assembly in LC molecules, novel

[1]Department of Electrical, Electronics and Information Engineering, Nagaoka University of Technology, Japan

LC alignment techniques that do not use surface alignment films have been demonstrated. In either technology, guest additives play a very important role as a coagulant agent. To introduce the new alignment technique, the flow alignment for the intrinsic self-assembly method is described and a slit coater method using reactive mesogen is demonstrated.

8.2 Shear flow alignment

8.2.1 LC alignment in flow

Recently, a technique has been employed in which LC substances are enclosed by one-drop fill (ODF) dispensers, rather than using capillary action. A continuous flow of LC molecules induces self-assembly of the deposited LC molecules, called flow alignment. Early studies on this flow phenomenon focused on the anisotropy of the viscosity rather than the orientation of the molecules under the flow field. In the 1960–70s, F. M. Leslie, J. L. Ericksen and O. Parodi established a sophisticated hydrodynamic theory for uniaxial nematic LCs. Their formulae and notations are regarded as standard (called LEP theory) and are detailed in major textbooks on LC science [3]. Here we follow the common expression of the flow properties. Among the six Leslie coefficients α_i ($i = 1$ to 6), which are inherent values for LC materials, α_2 and α_3 are important coefficients that depend on the anisotropy in the LC material. Negative values of α_2 and α_3 are the most frequent combination for thermotropic nematic LCs consisting of rod-like molecules. When the flow velocity gradient is set parallel to the LC flow, as shown in Figure 8.1, the flow alignment angle θ_0 can be expressed as

$$\tan^2\theta_0 = \alpha_3/\alpha_2. \tag{8.1}$$

Owing to the respective orders of magnitude of α_2 and α_3, θ_0 is usually small, on the order of 0.1 rad [4], and appears to have planar alignment. The surface alignment determined by the flow is easily destroyed by thermal fluctuation of the nematic phase with low viscosity. Therefore, it is necessary to ‘memorise the flow-induced LC alignment’ by adopting a method to maintain the alignment. A promising technique is the method of solidifying the surface LC alignment using additives, as described in Section 8.2.2.

8.2.2 Flow-coating method

To fabricate a field-effect transistor (FET), typically a semiconductor layer consisting of organic material is stacked by an evaporation method under vacuum. It has been reported that flow-coating can form highly oriented and fairly uniform organic semiconductor material films, such as 6,13-Bis(triisopropylsilylethynyl) pentacene (TIPSPEN) even without using vacuum environments. These organic semiconductor materials exhibit a viscous smectic phase and/or crystal phase at room temperature, allowing the molecular order imposed by an external field to be maintained. Sakamoto *et al.* [5] have provided an explanation of the flow-coating

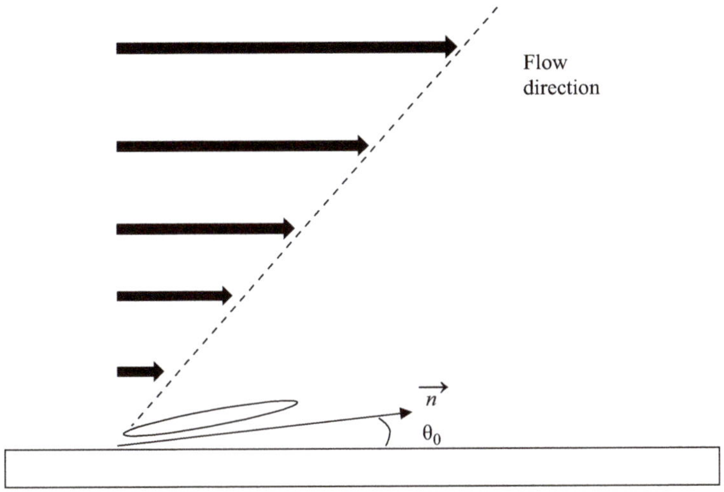

Figure 8.1 *Flow alignment induced by the gradient of flow velocity. The length of the thick arrows represents the flow velocity. The oval represents the LC molecule, $\vec{n}$ and θ_0 represent the LC director and the flow alignment angle*

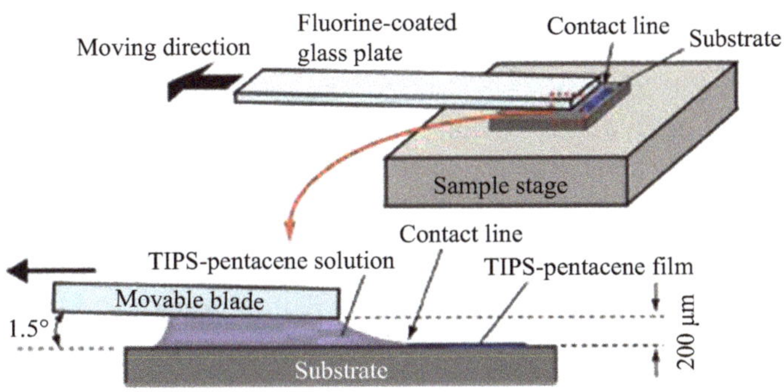

Figure 8.2 *Schematic illustration of flow-coating process with movable blade. Reprinted with permission from Reference [5]. Copyright © 2012 by the AIP (American Institute of Physics) Publishing LLC*

method, as illustrated in Figure 8.2. The flow coater consists of a movable blade, here a 13-mm-wide 2-mm-thick fluorine-coated glass plate, and a stationary sample stage. The movable blade is inclined at ~1.5° from the substrate surface plane, and the gap height between the blade and the substrate surface is set to 200 μm. After placing the leading edge of the movable blade above the start line of the coating, 16 μl of a 0.6 per cent by weight solution of TIPSPEN (Aldrich) in chloroform was poured into the blade/substrate gap by capillary action, and the flow-coating was immediately started with a constant blade speed of 200 μm/s. After the chloroform was volatilised, a TIPSPEN crystal layer was precipitated. The flow-coated films are composed of arrays of needle-shaped crystals whose long axes are aligned along the flow-coating direction. It was reported that the contact resistance, as well as the field-effect mobility, of bottom-contact/bottom-gate (BC/BG) type TIPSPEN FETs depends on the channel current direction with respect to the flow-coating direction.

8.3 LC alignment without using polyimide

8.3.1 Function of polyimide alignment film

In an LCD, the function of the alignment film is to maintain the LC alignment [1]. It has been found that even simply rubbing on the bare glass surface without a coated alignment film can assemble the LC molecules, though the rubbing-induced easy axis is easily destroyed by thermal fluctuations. PI alignment films are widely used due to the high quality of the surface LC alignment properties when they are used. The indices of the surface alignment quality are the surface anchoring strength and the surface order parameter. The surface anchoring strength expresses how strong the LC molecule is anchored on the alignment film, and strong surface anchoring is preferable. If the surface anchoring strength is very weak, no restoring force will act after the external driving voltage is turned off. In early LCD device development, PI alignment films with pre-rubbed surfaces were widely used. At present, replacement from classical rubbing method to UV photoalignment technique is progressing in production site for latest LCD panel. A UV photoalignment technique with a specially synthesised PI layer has been developed, though it has only recently started to replace the rubbing method because the surface anchoring strength and the surface order parameters have reached a satisfactory level for commercial products. Suitable PI materials for UV-photoalignment are being synthesised based on intermolecular interactions, though a unified explanation of the surface LC anchoring mechanism by PI has not been provided. LC alignment methods without using PI, such as the silicon dioxide vapour deposition [6] and the stretched polymer film technique [7,8], have also been developed and some part of these technologies are put into practice for LCD projector and/or sophisticated eyeglasses.

8.3.2 Reactive mesogen

A shortcoming of UV photoalignment PI is that the anchoring strength is relatively weak, especially for vertical alignment (VA). For the purpose of reinforcing the surface anchoring, the technique called polymer-sustained alignment (PSA) has been developed, where reactive mesogen (RM) doped with the host LC is applied near the surface alignment film by means of UV irradiation [9]. Figure 8.3 shows the fabrication process to form a polymer layer. An appropriate concentration of UV-curable monomer is added to a nematic LC with negative dielectric anisotropy and injected into an empty cell. Then, a voltage is applied to the cell while the cell is irradiated by UV light. This process can form polymer layers on the inside surface of the substrate. The polymer layer formed in this way can control the LC

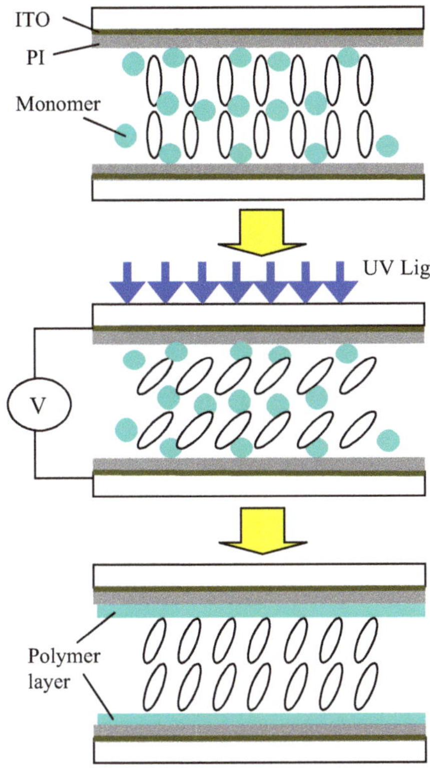

Figure 8.3 Fabrication process of polymer layer. Reprinted with permission from Reference [9]. Copyright © 2004 by John Wiley & Sons, Inc

alignment; in particular, the pre-tilt angle (polar angle) can be controlled by an applied voltage, monitoring the transmittance through the LCD. Additionally, azimuthal angle is controlled by adopting some skilful techniques such as rubbing method, protrusion method and diagonal electric field method. It is essential that the polymer network is locally segregated at the surface of the substrate, a technique known as phase separation. The segregated polymer network can solidify the LC molecules, which will generate the appropriate strong surface anchoring strength. The conditions of phase separation via spinodal decomposition determines whether the polymer network will be segregated in the bulk or near the substrate, which depends on the kind of RM, LC substance and UV irradiation conditions, such as power, duration and wavelength. For these purposes, acrylate monomer derivatives with a mesogenic structure are widely used (for example, UV-curable LC series provided by DIC, Japan, and RM257 provided by Merck, Germany).

8.3.3 PI-less alignment technique

If PI could be omitted in LCD production, the manufacturing cost would be significantly reduced and the thermal curing process for PI, which distorts the plastic

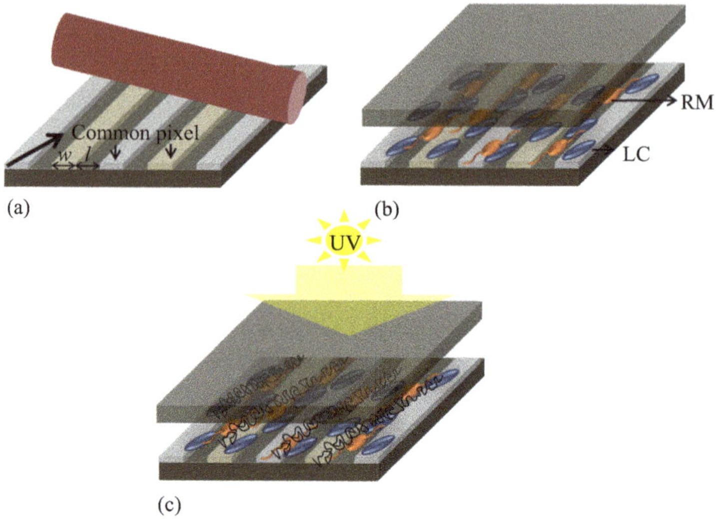

Figure 8.4 Fabrication method to achieve a homogenous alignment of LC: (a) Strong rubbing process is applied on the substrate, (b) The LC/RM mixture is homogeneously aligned, and (c) UV is exposed to the IPS cell to fix the LC molecules at both surfaces of the substrates. Reprinted with permission from Reference [11]. Copyright © 2014 by John Wiley & Sons, Inc

substrate for flexible LCDs, could be avoided. One promising PI-free technique involves dissolving dendrimers into a host nematic LC [10]. One of the attractive properties of dendrimers is that the spontaneous formation of VA on a glass substrate is realised by a uniformly adsorbed dendrimer layer that acts as an LC alignment layer after the surface LC alignment is solidified with additives after filling the host LC/dendrimer mixture into the LCD panel.

This solidifying additive technique is also applicable for planar alignment [11]. Initially, strong rubbing is performed on bare glass without a coated PI alignment layer. The two substrates are then attached to form an LCD cell. After filling the LC and RM mixture into the LCD cell via capillary force, homogeneous alignment of the LC and phase separation between the LC and RM molecules are expected to be obtained before UV exposure, as shown in Figure 8.4(b). Finally, UV is exposed to polymerise the RM at both substrates and thus fix the alignment in the formed polymer. Two kinds of RM (Merck, Germany) are used: powder form RM 257 and RM solution in a solvent, RMS03-013C. The weight ratio of LC and RM used is 99.6:0.4. The exposed UV light is 365 nm at 1.5 mW/cm^2, for 30 min. In the literature [11], it has been demonstrated that the electrooptical characteristics for cells fabricated by this method are comparable to those for conventional cells with rubbed PI.

Recently, a new approach to obtain homogeneous self-alignment of LC without using conventional PI alignment material was proposed that uses neither a PI coating process nor a rubbing process in the LCD fabrication. Details of the manufacturing process and the excellent electrooptical characteristics obtained are described in Section 8.4.

8.4 Slit coater method

8.4.1 Basic concept

In recent years, printed electronics and related technologies have attracted considerable attention. Among them, inkjet technology has been extensively studied. Both ODF and inkjet method are the same in terms of the materials transfer technology, namely an LC substance is applied to one substrate as drops before sealing the two substrates. It is difficult to arrange the LC alignment by these methods however. The 'slit coater' was developed as a way to control the molecular alignment exactly when the LC is dropped onto the substrate [12–14]. The slit coater is also called a die coater or a lip coater and is widely used for painting or spreading a fluid with uniform thickness. It should be noted that during slit coating, an LC bead discharged between the slit nozzle and substrate is subject to a shear flow force. Therefore, it would be possible to arrange the LC molecular alignment by means of the slit coater. It is unclear whether an insulator film would be still necessary if the PI film is omitted. This is a common problem not only for the slit coater method but for all PI-less techniques. Even if an insulator film is still required, it is only necessary to spread an insulating film between glass and ITO film beforehand.

8.4.2 Mechanism of slit coater method

The slit coater method is a technique that integrates the above-mentioned elemental techniques. Figure 8.5 shows the slit coater system and its schematic diagram, including a moving stage, slit nozzle and UV light source. The substrate on the moving stage is moved horizontally so that the LC molecule discharged from the nozzle coats the substrate surface, and simultaneously the LC molecular alignment can be self-organised by the shear flow force. A photograph of the experimental machine is shown in Figure 8.5(b). An enlarged side view around the nozzle is illustrated in Figure 8.5(c). UV irradiation from the bottom side of the substrate polymerises a thin LC layer adjacent to the substrate. A locally polymerised LC region acts as a molecular alignment film whose thickness is expected to be several tens of nanometres. Consequently, the slit coater can be used to achieve both LC enclosing and molecular alignment control. Note that after the LC coating and UV irradiation process, the top surface of the LC layer remains in the nematic phase rather than the polymerised phase, because the oxygen in the air inhibits a radical polymerisation reaction. Therefore, by combining these two substrates with LC layers, an LCD panel can be fabricated without injecting additional LC material. The LC alignment is brought about by the shear flow between the slit nozzle and the substrate. The process parameters for controlling the LC alignment are the slit width, slit length, coating gap, delivery rate of LC materials from the slit nozzle, stage movement velocity and UV irradiation intensity. In the preliminary

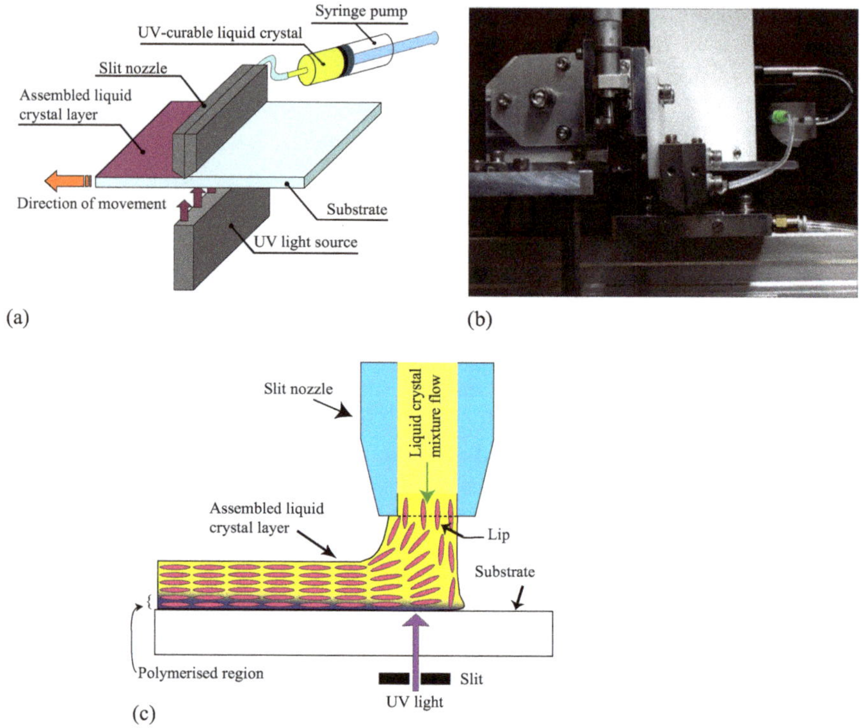

Figure 8.5 (a) Schematic illustration of slit coater system; (b) photograph of the experimental machine; and (c) illustration of shear flow LC alignment during the coating process. Reprinted with permission from Reference [12]. Copyright © 2012 by John Wiley & Sons, Inc

experiment, the slit width was fixed at 20 μm, the slit length at 13 mm and the coating gap at approximately 100 μm.

8.4.3 LC alignment and electrooptical properties

In the preliminary experiment, the host nematic LC used was ZLI-2293 (Merck) and the guest reactive mesogen used was UCL-011-K1 (DIC), called 'UV-curable LC'. The mixing rate of the reactive mesogen was 1–5 per cent by weight. The LC mixture was delivered using a syringe pump to the slit nozzle at a fixed rate. The stage was also moved at a fixed rate using a controlled stepper motor. As a UV light source, a high-pressure mercury lamp with a central wavelength of 365 nm was employed (SP-9, Ushio). The stage moving velocity was 0.50 mm/s and the UV irradiation intensity was 100 mW/cm^2. Figure 8.6 shows photographs of the coated LC films, where the coating direction is parallel to the crossed polariser and is 45° with respect to the polarisation axis of one polariser. From these photographs, a uniform and uniaxial alignment is observed. It is found that the extinction direction

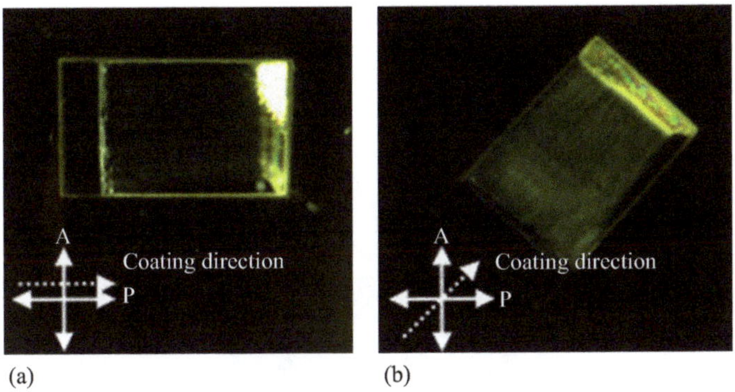

Figure 8.6 Photographs of the LC-coated substrate: (a) coating direction parallel to the polariser under crossed nicols and (b) coating direction is 45° with respect to the polariser under crossed nicols. Reprinted with permission from Reference [13]. Copyright © 2011 by the Japan Society of Applied Physics

is parallel to the coating direction of the LC layer, which implies that a uniaxial planar alignment of the LC director is formed.

Figure 8.7 shows a numerical simulation of the liquid flow around the nozzle based on the Navier–Stokes equations, examined with commercially available software (ANSYS Fluent, ANSYS Inc., USA) [12]. The Navier–Stokes equations are discretised using the finite volume method. In this simulation, to simplify the calculation, an isotropic liquid with a viscosity of 160 mPa.s was assumed, whereas the LC is actually an anisotropic fluid. The calculation parameters assumed were as follows: the thickness of the LC layer was 2 μm, the gap between the lip and substrate was 100 μm, the contact angle between the substrate and fluid was 30°, the contact angle between the lip and fluid was 60°, the surface tension was 30 mN/m and the moving velocity of the substrate was 10 mm/min. Figure 8.7(a) shows the shape of the fluid bead between the slit coater lip and the substrate. It has been found that the meniscus radii in the front and rear are approximately 0.2 and 0.3 mm, respectively. Figure 8.7(b) shows the pathlines of the fluid, which represent the trace of the fluid in a certain duration of the reduced time. The pathline following the nozzle depicts the fluid flow of the upper LC layer, which shows that the LC molecule reaches the surface of the bead (interface between the air and fluid) and the flow velocity is relatively low. The line directly running toward the substrate depicts the fluid flow of the lower LC layer, which shows that the flow velocity is relatively high. It is evident that there is a flow velocity gradient between the lip and substrate, which induces the shear flow described in the previous subsection 8.2. Figure 8.7(c) shows a snapshot around the lip of the slit coater taken by a high-speed camera (Fastcam MC24, Photron), in which a nematic mixture composed of a fluorinated derivative was used. Comparing Figure 8.7(a) and (c), it is clearly seen that the shape of the bead taken by the high-speed camera is similar to that depicted by the numerical calculation. Although the fluid dynamics of an anisotropic fluid such as

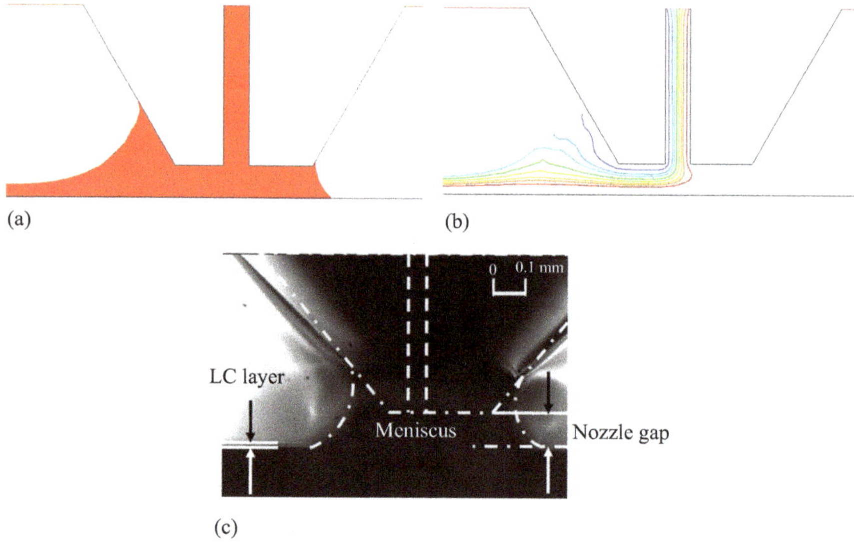

Figure 8.7 Aspect of shear flow: (a) shape of the fluid bead between the slit coater lip and the substrate by numerical simulation; (b) pathlines of the fluid, which represent the flow velocity between the slit coater lip and the substrate; and (c) snapshot around the lip of the slit coater taken by high-speed camera. Reprinted with permission from Reference [12]. Copyright © 2012 by John Wiley & Sons, Inc

the nematic LC is a complicated subject, the flow velocity gradient in the numerical simulation suggests the existence of shear flow near the substrate, which induces the LC molecular alignment.

From the viewpoint of productivity, the coating speed should be as fast as possible. Figure 8.8 shows microphotographs of LC alignment using a polarised optical microscope (POM), where the nominal coated LC thickness was 2 μm and the coating velocity in the range 1.0 to 3.0 mm/s. As shown in Figure 8.8(a), when the coating velocity is 1.0 mm/s, fairly homogeneous LC alignment is obtained and the order of uniformity is as good as an LC cell using photo-alignment PI [14]. When the coating velocity is increased up to 3.0 mm/s, patterns like cursive script appear periodically, as shown in Figure 8.8(b) and (c). This periodic pattern is called a loop defect. It is found that the pattern cycle becomes shorter as the LC coating velocity is increased. It is assumed that this loop defect is caused by turbulence induced by excessive shear flow, and the increase in coating velocity increases LC turbulence and the contrast ratio of the LCD deteriorates due to light scattering around these loop defects.

Also of interest is whether the origin of this pattern is in the bulk LC or a boundary layer in the vicinity of the substrate surface and whether the LC alignment is retained by a UV photo-polymerised layer. To verify this question, a thermal cycle was applied to the samples. Figure 8.9 shows microphotographs of

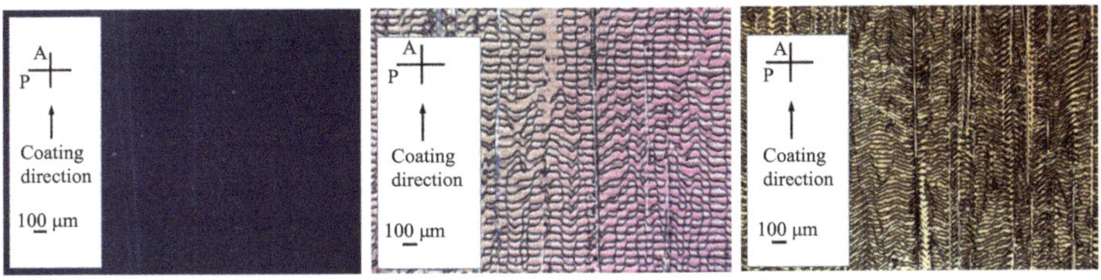

(a) (b) (c)

Figure 8.8 POM image of LC alignment fabricated by slit coater. In these photographs, A and P represent the optical axes of the analyser and polariser, respectively. The coating direction is parallel to the polariser: (a) coating velocity of 1.0 mm/s; (b) coating velocity of 2.0 mm/s; and (c) coating velocity of 3.0 mm/s. Reprinted with permission from Reference [14]. Copyright © 2017 by the Society of Photopolymer Science and Technology

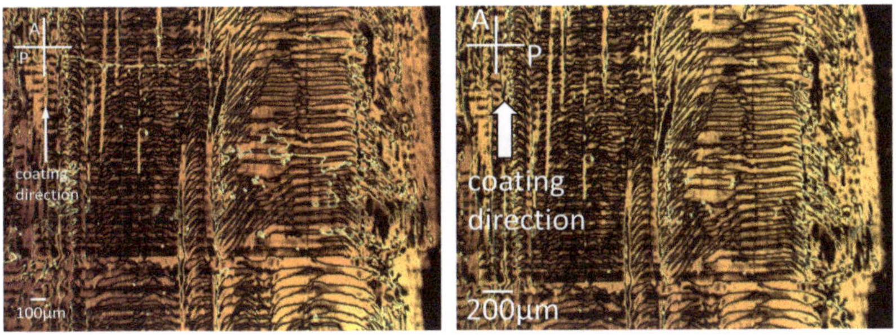

(a) (b)

Figure 8.9 POM image of LC alignment fabricated by slit coater: (a) right after fabrication and (b) once heated above nematic–isotropic phase transition temperature (114 °C), then cooled to room temperature. Reprinted with permission from Reference [14]. Copyright © 2017 by the Society of Photopolymer Science and Technology

LC alignment fabricated by slit coating before and after heating above the nematic–isotropic phase transition temperature ($T_{NI} = 114$ °C), where the nominal coated LC thickness was 2 μm and the coating velocity was 2.0 mm/s. Before heating, many loop defects were evident in the LC texture, as shown in Figure 8.9(a), because the coating velocity was too fast. After the sample was heated above T_{NI}, the nematic LC alignment melted and the loop defects disappeared. After the temperature was decreased from T_{NI}, almost the same texture of loop defects was reproduced, as shown in Figure 8.9(b). These results suggest that the LC alignment is memorised by the polymerised RM.

For LCDs, the pretilt angle and surface anchoring energy are important device parameters. Figure 8.10 shows an experimental result for the dependence of the

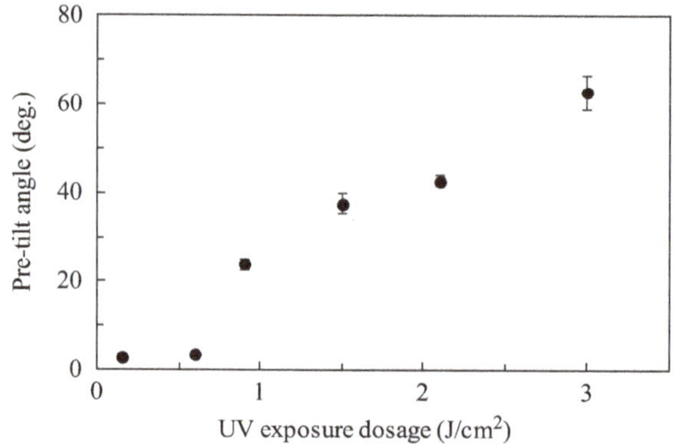

Figure 8.10 Dependence of pretilt angle on UV exposure dosage. Reprinted with permission from Reference [15]. Copyright © 2016 by the Institute of Image Information and Television Engineers

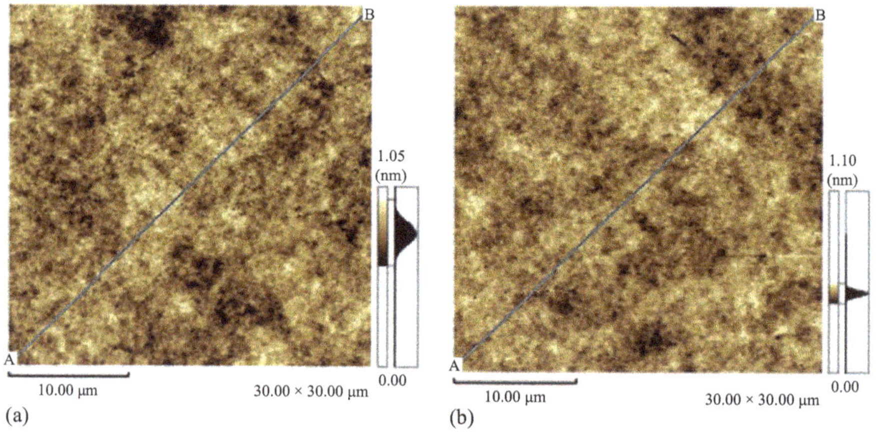

Figure 8.11 Surface topology of RM257: (a) before UV irradiation and (b) after UV radiation dosage of 3 J/cm^2. Reprinted with permission from Reference [15]. Copyright © 2016 by the Institute of Image Information and Television Engineers

pretilt angle on UV exposure dosage [15], where the host nematic LC was MLC-14100-000 (Merck) and the dopant was 1.0 per cent by weight of RM-257 (Merck). It is found that the pretilt angle increases with increasing UV dosage. Comparable results were obtained using different additive mesogen (UCL-011-K1, DIC). According to research procedures on photo alignment films, studying the surface roughness and surface energy may shed light on this phenomenon. Figure 8.11 shows atomic force microscopy (AFM) observations, where the RM doped was RM257 (Merck). After the RM and NLC (MLC-14100-000, Merck) mixture was

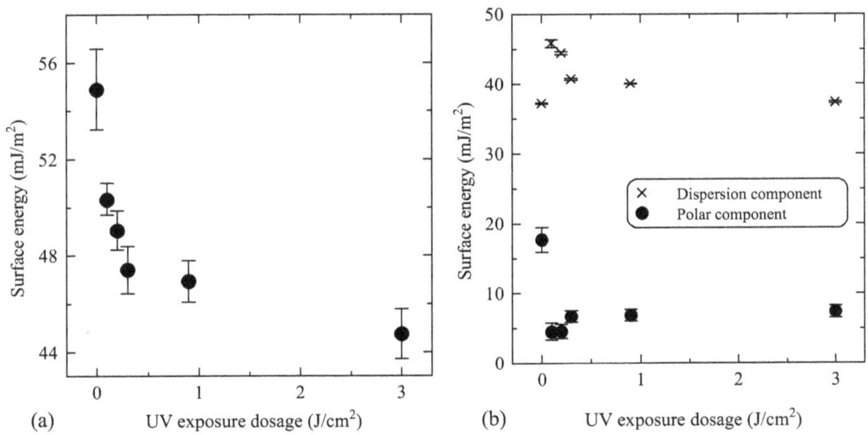

Figure 8.12 (a) Surface energy of RM257 and (b) dispersive and polar components of surface energy. Reprinted with permission from Reference [15]. Copyright © 2016 by the Institute of Image Information and Television Engineers

coated on the glass substrate, UV light of 100 mW/cm^2 was irradiated for 30 s (Figure 8.11(b)). Note that the LC mixture was not removed, to avoid any damage due to washing. The true boundary between the RM and LC could then be observed by a special technique called 'in-LC AFM'. As shown in Figure 8.11, no significant difference before and after UV irradiation could be confirmed.

Figure 8.12(a) shows the dependence of the surface free energy on the irradiation UV dosage. Figure 8.12(b) shows the UV dosage dependence of the dispersive and polar components of surface energy. It is deduced that a high pretilt angle is induced by the surface energy rather than by the surface topology. However, it is not clear yet whether the cause of the high pretilt angle is the dispersive or polar components of the surface energy, because the changes in surface energy due to UV irradiation are not monotonic. To realise precise pretilt angle control, it is necessary to better determine the mechanism of surface LC alignment.

8.5 Future applications

This method is suitable for flexible devices, because no thermal process is required that might damage a plastic substrate. Figure 8.13 shows a photograph of a coated LC film using the prototype slit coater, where the slit length was 100 mm, slit width was 20 μm, coating gap was approximately 100 μm and the thickness of the coated LC layer was approximately 1 μm. It is found that the point at which the LC coating alignment started to be lost, where the slit nozzle initially approaches the substrate and the LC molecular alignment firstly discharges from the lip, cannot be controlled. This is unavoidable, and the same phenomenon occurs at the terminal point of the LC coating because no shear flow force is generated around the initial and

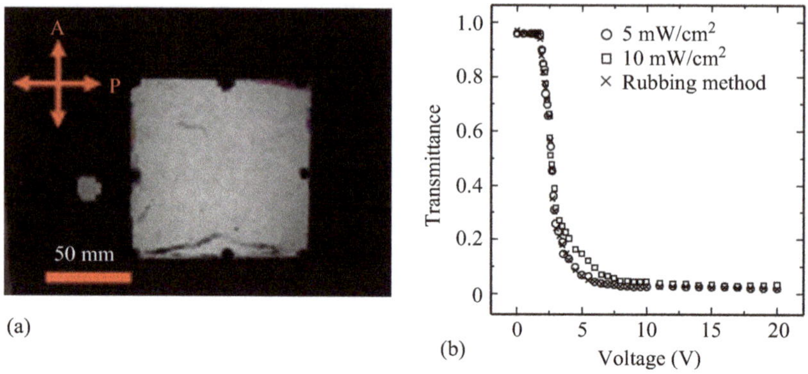

Figure 8.13 (a) Photograph of the aspect of TN-LCD and (b) electrooptical properties of TN-LCD, where the black circles represent points for UV irradiation of 5 mW/cm^2 and the squares represent UV irradiation of 10 mW/cm^2. The crosses represent results for a conventional rubbing cell. Reprinted with permission from Reference [12]. Copyright © 2012 by John Wiley & Sons, Inc

terminal portions. However, this issue can be addressed. For example, in the case of roll to roll process for long films, the initial and terminal portions of the substrate film can be cut off, and such removal of both ends is commonly performed in the process of manufacturing ordinary optical films.

Another interesting application is achieving a uniform lying helix (ULH) mode, which exhibits a fast response speed. Generally, it is very hard to uniformly align short-pitch cholesteric LC. PI alignment films alone are inadequate for this purpose and a supplemental field must be applied to the sandwiched LC cell to induce the ULH alignment after filling the short-pitch cholesteric LC. The slit coater method can be used to achieve this [16,17]. During coating of the short-pitch cholesteric LC on a substrate, an electric field of 400 V/100 μm, 1 kHz was applied between the slit nozzle and an ITO electrode on the substrate to align the helical axis parallel to the substrate to avoid the generation of a uniform standing helix (USH), which is disadvantageous in terms of the bulk free energy. Figure 8.14 shows photographs of a ULH mode cell made by slit coating an LC consisting of a mixture of short-pitch cholesteric LC ($\Delta\varepsilon > 0$, fluorine-based, Merck) and RM (UCL-011-K1, DIC). It was experimentally found that the helical axis is perpendicular to the coating direction. As described above, a feature of the slit coater method is that it allows modification of the LC alignment during the manufacturing process at room temperature.

The scope of applications of the slit coater method is not limited to LCD fabrication. For example, quantum nano-rods have attracted engineering interest for their potential as polarised-light sources, which can be a substitute for polariser films. To date, it has been difficult to fabricate large quantum nano-rod sheets [18]. However, as described before, slit coating can induce an anisotropic

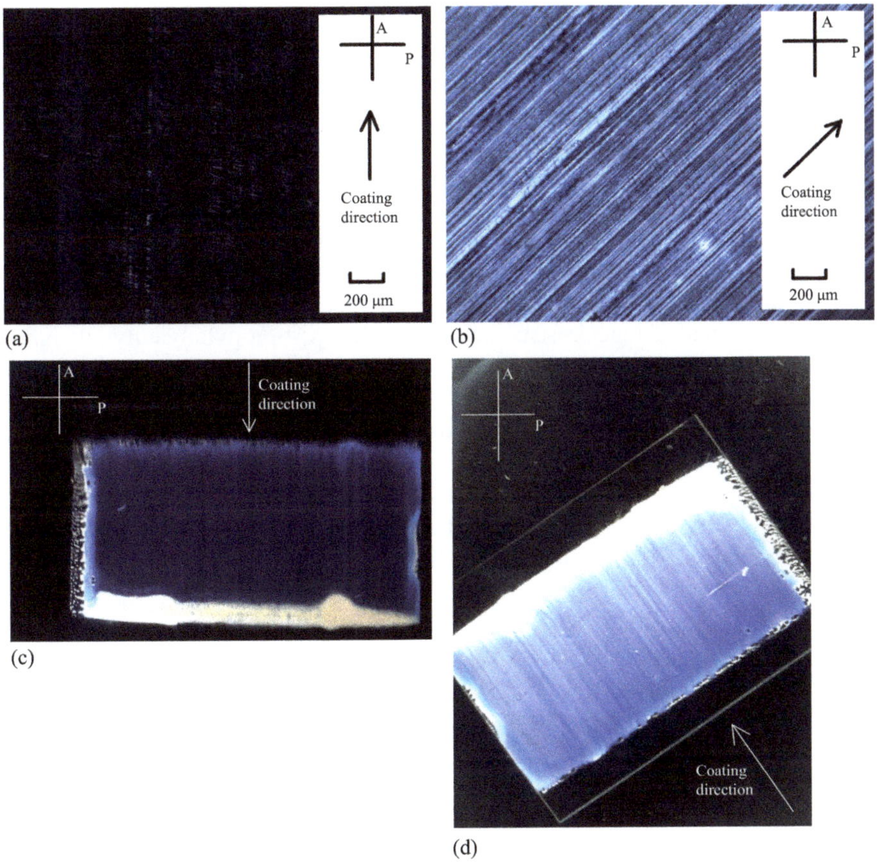

Figure 8.14 Microphotographs and aspect of ULH-LCD under crossed nicols: (a) and (c) coating direction parallel to the analyser; (b) and (d) coating direction 45° with respect to the analyser. The longer side width of the substrate is 100 mm. Reprinted with permission from Reference [17]. Copyright © 2016 by the Institute of Image Information and Television Engineers

flow field, which has the potential to align quantum nano-rods. If the problem of poor dispersibility of quantum nano-rods in an LC substance can be solved, the fabrication of quantum nano-rod sheets by the slit coater method will be realised in the near future.

References

[1] Nishikawa M. 'Design of polyimides for liquid crystal alignment films'. *Polym. Adv. Tech*. 2000, vol. 11(8–12), pp. 404–12.

[2] Kawatsuki N., Arita T., Kawakami Y. and Yamamoto T. 'Photoalignment of liquid crystals on photo-cross-linkable polymer films with anthracene at the end of the side group'. *Jpn. J. Appl. Phys.* 2000, vol. 39(10), pp. 5943–46.

[3] Demus D., Goodby J., Gray G. W., Spiess H.-W. and Vill V. (eds.). *Physical Properties of Liquid Crystals. London*: Weinheim: Wiley-VCH; 1999. pp. 356–58.

[4] Buka A. and Kramer L. (eds.). *Pattern Formation in Liquid Crystals.* New York: Springer; 1996. pp. 97–99.

[5] Sakamoto K., Ueno J., Bulgarevich K. and Miki K. 'Anisotropic charge transport and contact resistance of 6,13-bis(triisopropylsilylethynyl) pentacene field-effect transistors fabricated by a modified flow-coating method'. *Appl. Phys. Lett.* 2012, vol. 100(12), 123301.

[6] Cuypers, D., De Smet, H. and Van Calster, A. 'VAN LCOS microdisplays: A decade of technological evolution'. *J. Disp. Technol.* 2011, vol. 7(3), pp.127–34.

[7] Aoyama H., Yamazaki Y., Matsuura N., Mada H. and Kobayashi S. 'Alignment of liquid crystals on the stretched polymer films'. *Mol. Cryst. Liq. Cryst.* 1981, vol. 72(4), pp. 127–32.

[8] Araishi T., Yoshida Y., Kimura Y. and Ishihara S., 'Film substrate which needs no alignment layer for LC molecules'. IDW'15. 2015 pp.102–03.

[9] Hanaoka K., Nakanishi Y., Inoue Y., Tanuma S. and Koike Y. 'A new MVA-LCD by polymer sustained alignment technology'. SID'04 Digest. 2004, New York: Wiley; pp. 1200–3.

[10] Momoi Y., Kwak M., Choi D., *et al.* 'Polyimide-free LCD by dissolving dendrimers'. *J. SID.* 2012, vol. 20(9), pp. 486–92.

[11] Jeong I. H., Yu J. H., Jin H.-S., Song K. H., Kimura M. and Lee S. H. 'Novel approach to achieve conventional polyimide-less IPS/FFS LCDs'. SID'14 *Digest.* 2014, New York: Wiley; pp. 1418–20.

[12] Kimura M., Honda K., Yodogawa S., *et al.* 'Flexible LCDs fabricated with a slit coater: Not requiring an alignment film'. *J. SID*, 2012, vol. 20(12), New York: Wiley; pp. 633–39.

[13] Sato H., Miyashita K., Kimura M. and Akahane T. 'Study of liquid crystal alignment formed using slit coater'. *Jpn. J. Appl. Phys.* 2011, vol. 50(1), 01BC16.

[14] Kimura M. and de la Mora D. 'Study of liquid crystal alignment using slit coater under different application velocities'. *J. Photopolym. Sci. Technol.* 2017. vol. 30(4), pp. 457–61.

[15] Takahashi S., Kobayashi Y. and Kimura M. 'Continuous change of dispersive and polar components of surface energy: Irradiation on reactive mesogen'. IDW'16. 2016, Japan; pp.168–69.

[16] Kimura M. and Endo N. 'Uniform lying helix of cholesteric liquid crystal aligned by means of slit coater method with electric treatment'. *IEICE Trans. Electron.* 2016. vol. 99-C pp. 1240–43.

[17] Kouno T. and Kimura M., 'Uniform lying helix of cholesteric liquid crystal aligned by means of slit coat technique with electric treatment'. IDW'16. 2016, pp. 1432–33.

[18] Sitt A., Salant A., Menagen G. and Banin U. 'Highly emissive nano rod-in-rod heterostructures with strong linear polarization'. *Nano Lett.* 2011. vol. 11(4) pp. 2054–60.

Chapter 9

Liquid crystal self-alignment technology

Masanobu Mizusaki[1]

The preparation method for uniaxially homogeneous and vertical alignments of liquid crystal (LC) molecules was proposed without using a conventional alignment layer such as polyimide film. The method produces the self-alignment layer (SAL) by the polymerisation of the monomers in the LC layer above the T_{NI} of the LC material. The fringe-field-switching (FFS) and the vertical alignment (VA) modes LC cells with the SALs (self-FFS or self-VA cells) produced from the monomers exhibited equivalent level of alignment state and electro-optical properties compared with the conventional LC cells having the polyimide-film alignment layers. The self-FFS and the self-VA cells would be useful for next-generation displays such as flexible liquid crystal displays (LCDs) because the method does not need high-temperature process over 200 °C.

9.1 Introduction

Alignment control of LC molecules is important for several applications such as displays, sensors and transistors [1,2]. In most cases, the alignment film such as a polyimide film with rubbing process is used for the display application [2–5]. However, the preparation of the polyimide film usually requires large quantities of solvent, high-temperature process over 200 °C for imidisation, and rubbing process [5]. The rubbing process sometimes generates defects of the LC alignment. Hence, new alignment technologies with comparably low temperature process and without any alignment defects are desired, and conventional alignment layer-free technologies are expected to solve these problems.

Concerning about the application of the LC display (LCD), the market has been growing with many fields such as mobile phones, note PCs, tablets, car navigations and TVs, and then high image qualities are demanded with high contrast ratio, wide viewing angle and high transmittance. Therefore, several types of the LC alignment modes such as in-plane switching (IPS) [6], FFS [7] and vertical alignment (VA) [8,9] modes have been developed in late years. Above all, homogeneously aligned LC modes such as IPS and FFS are known to exhibit intrinsically

[1]Display Device Company, Sharp Corporation, Japan

wide viewing angle and high transmittance [6,7]. In contrast, the VA mode shows markedly high contrast ratio over 3,000:1 [9].

On the basis of the above background, the new alignment technologies without using the conventional alignment layer materials have been developed for both the homogeneous and the vertical LC alignments. The processes of the new alignment technologies will be explained in this chapter. The new alignment technologies introduced here are named 'LC self-alignment technologies' because they do not use the conventional alignment layer materials.

9.2 Fabrication process of the self-alignment LC cells

9.2.1 Homogeneous self-alignment

The process for fabrication of the homogeneous self-alignment FFS-mode LC cell (self-FFS cell) without forming the conventional alignment layer is illustrated in Figure 9.1. First of all, a pair of substrate is assembled with each other; one side of the substrate has pixel and common electrodes and insulating layer between the pixel and the common electrodes. Both the pixel and the common electrodes are formed from transparent indium-tin-oxide (ITO) electrodes. A mixture of LC material and an appropriate concentration of an additive, which induces the homogeneous alignment, is injected into an empty cell. Then, polarised ultraviolet (UV) light is exposed to the cell from the normal direction of the substrate having the electrodes in order to induce the homogeneous alignment with the SAL. During the exposure of the polarised UV light, the temperature of the cell should be optimised over the nematic-to-isotropic transition temperature (T_{NI}) of the LC material because the polarised UV light is shifted to depolarised one in the LC layer due to scattering.

The molecular design of the additive agent solubilised in the LC material is significantly important for the homogeneous self-alignment. The additive agents will be separated from the LC layer after exposing polarised UV light, and then produce the SAL. Figure 9.2(a) indicates the concept for the molecular design of the additive agent. The additive agent consists of two parts; one is a polarised UV responsive unit and the other is a polymerisable unit. The polarised UV responsive

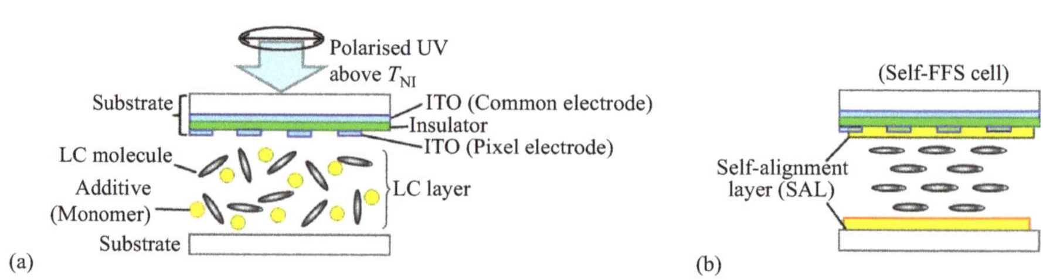

Figure 9.1 Schematic illustration of the fabrication process for self-alignment FFS-mode LC cell (self-FFS cell): (a) before preparation of the SAL and (b) after preparation of the SAL and obtaining the homogeneous alignment

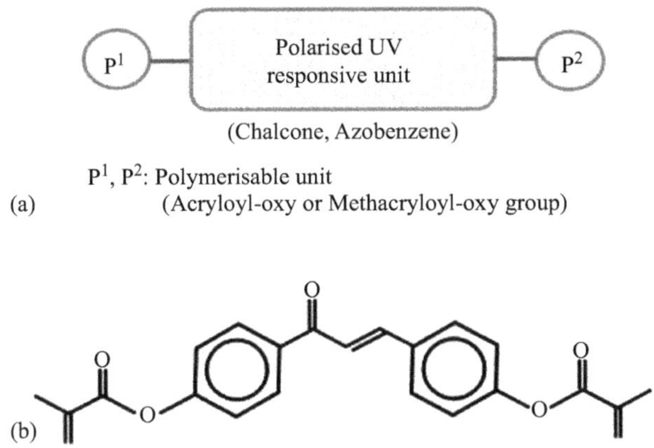

Figure 9.2 (a) Molecular design concept of the additive agents and (b) chemical structure of the monomer 4,4′-DMOCh

unit would be selected based on the photo-alignment technology. In the technology, polarised UV light is usually exposed to the alignment layer carrying photo-functional units [10,11]. It is generally known that there are three types of the conventional photo-alignment layer materials, i.e., photo-isomerisation, photo-crosslink and photo-decomposition [11]. On the basis of this information, the photo-isomerisation and the photo-cross-linked types were selected as the polarised UV responsive unit. In this work, the monomer carrying a chalcone unit 4,4′-di-methacryloyloxy chalcone (4,4′-DMOCh) was designed, which is shown in Figure 9.2(b). The uniaxially aligned chalcone unit would induce the LC alignment [10]. Moreover, the monomer would be polymerised during the exposure of the polarised UV light, and the polymerised compounds would be separated from the LC layer, inducing the formation of the SAL.

9.2.2 Vertical self-alignment

The self-alignment VA-mode LC cell (self-VA cell) is fabricated in a manner almost similar to that of the self-FFS cell, which is shown in Figure 9.3. First, a pair of substrate is assembled with each other to produce the empty cell. Each substrate is coated with ITO electrode having small slits. Then, a mixture of LC material and appropriate concentrations of a monomer and a photo-initiator is injected into an empty cell (Figure 9.3(a)). The monomer should carry the side chain such as a LC group for inducing the vertical alignment of the LC molecules. Finally, unpolarised UV light is exposed to the cell from the normal direction of the substrate in order to prepare the SAL. During the exposure of the unpolarised UV light, polymerisation would occur and the polymerised compounds would be separated from the LC layer, which induce the formation of the SAL. The temperature should be over the T_{NI} of the LC material for increasing the rate constant of the polymerisation. The monomer carrying the side LC group used in this work is shown in Figure 9.4, which is acrylic acid 4-(4′-octyloxy-biphenyl-4-yloxy) butyl ester (AOBBE).

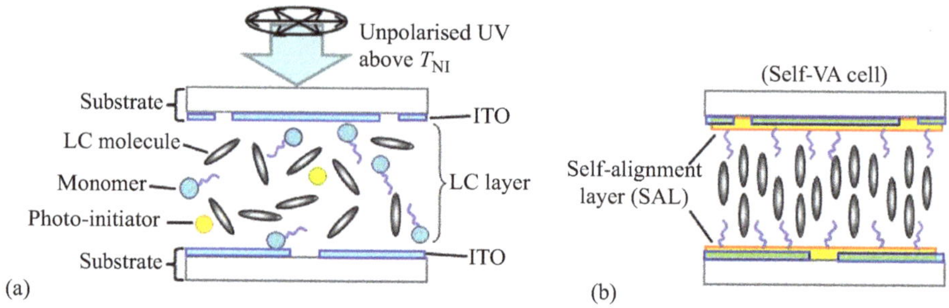

Figure 9.3 Schematic illustration of the fabrication process for self-alignment VA-mode LC cell (self-VA cell): (a) before preparation of the SAL and (b) after preparation of the SAL and obtaining the vertical alignment

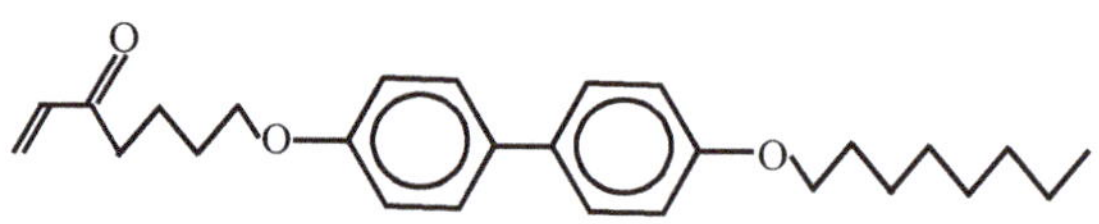

Figure 9.4 Chemical structure of the monomer AOBBE

9.3 Characteristics of the self-alignment LC cells

9.3.1 FFS mode self-alignment (self-FFS cell)

9.3.1.1 Fabrication of the self-FFS cells

An example for fabrication of the self-FFS cell is described here. First, a mixture of the LC material and the monomer 4,4′-DMOCh was prepared in the following manner: 0.05 g 4,4′-DMOCh was introduced into 5 g LC material with negative dielectric anisotropy (T_{NI} 80 °C, Δn 0.09, $\Delta\varepsilon$ −4.0) purchased from Merck. The mixture was kept in an ultrasonic bath for 3 min, and then kept in a 90 °C oven for 1 h for solubilising the monomer to the LC material. In the next step, the cell was fabricated in the followings.

Two glass substrates were assembled together with seal adhesive to produce the cell without the LC material; one side of the substrates carries the pixel and the common ITO electrodes through the insulating SiN layer, and the other side of the substrates does not have the electrodes and the SiN layer. The cell gap was maintained at 3.5 μm with the glass fibre. The mixture of the monomer 4,4′-DMOCh and the LC material was injected into the cell by vacuum filling method. Finally, the cell including the mixture of the monomer and the LC material was exposed the polarised UV light in the normal direction of the cell while keeping the temperature of the cell at about 95 °C for producing the SAL. The UV light source of UXM-501MA from Ushio Electronics Corp, Yokohama, Japan and a cut filter that blocks light below 280 nm from Asahi Spectra Corp, Tochigi Japan were used. The exposure energies for the polarised UV light were optimised to be 0, 2 and 5 J/cm^2.

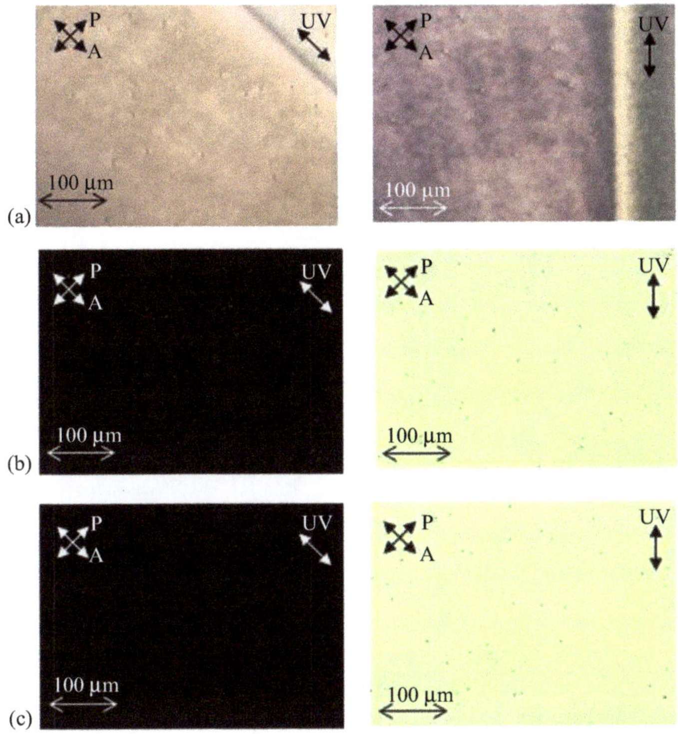

Figure 9.5 POM images of the self-FFS cell with the exposure energies of polarised UV light for (a) 0 J/cm^2; (b) 2 J/cm^2; and (c) 5 J/cm^2 at 95 °C; the angles between crossed polarisers and polarised UV light are 0° and 45°

To compare the electro-optical property of the self-FFS cell with the conventional FFS-LC cell having the polyimide-film alignment layer (PI-FFS cell), the PI-FFS cell was fabricated in the following way. First, homogeneous alignment polyimide material was coated on the pair of glass substrates and rubbed to obtain the antiparallel alignment direction. Two glass substrates having the polyimide alignment layers were assembled together with seal adhesive. The cell gap was maintained at 3.5 μm with the glass fibre. Finally, the LC material without monomers was injected into the empty cell by the vacuum-filling method, being obtained the PI-FFS cell.

9.3.1.2 Alignment state of the self-FFS cells

The alignment state of the self-FFS cell before and after exposure the polarised UV light was investigated by polarised optical microscopic (POM) observation at room temperature. Figure 9.5 shows the POM images of the self-FFS cells. The exposure energies of the polarised UV light for formation of the SAL was in the range from 0 to 5 J/cm^2. The observation was carried out under crossed polarisers, and alignment

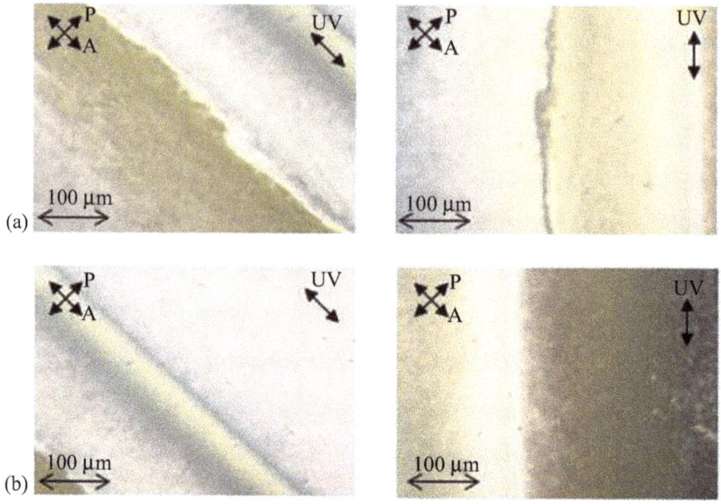

Figure 9.6 POM images of the self-FFS cell with the exposure energies of polarised UV light for (a) 0 J/cm^2 and (b) 5 J/cm^2 at 25 °C

directions of the self-FFS cells were set to be 0° and 45° to the crossed polarisers. The alignment direction indicates an axis of the polarised UV light, as shown in the figures. In the case of 0 J/cm^2 (Figure 9.5(a)), the transmittance of the alignment directions 0° and 45° to the crossed polarisers are comparably equal and not uniform, indicating that the uniform alignment of the LC molecules was not shown for this cell. In the cases of 2 and 5 J/cm^2 exposures (Figure 9.5(b) and (c)), however, dark states were obtained at the alignment direction of 0° to the crossed polarisers, whereas high transmittance was observed at the alignment direction of 45°. In addition, the dark level of the self-FFS cell with the exposure energy of 5 J/cm^2 is slightly higher than that of 2 J/cm^2. The result indicates that the uniaxially homogeneous alignment was obtained by exposing the polarised UV light over the T_{NI} of the LC material. In contrast to the fact, uniaxially homogeneous alignment was not observed even when 5 J/cm^2 polarised UV light was exposed at 25 °C, which is lower than the T_{NI} of the LC material. The POM images of the self-FFS cells before and after exposing 5 J/cm^2 polarised UV light at 25 °C are shown in Figure 9.6(a) and (b), respectively. This implies that the optimising the temperature is significantly important for obtaining the homogeneous alignment during the polarised UV light exposure. These results suggest that the uniaxially homogeneous alignment was obtained by exposing the polarised UV light over the T_{NI} of the LC material due to formation of the SAL.

9.3.1.3 Electro-optical property of the self-FFS cells

The photographs and POM images of the self-FFS cell with the exposure energy of 5 J/cm^2 polarised UV under application of voltage are shown in Figures 9.7 and 9.8, respectively. According to both the figures, the transmittance was increased with

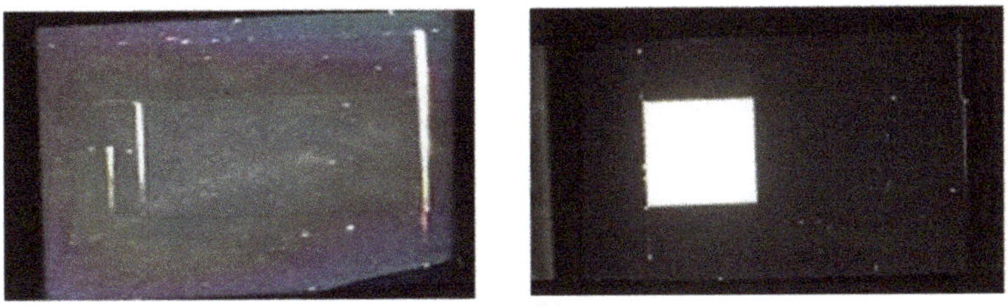

(a) (b)

Figure 9.7 Voltage dependence of (a) 0 V and (b) 5 V of the photographs for the self-FFS cell with the exposure energy of polarised UV light for 5 J/cm^2 at 95 °C; the angle between crossed polarisers and polarised UV light is 0°

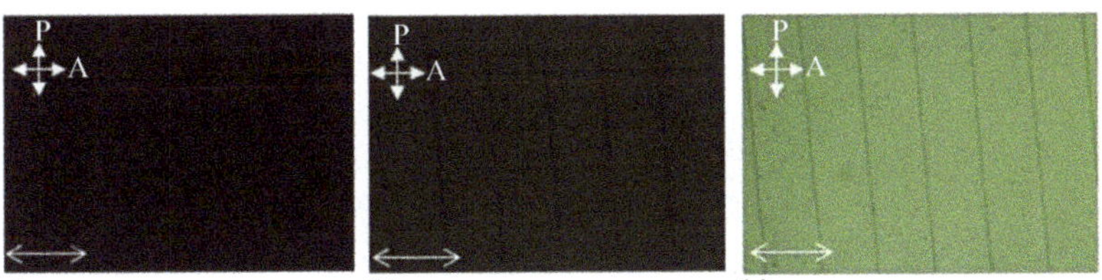

(a) (b) (c)

Figure 9.8 Voltage dependence of (a) 0 V; (b) 2 V; and (c) 5 V of the POM images for the self-FFS cell with the exposure energy of polarised UV light for 5 J/cm^2 at 95 °C; the angle between crossed polarisers and polarised UV light is 0°

increasing the application voltage from 0 to 5 V, confirming that proper electro-optical switching was shown for the self-FFS cell. Figure 9.9 presents voltage-dependent transmittance (*V–T*) curves for the self-FFS and the PI-FFS cells. Here, 100 per cent is defined to be the transmittance that the polariser and analyser are set to be parallel without placing any LC cell. The *V–T* curves for the two LC cells are well coincident with each other, anticipating that the alignment state of the self-FFS cell is at the same level as the PI-FFS cell.

The response times and the voltage holding ratios (VHRs) of the self-FFS and the PI-FFS cells were evaluated, as listed in Table 9.1. The response times of τ_{on} (turn-on time) and τ_{off} (turn-off time) are defined as the transient times from 10 to 90 per cent and from 90 to 10 per cent of the maximum transmittance, respectively.

Both τ_{on} and τ_{off} are almost equal between the self-FFS and the PI-FFS cells. The results also support that the alignment state of the self-FFS cell is at the same level as the PI-FFS cell. On the other hand, the VHR of the self-FFS cell is a little lower than that of the PI-FFS cell. One can consider that the VHR of the self-FFS cell was decreased due to the exposure of the polarised UV light. However, the value is still over 98.0 per cent, implying that the amount of impurity ion and/or radical included in the self-FFS cell was not so much that flicker would not be

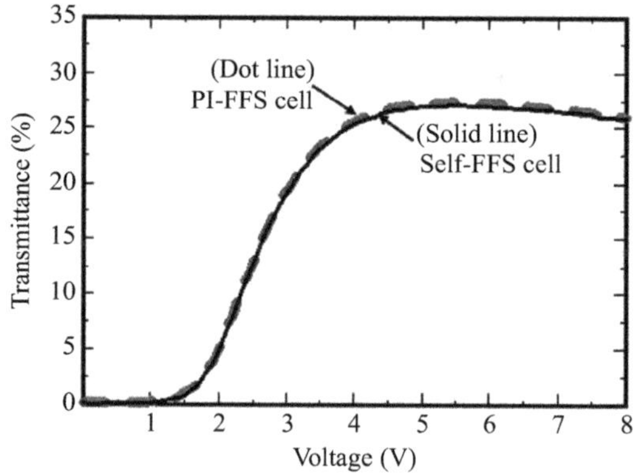

Figure 9.9 V–T curves for the self-FFS and the PI-FFS cells; the measurements were carried out with 60 Hz square voltage application at 25 °C

Table 9.1 Response times (τ_{on} and τ_{off}) and VHRs of the self-FFS and the PI-FFS cells

LC cell	τ_{on} (ms)	τ_{off} (ms)	VHR (%)[1]
Self-FFS	11.8	17.9	98.2
PI-FFS	12.5	17.4	98.8

[1]VHRs were measured with a 6254-type VHR meter developed by Toyo Corp. at 70 °C. The VHR was measured during an open-circuit period of 16.61 ms after the application of 5 V for 60 μs.

visible [10]. The results indicate that the SAL produced from the monomer 4,4′-DMOCh in the LC layer can align the LC molecules horizontally, and the alignment state and the electro-optical properties of the self-FFS cell are almost equal to those of the PI-FFS cell.

9.3.2 VA mode self-alignment (self-VA cell)

9.3.2.1 Fabrication of the self-VA cells

The self-VA cells were fabricated in a manner almost similar to the fabrication of the self-FFS cell. First, a mixture of the LC material, the monomer AOBBE, and the photo-initiator was prepared; 0.06 g AOBBE and 0.015 g photo-initiator were introduced into 5 g LC material with negative dielectric anisotropy (T_{NI} 80 °C, Δn 0.10, $\Delta\varepsilon$ −3.1) purchased from Merck. As the photo-initiator, 1,2-bis-(4-methacryloyloxy-phenyl)-2,2-dimethoxy-ethane-1-one was used here [12]. The mixture was kept in an ultrasonic bath, and then kept in a 90 °C oven for solubilising the monomer and the photo-initiator to the LC material. In the next step, two glass substrates with the ITO electrodes were assembled together. The cell gap was maintained at 3.5 μm with the glass fibre. The mixture was injected into the cell,

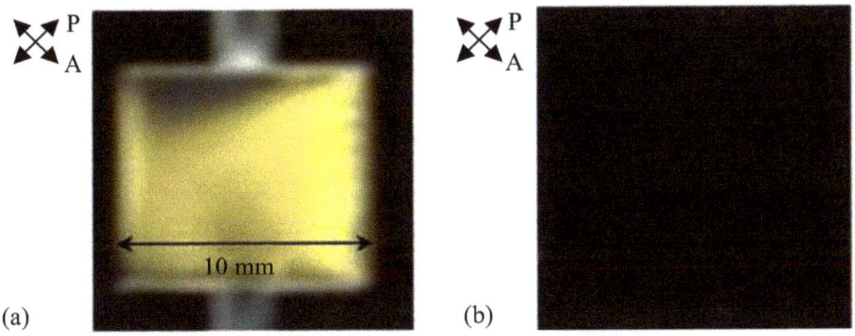

Figure 9.10 Photographs of the self-VA cell before (a) and after (b) 792 mJ/cm^2 UV light exposure at 95 °C; the self-VA cell was placed between crossed polarisers

Table 9.2 Pre-tilt angles of the self-VA and the PI-VA cells before and after exposure the UV light

LC cell	Before UV light exposure	UV light exposure[1] at 95 °C	UV light Exposure[1] at 35 °C
Self-VA	(Impossible to determine)	90	(Impossible to determine)
PI-VA	90	–	–

[1]Exposure energy of the UV light was 792 mJ/cm^2 from normal direction of the self-VA cell.

and finally, the unpolarised UV light was exposed from the normal direction of the cell at 95 °C and 35 °C in order to prepare the SAL for vertical alignment. During the exposure of the UV light, the cut filter that blocks light below 280 nm was used, and the exposure energies of the unpolarised UV light were optimised in the range from 70 to 792 mJ/cm^2. For comparison of the electro-optical property with the conventional VA-LC cell having the VA-type polyimide layer (PI-VA cell), the PI-VA cell was fabricated in a manner similar to that of the PI-FFS cell. As the VA-type polyimide layer material, AL60101 from JSR was used.

9.3.2.2 Alignment state of the self-VA cells

The vertical alignment with enough level of alignment state was obtained when the 792 mJ/cm^2 unpolarised UV light (UV light) was exposed at 95 °C. The photographs of the alignment state for the self-VA cell placed under crossed polarisers are presented in Figure 9.10 before and after UV light exposure with the exposure energy of 792 mJ/cm^2 at 95 °C. Before the UV light exposure, the cell shows non-uniform bright state because the alignment state of the LC molecules is random. After the UV light exposure, however, the dark state is observed, implying that the vertical alignment was obtained for the self-VA cell. For more confirmation that the vertical alignment was obtained, pre-tilt angles were measured before and after exposure the UV light. The pre-tilt angle was determined by crystal rotation technique with an OMS-AF2 developed by Chuo Seiki [13], and the results are listed in Table 9.2.

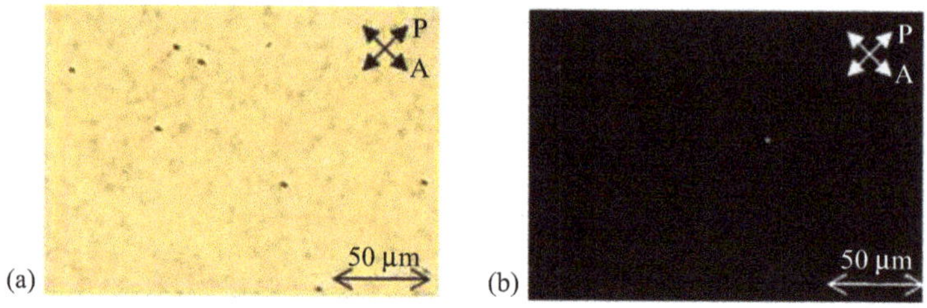

Figure 9.11 POM images of the self-VA cell before (a) and after (b) 792 mJ/cm^2 UV light exposure at 95 °C

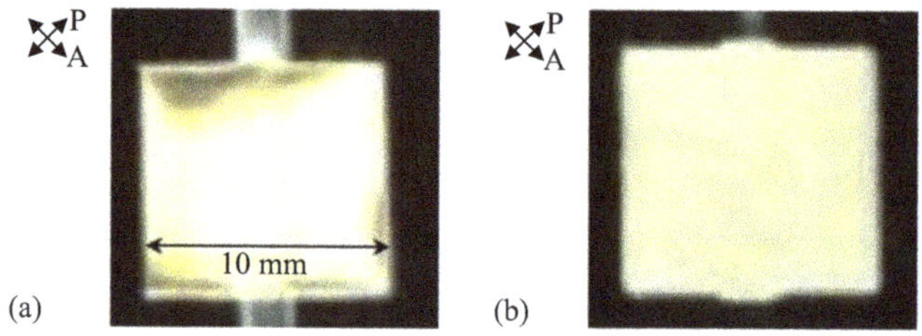

Figure 9.12 Photographs of the self-VA cell before (a) and after (b) 792 mJ/cm^2 UV light exposure at 35 °C; the self-VA cell was placed between crossed polarisers

The pre-tilt angle of the self-VA cell after exposure the UV light was 90°, whereas that before exposure the UV light was not able to measure. The fact that the pre-tilt angle could not be measured indicates that uniform alignment was not obtained for the LC molecules. The POM images of the self-VA cell are shown in Figure 9.11 before and after UV light exposure with the exposure energy of 792 mJ/cm^2 at 95 °C. One cannot observe any alignment defects after the UV light exposure. The small-sized dots, which are observed in the images, are originated from spacer beads for maintaining the cell gap. The POM image indicates that the uniform vertical alignment was obtained for the self-VA cell after the UV light exposure at 95 °C (Figure 9.11(b)). In contrast to the result, the vertical alignment was not obtained when the 792 mJ/cm^2 UV light was exposed at 35 °C, which is shown in Figure 9.12. One can predict two possibilities for not obtaining the vertical alignment; one is that the rate constant of the polymerisation was too small to produce the SAL, the other reason is that the random alignment of the LC molecules was maintained even the polymerisation was occurred.

9.3.2.3 Electro-optical property of the self-VA cells

The electro-optical and response properties are compared between the self-VA and the PI-VA cells. The *V–T* curves for the self-VA and the PI-VA cells are presented in Figure 9.13. The *V–T* curves for the two LC cells are well coincident with each other, inferring that the vertical alignment state of the self-VA cell is at the same level as the PI-VA cell.

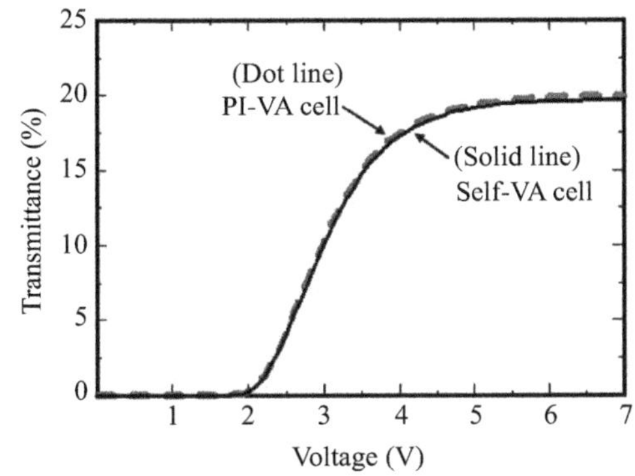

Figure 9.13 V–T curves for the self-VA and the PI-VA cells; the measurements were carried out with 60 Hz square voltage application at 25 °C

Table 9.3 Contrast ratios, response times (τ_{on} and τ_{off}), and VHRs of the self-VA and the PI-VA cells

LC cell	Contrast ratio[1]	τ_{on} (ms)	τ_{off} (ms)	VHR (%)[1]
Self-VA	1,260	13.8	7.2	99.1
PI-VA	1,220	14.4	6.9	99.3

[1]Contrast ratio was defined from the ratio of the transmittance at 7.0 V and 0.2 V of the *V–T* curve.

Contrast ratios, response times (τ_{on} and τ_{off}), and VHRs of the self-VA and the PI-VA cells are listed in Table 9.3. The contrast ratio is defined from the ratio of the transmittance at 7.0 V and 0.2 V. The results of the contrast ratio and the response times (τ_{on} and τ_{off}) between the self-VA and the PI-VA cells are almost equal, inferring that the alignment state of the self-VA cell is at the same level as the PI-VA cell. On the other hand, the VHR of the self-VA cell is slightly lower than that of the PI-VA cell. This is almost the same tendency with that between the self-FFS and the PI-FFS cells, and one can estimate that the VHR of the self-VA cell was decreased due to the exposure of the UV light. However, the value is over 99.0 per cent, which would be high enough that image sticking would not be visible in the initial stage [14]. The results indicate that the SAL produced from the monomer AOBBE in the LC layer can induce the vertical alignment of the LC molecules without sacrificing the image quality of the self-alignment LCDs.

9.4 Reliability

With increasing the market of the LCDs, the demand for the reliability has been increasing because the LCDs are being utilised in various fields. In particular, the

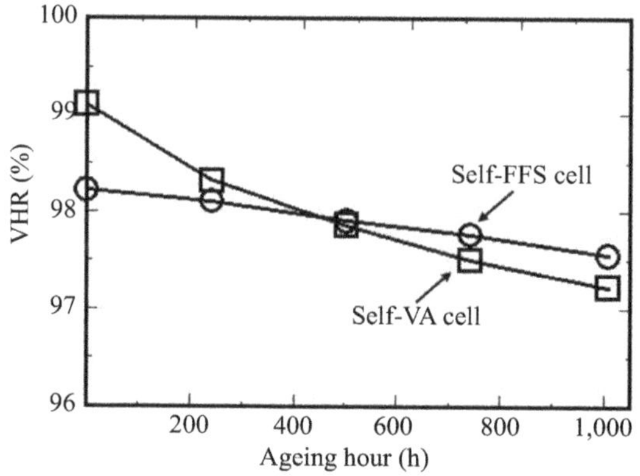

Figure 9.14 VHR as a function of ageing hour; ageing condition of the preservation test is keeping in a 70 °C oven

LCDs for mobile applications are necessary to have high reliability under high temperature because they are used outdoors.

By the way, there is a tendency that the image sticking becomes observable for the LCDs with low VHR when they are used for a long period [15]. The image sticking is the phenomenon which the previous pattern is visible when the next pattern is addressing [16–18]. One of the reasons for the image sticking of the LCDs is the generation of residual direct current voltage due to existence of impurity ion and/or radical in the LC layer [17,18]. Moreover, it is well known that the VHR decreases with increasing the density of the impurity ion and/or radical in the LC layer [15,19,20]. Therefore, evaluation of the VHR as a function of ageing period under some stress is crucially important because both the self-FFS and the self-VA cells, which are fabricated above, include the SAL produced from the monomers by the polymerisation. It means that there is a possibility that the radicals remain in the LC layers of the cells.

The self-FFS and the self-VA cells were preserved in a 70 °C oven for 1,000 h, and the VHR was measured as a function of ageing period. Figure 9.14 plots the VHR as a function of ageing hour. The VHRs for both the self-FFS and the self-VA cells were slightly decreased, and the values became below 98.0 per cent after 1,000 ageing hours. In addition, the degree of the decrease for the self-VA cell is larger than that for the self-FFS cell. The reason for the decrease of the VHR under the preservation test is now under consideration in order to improve the reliability of the self-alignment technology.

9.5 Future LCDs fabricated by self-alignment technologies

Most of the latest LCDs have the alignment films produced from the polyimides. The preparation of the polyimide-film alignment layer requires not only large quantities of solvents but also heating process. In contrast, the self-alignment technology does not need any solvents and heating process over 200 °C. Therefore, the self-alignment technology would be environmentally friendly. Moreover, deformations would be generated for plastic substrates under high temperature over 200 °C. Hence, it would be difficult to fabricate the flexible LCDs with the conventional process. On the other hand, the self-alignment technology is useful for the development of the flexible LCDs since the technology needs the process below 100 °C for preparation of the SAL. In addition, it is generally known that adhesive force between the polyimide film and the sealant is considerably weak [21]. The self-alignment technology can solve the problem since the sealant directly contacts with the substrates. Therefore, it would be useful for the development of narrow-frame LCDs. The self-alignment technology is the promising technology for the development of the flexible and/or the narrow-frame LCDs.

9.6 Conclusions

In this chapter, the self-alignment technology and its materials were introduced as the novel preparation method of the LCDs. The method forms the SAL in order to induce both the homogeneous and the vertical alignments of the LC molecules. The SAL is produced by polymerisation of the monomers solubilised in the LC layers over the T_{NI} of the LC materials. Both the FFS and the VA modes LC cells were fabricated with the self-alignment technology. The cells exhibited enough level of alignment states and electro-optical properties compared with the conventional LC cells having the polyimide-film alignment layers. However, further improvement related to the reliability is necessary for the self-alignment cell.

The self-alignment technology would be quite useful for the development of the flexible and/or the narrow frame LCDs because the technology does not need the conventional alignment layer materials and high temperature process over 200 °C.

References

[1] Schadt M. 'Nematic liquid crystals and twisted-nematic LCDs'. *Liquid Crystals*. 2015, vol. 42(5), pp. 646–52.

[2] Chigrinov V.G., Kozenkov V.M. and Kwok H.-S. *Photoalignment of Liquid Crystalline Materials: Physics and Applications*. Chichester: Wiley. SID; 2008. Chap. 4.

[3] Seo D.-S., Muroi K. and Kobayashi S. 'Generation of pretilt angles in nematic liquid crystal, 5CB, media aligned on polyimide films prepared by spin-coating and LB techniques: effect of rubbing'. *Molecular Crystals and Liquid Crystals*. 1992, vol. 213(1), pp. 223–28.

[4] Samant M.G., Stohr J., Brown H.R., Russell T.P., Sands J.M. and Kumar S.K. 'NESAFS studies on the surface orientation of rubbed polyimides'. *Macromolecules*. 1996, vol. 29(26), pp. 8334–42.

[5] Haba O., Hiratsuka D., Shirakawa, T., *et al*. 'Homeotropic orientation of nematic liquid crystals induced by dissolving polypropyleneimine dendrimer

having peripheral mesogens'. *Optical Materials Express*. 2014, vol. 4(5), pp. 934–43.

[6] Oh-e M. and Kondo K. 'Quantitative analysis of cell gap margin for uniform optical properties using in-plane switching of liquid crystals'. *Japanese Journal of Applied Physics*. 1997, vol. 36(11), pp. 6798–803.

[7] Lee S.H., Lee S.L. and Kim H.Y. 'Electro-optic characteristics and switching principle of a nematic liquid crystal cell controlled by fringe-field switching'. *Applied Physics Letters*. 1998, vol. 73(20), pp. 2881–83.

[8] Takeda A., Kataoka S., Sasaki T., *et al*. 'A super-high image quality multi-domain vertical alignment LCD by new rubbing-less technology'. 1998 SID International Symposium Digest of Technical Papers. 1998, vol. 29(1), pp. 1077–80.

[9] Lee G.-D., Son J.-H., Choi Y.-H., Lju J.-J., Kim K.H. and Lee S.H. 'Control of the defect in the liquid crystal director field on the slit of the patterned vertical alignment cell'. *Applied Physics Letters*. 2007, vol. 90(3), 033509.

[10] Miyake I., Asagi H., Shimizu M., *et al*. 'Photo-alignment and n-FFS LCD technologies with IGZO-TFT applied to generation eight factory'. 2016 SID International Symposium Digest of Technical Papers. 2016, vol. 47(1), pp. 592–95.

[11] O' Neill M. and Kelly S.M. 'Photoinduced surface alignment for liquid crystal displays'. *Journal of Physics D: Applied Physics*. 2000, vol. 33(10), pp. R67–84.

[12] Mizusaki M., Nakanishi Y. and Enomoto S. 'Fabrication of vertically aligned liquid crystal cell without using a conventional alignment layer'. *Liquid Crystals*. 2018, vol. 45(2), pp. 270–78.

[13] Han K.-Y., Miyashita T. and Uchida T. 'Accurate determination and measurement error of pretilt angle in liquid crystal cell'. *Japanese Journal of Applied Physics*. 1993, vol. 32(2B), pp. L277–79.

[14] Mizusaki M., Enomoto S. and Hara Y. 'Generation mechanism of residual direct current voltage for liquid crystal cell with polymer layers produced from monomers'. *Liquid Crystals*. 2017, vol. 44(4), pp. 609–17.

[15] Nakanishi Y. and Okamoto K. 'Relation between concentration of initiator and image-sticking phenomenon of polymer-sustained-alignment liquid crystal displays'. *Japanese Journal of Applied Physics*. 2012, vol. 51(4R), 041701.

[16] Park Y., Kim S. and Lee E. 'A study on reducing image-sticking artifacts in wide-screen TFT-LCD monitors'. *Journal of the SID*. 2007, vol. 15(11), pp. 969–73.

[17] Mizusaki M., Miyashita T. and Uchida T. 'Behavior of ion affecting image sticking on liquid crystal displays under application of direct current voltage'. *Journal of Applied Physics*. 2010, vol. 108(10), 104903.

[18] Xu D., Peng F., Chen H., *et al*. 'Image sticking in liquid crystal displays with lateral electric field'. *Journal of Applied Physics*. 2014, vol. 116 (19), 193102.

[19] Mizusaki M., Yoshimura Y., Yamada Y. and Okamoto K. 'Analysis of ion behavior affecting voltage holding property of liquid crystal displays'. *Japanese Journal of Applied Physics*. 2012, vol. 51(1R), 014102.

[20] Mizusaki M., Nakanishi Y., Enomoto S. and Hara Y. 'Evaluation of image sticking property on liquid crystal displays with polymer layers produced from phenanthrene-carrying monomers'. *Liquid Crystals.* 2016, vol. 43(9), pp. 1208–14.

[21] Mizusaki M., Tsuchiya H. and Minoura K. 'Homogeneous self-alignment technology without using conventional alignment layer materials'. *Proceedings of the IDW'17*; Dec. 2017. Sendai: vol. 24(1), pp. 171–74.

Chapter 10

Color-reflective LCD based on cholesteric liquid crystals

Deng-Ke Yang[1] *and Quan Li*[1]

Cholesteric liquid crystals (hereafter Ch LCs) are self-assembled systems consisting of elongated chiral organic molecules. They possess a helical structure where the local average direction of the molecules twists spatially around an orthogonal helical axis. Their refractive index varies periodically, and thus exhibits a Bragg reflection band centered at the wavelength $\lambda = [(n_e + n_o)/2]P$ and with the bandwidth $\Delta\lambda = (n_e - n_o)P$, where n_e and n_o are the extraordinary and ordinary refractive indices of the LC, respectively, and P is the helical pitch. They can be used to make reflective displays which do not need polarizers and have high reflectance. In this chapter, various mechanisms, which can be used to modify the reflection of the displays, are described. The performance of the displays is also discussed.

10.1 Introduction to cholesteric LCs

LCs are a class of organic materials consisting of elongated rod-like molecules or flat disk-like molecules [1–4]. In this chapter, we only discuss rod-like molecules. The most common LC is nematic LC if the constituent molecules are achiral (the mirror image of the molecule is the same as the molecule itself). The chemical structure of a typical nematic LC molecule is shown in Figure 10.1(a1). Because of the rapid spinning and flipping motion, the molecule can be considered as a rod as shown in Figure 10.1(a2). The length of the rod is about 2 nm and the diameter is about 0.5 nm. A nematic LC has orientational order that the average direction of the long molecular axes of the constituent molecules are aligned along a common direction known as the LC director and denoted by $\vec{n}$, as shown in Figure 10.1(a3). Therefore, it exhibits anisotropic optical properties as crystals. The position of the constituent molecules, however, has no positional order. Therefore, it can flow as a regular liquid.

If the constituent molecules are chiral (the mirror image of the molecule is different from the molecule itself), they form a chiral phase called cholesteric liquid

[1]Liquid Crystal Institute, Kent State University, USA

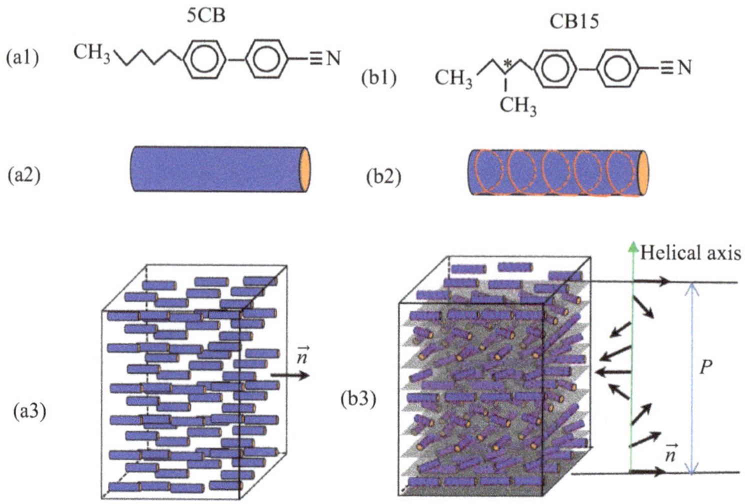

Figure 10.1 Schematic diagrams of molecular structure and orientational order of nematic and cholesteric LCs

crystal (also referred as chiral nematic liquid crystal). The chemical structure of a typical chiral LC molecule is shown in Figure 10.1(b1). The molecule can be considered as a screw as shown in Figure 10.1(b2). The LC director is no longer uniform in space but twists periodically around an orthogonal helical axis, as shown in Figure 10.1(b3). The distance over which for the LC director to twist 360° is the pitch, denoted by P. Because the physical properties of the LC are the same along $\vec{n}$ and $-\vec{n}$, the periodicity is $P/2$.

In most electro-optical applications, cholesteric (Ch) LCs are mixtures of a nematic host and a chiral dopant. The pitch P of the Ch LC is approximately related to the concentration x of the chiral dopant by [1]:

$$P = \frac{1}{(HTP) \cdot x} \tag{10.1}$$

where HTP is the helical twisting power of the chiral dopant, which mainly depends on its chemical structure and slightly depends on the nematic host. The unit of HTP is μm^{-1}. For commercially available chiral dopants, the HTP is in the region of 1 μm^{-1} to 150 μm^{-1}. For example, chiral dopant CB15 has the HTP about 8 μm^{-1}. If its concentration is 30 percent, the pitch is $P = 1/[8\ \mu m^{-1} \times 0.3] = 0.416\ \mu m$. The pitch of a Ch LC can be adjusted by varying either the concentration or the chemical structure of the chiral dopant.

10.2 Reflection of Ch LCs

Ch LCs selectively reflect light because of their periodical helical structure. A Ch LC exhibits birefringence. It has the ordinary refractive index n_o for light linearly

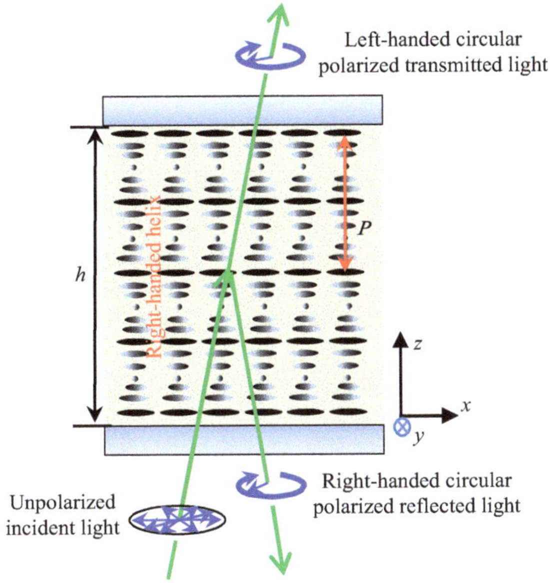

Figure 10.2 Schematic diagrams of the selective reflection of Ch LC

polarized with the electric field perpendicular to the LC director, and has the extraordinary refractive index n_e for light linearly polarized with the electric field parallel to the LC director. Because the LC director changes direction along the helical axis, the effective refractive index changes periodically along the helical axis. Bragg reflects light peaked at the wavelength $2 \cdot [(n_o + n_e)/2] \cdot (P/2)$ [1,4–8]. Many Ch LCs have pitches comparable to the wavelength of visible light. Therefore they reflect visible light and can be used to make reflective displays.

For light incident normally on a Ch LC layer, the reflection can be calculated analytically. We consider a Ch LC with right-handed helical structure as shown in Figure 10.2, the LC director is given by

$$n_x = \cos(qz), \quad n_y = \sin(qz), \quad n_z = 0, \tag{10.2}$$

where $q = 2\pi/P$ is the helical wave number of the LC. The dielectric tensor in the xy plane is given by

$$\overleftrightarrow{\varepsilon}(z) = \begin{pmatrix} \bar{\varepsilon} + \delta\cos(2qz) & \delta\sin(2qz) \\ \delta\sin(2qz) & \bar{\varepsilon} - \delta\cos(2qz) \end{pmatrix}. \tag{10.3}$$

where $\delta = (\varepsilon_{//} - \varepsilon_{\perp})/2$, $\bar{\varepsilon} = (\varepsilon_{//} + \varepsilon_{\perp})/2$, $\varepsilon_{//} = n_e^2$ and $\varepsilon_{\perp} = n_o^2$. Therefore $n_e^2 = \varepsilon_{//} = \bar{\varepsilon} + \delta$ and $n_o^2 = \varepsilon_{\perp} = \bar{\varepsilon} - \delta$. Consider a polarized light propagating along the z-axis (cell normal direction) with the wavelength λ, its electric field vector is described by

$$\vec{E}(z,t) = \vec{A}(z)e^{i\omega t}. \tag{10.4}$$

The electric field vector $\vec{A}(z)$ of the light is governed by the wave equation:

$$\frac{\partial^2 \vec{A}(z)}{\partial z^2} = -k_o^2 \overleftrightarrow{\varepsilon}(z) \cdot \vec{A}(z), \tag{10.5}$$

where $k_o = \omega/c = 2\pi/\lambda$ (λ is the wavelength of the light in vacuum). We consider the light propagation eigenmodes whose electric field vector $\vec{A}$ does not change direction with position when propagating through the medium [4,9]. In the lab frame, because the dielectric tensor varies with position, unfortunately there is no solution where $\vec{A}$ is a constant vector. We have to consider the problem in the local frame whose x' axis is parallel to the LC director, y' axis is perpendicular to the LC director and z' axis is still parallel to the cell normal. We assume there are eigenmodes in the local frame, whose electric field vector is given by

$$\vec{A}'(z) = \vec{A}'_o e^{-ikz} = \left(A'_{ox}\hat{x}' + A'_{oy}\hat{y}'\right)e^{-ink_o z}, \tag{10.6}$$

where A'_{ox} and A'_{oy} are constants and n is the effective refractive index. The electric field vectors in the lab and local frames are related to each other by

$$\vec{A} = \begin{pmatrix} A_x \\ A_y \end{pmatrix} = \begin{pmatrix} \cos(qz) & -\sin(qz) \\ \sin(qz) & \cos(qz) \end{pmatrix} \begin{pmatrix} A'_x \\ A'_y \end{pmatrix} \equiv \overleftrightarrow{S}(qz)\vec{A}' = \overleftrightarrow{S}(qz)\vec{A}'_o e^{-ink_o z}, \tag{10.7}$$

where $\overleftrightarrow{S}(qz) = \begin{pmatrix} \cos(qz) & -\sin(qz) \\ \sin(qz) & \cos(qz) \end{pmatrix}$ is the rotation matrix. The dielectric tensor in the local frame is

$$\overleftrightarrow{\varepsilon}' = \overrightarrow{S}^{-1}(qz) \cdot \overleftrightarrow{\varepsilon}(z) \cdot \vec{S}(qz) = \begin{pmatrix} \varepsilon_{//} & 0 \\ 0 & \varepsilon_{\perp} \end{pmatrix} \tag{10.8}$$

The wave equation (10.5) becomes

$$\frac{\partial^2 \left[\overleftrightarrow{S}(qz)\vec{A}'_o e^{-ink_o z}\right]}{\partial z^2} = -k_o^2 \overleftrightarrow{S}(qz) \cdot \overleftrightarrow{\varepsilon}'(z) \cdot \overleftrightarrow{S}^{-1}(qz) \cdot \overleftrightarrow{S}(qz) \cdot \vec{A}'(z) = -k_o^2 \overleftrightarrow{\varepsilon}'(z) \cdot \vec{A}'(z) \tag{10.9}$$

When expressed in terms of components, the above equation becomes

$$\begin{pmatrix} n_e^2 k_o^2 - n^2 k_o^2 - q^2 & i2qnk_o \\ -i2qk_o & n_o^2 k_o^2 - n^2 k_o^2 - q^2 \end{pmatrix} \begin{pmatrix} A'_{ox} \\ A'_{oy} \end{pmatrix} = 0 \tag{10.10}$$

For nonzero solutions, it is required that

$$\begin{vmatrix} n_e^2 k_o^2 - n^2 k_o^2 - q^2 & i2qnk_o \\ -i2qk_o & n_o^2 k_o^2 - n^2 k_o^2 - q^2 \end{vmatrix} = 0 \tag{10.11}$$

The solutions for the eigenvalue n are

$$n_{1\pm} = \pm\left\{(\lambda/P)^2 + \bar{\varepsilon} + \left[4(\lambda/P)^2\bar{\varepsilon} + \delta^2\right]^{1/2}\right\}^{1/2}, \tag{10.12}$$

$$n_{2\pm} = \pm\left\{(\lambda/P)^2 + \bar{\varepsilon} - \left[4(\lambda/P)^2\bar{\varepsilon} + \delta^2\right]^{1/2}\right\}^{1/2}, \tag{10.13}$$

where the sign + is for eigenmodes propagating in the $+z$ direction and the sign − is for eigenmodes propagating in the $-z$ direction. There are two eigenmodes in each propagation direction. $n_{1\pm}$ is always real for any wavelength λ. $n_{2\pm}$ can be real or imaginary depending on the wavelength λ and the pitch P. Substituting the eigen values back to the wave equation, the corresponding eigen vectors, respectively, can be obtained:

$$\vec{A}'_{o1\pm} = a\begin{pmatrix} 1 \\ i\left[n_e^2 - n_{1\pm}^2 - (\lambda/P)^2\right]/2(\lambda/P)n_{1\pm} \end{pmatrix}, \tag{10.14}$$

$$\vec{A}'_{o2\pm} = b\begin{pmatrix} 1 \\ i\left[n_e^2 - n_{2\pm}^2 - (\lambda/P)^2\right]/2(\lambda/P)n_{2\pm} \end{pmatrix}. \tag{10.15}$$

where a and b are the normalization constants. Generally, the eigen modes are elliptically polarized.

It is well known that only light with wavelength λ close to $[(n_e + n_o)/2]P$ is reflected. In this chapter, we are only interested in the reflection of Ch LCs, therefore, we consider the case where the wavelength is in the region $n_oP \le \lambda \le n_eP$. Furthermore, most Ch LCs have $\Delta n \le 0.3$ and $(n_e + n_o)/2 \sim 1.6$, and therefore $\delta/\bar{\varepsilon} = \left[(\varepsilon_{//} - \varepsilon_\perp)/2\right]/\left[(\varepsilon_{//} + \varepsilon_\perp)/2\right] = (n_e^2 - n_o^2)/(n_e^2 + n_o^2) \sim 1/6 << 1$. Define $\lambda/P = \sqrt{\bar{\varepsilon} + \gamma}$. When the wavelength is in the region $\sqrt{\varepsilon_\perp}P \le \lambda \le \sqrt{\varepsilon_{//}}P$, then $-\delta \le \gamma \le \delta$ and thus $|\gamma|/\bar{\varepsilon} \le \delta/\bar{\varepsilon} << 1$. In the following derivations, the lowest non-zero terms are kept. From (10.12), we have

$$n_{1\pm} = \pm\left[(\bar{\varepsilon} + \gamma) + \bar{\varepsilon} + 2\sqrt{\bar{\varepsilon}}\left(\bar{\varepsilon} + \gamma + \delta^2/4\bar{\varepsilon}\right)^{1/2}\right]^{1/2} \approx \pm 2\sqrt{\bar{\varepsilon}}(1 + \gamma/4\bar{\varepsilon}) \approx \pm 2\sqrt{\bar{\varepsilon}}. \tag{10.16}$$

The corresponding eigenmodes have the polarization

$$\vec{A}'_{o1\pm} = a\begin{pmatrix} 1 \\ i[(\bar{\varepsilon} + \delta) - 4\bar{\varepsilon} - (\bar{\varepsilon} + \gamma)]/2(\sqrt{\bar{\varepsilon} + \gamma})\left[\pm(2\sqrt{\bar{\varepsilon}})\right] \end{pmatrix} \approx \begin{pmatrix} 1 \\ \mp i \end{pmatrix}. \tag{10.17}$$

In the lab frame, the polarization is

$$\vec{A}_{1\pm} = \frac{1}{\sqrt{2}}\begin{pmatrix} 1 \\ \mp i \end{pmatrix} e^{-i(\pm k_o)\sqrt{\bar{\varepsilon}}z}. \tag{10.18}$$

Eigen mode 1 is left-handed circularly polarized and propagates in the $+z$ direction with the effective reactive index $\sqrt{\bar{\varepsilon}} = \sqrt{(n_e^2 + n_o^2)/2} \sim (n_e + n_o)/2$. Eigen mode 2 is also left-handed circularly polarized but propagates in the $-z$ direction with the same effective reactive index $\sqrt{\bar{\varepsilon}}$. The instantaneous electric field pattern of these eigenmodes is of opposite sense to the cholesteric helix which is right-handed. From (10.13), we have

$$n_{2\pm} = \pm\left\{(\bar{\varepsilon} + \gamma) + \bar{\varepsilon} - \left[4(\bar{\varepsilon} + \gamma)\bar{\varepsilon} + \delta^2\right]^{1/2}\right\}^{1/2} = \pm\frac{1}{2\sqrt{\bar{\varepsilon}}}\left(\gamma^2 - \delta^2\right)^{1/2}. \quad (10.19)$$

Because $-\delta \leq \gamma \leq \delta$, $n_{2\pm}$ is imaginary. The corresponding polarization is

$$\vec{A}'_{o2\pm} = b\begin{pmatrix} 1 \\ \left[(\delta - \gamma) + (\delta^2 - \gamma^2)/4\bar{\varepsilon}\right]/\left[\pm(1 + \gamma/2\bar{\varepsilon})(\delta^2 - \gamma^2)^{1/2}\right] \end{pmatrix} \quad (10.20)$$

In order to see the physical meaning, we define

$$\lambda/P = \sqrt{\bar{\varepsilon} + \gamma} = n_e n_o/\sqrt{n_o^2\cos^2\alpha + n_e^2\sin^2\alpha}, \quad (10.21)$$

where $0 \leq \alpha \leq \pi/2$. When $\alpha = 0$, $\lambda/P = n_e$ and $\gamma = \delta$. When $\alpha = \pi/2$, $\lambda/P = n_o$ and $\gamma = -\delta$. Because $n_e^2 = \varepsilon_{//} = \bar{\varepsilon} + \delta$ and $n_o^2 = \varepsilon_\perp = \bar{\varepsilon} - \delta$, $\bar{\varepsilon} + \gamma = (\bar{\varepsilon}^2 - \delta^2)/\left[\bar{\varepsilon} + \delta(\sin^2\alpha - \cos^2\alpha)\right]$. As mentioned before $\delta/\bar{\varepsilon} << 1$, therefore

$$\gamma = \delta(\cos^2\alpha - \sin^2\alpha). \quad (10.22)$$

Substituting the above equation into (10.20), we have

$$\begin{aligned} \vec{A}'_{o2\pm} &= b\begin{pmatrix} 1 \\ \left[2\sin^2\alpha + 4\delta\sin^2\alpha\cos^2\alpha/4\bar{\varepsilon}\right]/\left[\pm(1 + \gamma/2\bar{\varepsilon})2\sin\alpha\cos\alpha\right] \end{pmatrix} \\ &\approx b\begin{pmatrix} 1 \\ \pm\sin\alpha/\cos\alpha \end{pmatrix} \end{aligned} \quad (10.23)$$

After renormalization, we obtain

$$\vec{A}'_{o2\pm} = \begin{pmatrix} \cos\alpha \\ \pm\sin\alpha \end{pmatrix} \quad (10.24)$$

In the local frame, the eigen vector (polarization) makes the angle α with the LC director. We also know that when the electric field makes the angle α with the LC director, the effective refractive index is $n_e n_o/\sqrt{n_o^2\cos^2\alpha + n_e^2\sin^2\alpha}$. Therefore, if the polarization of the incident light makes the angle α with the LC director on the entrance plane, when it propagates through the Ch LC, its polarization will remain the same angle with respect to the LC director. In the lab frame, the polarization is

$$\vec{A}_{2\pm} = \begin{pmatrix} \cos(qz \mp \alpha) \\ \sin(qz \mp \alpha) \end{pmatrix} e^{\pm k_o \sigma z}, \tag{10.25}$$

where $\sigma = \left(\delta^2 - \gamma^2\right)^{1/2} / 2\sqrt{\bar{\varepsilon}}$ is real. Note these eigenmodes have linear polarization. Because the effective refractive index is imaginary, these eigenmodes are a nonpropagating wave. The instantaneous electric field pattern of these eigenmodes varies in space in the same way as the cholesteric helix. The light intensity decays as these eigenmodes propagate into the LC. This means that the Ch LC reflects circularly polarized light with the same handedness as the helical structure of the Ch LC. The reflection bandwidth is

$$\Delta\lambda = P(\sqrt{\bar{\varepsilon} + \delta} - \sqrt{\bar{\varepsilon} - \delta}) = P(\sqrt{\varepsilon_{//}} - \sqrt{\varepsilon_{\perp}}) = P(n_e - n_o) = \Delta n P. \tag{10.26}$$

When the incident light has the polarization making the angle α with respect to the LC director, the wavelength of the reflected light is $\lambda = \left[n_e n_o / \sqrt{n_o^2 \cos^2 \alpha + n_e^2 \sin^2 \alpha} \right] P$. When $\alpha = 0$ (polarized parallel to the LC director), the wavelength of the reflected light is $\lambda = n_e P$. When $\alpha = 90^\circ$ $\alpha = 0$ (polarized perpendicular to the LC director), the wavelength of the reflected light is $\lambda = n_o P$.

Now, we calculate the reflection of the Ch LC cell with the cell thickness h. Because the reflection of the (right-handed) Ch LC for left-handed circularly polarized incident light is 0, we only have to consider the reflection for right-handed circular polarized light. For incident light with other polarization, it can always be decomposed into a component with left-handed circular polarization and another component with right-handed circular polarization. For the purpose of simplicity, we assume that the media above and below the Ch LC have the refractive index $\bar{n} = (n_e + n_o)/2$, and $h/(P/2)$ is an integer. On the entrance side, there are incident light and reflected light, both of them right-handed circularly polarized. The incident light has the field amplitude u and propagates in the $+z$ direction. The electric field is

$$\vec{E}_i = \frac{u}{\sqrt{2}} \begin{pmatrix} 1 \\ i \end{pmatrix} e^{-ik_o \bar{n} z} \tag{10.27}$$

The reflected light has the amplitude r and propagates in the $-z$ direction. The field is

$$\vec{E}_r = \frac{r}{\sqrt{2}} \begin{pmatrix} 1 \\ -i \end{pmatrix} e^{ik_o \bar{n} z} \tag{10.28}$$

On the exit side, there is only the transmitted light which also is right-handed circularly polarized and propagates in the $+z$ direction, and has the field:

$$\vec{E}_t = \frac{t}{\sqrt{2}}\begin{pmatrix}1\\ i\end{pmatrix}e^{-ik_o\bar{n}z} \tag{10.29}$$

where t is the amplitude of the wave. Inside the Ch LC, there are two eigenmodes which are linearly polarized (one mode propagating in the $+z$ direction and the other mode propagating in the $-z$ direction). From (10.24), we know the electric field inside the Ch film is

$$\vec{E}_{ch} = \frac{v_1}{\sqrt{2}}\begin{pmatrix}\cos\alpha\\ \sin\alpha\end{pmatrix}e^{-ik_on_2z} + \frac{v_2}{\sqrt{2}}\begin{pmatrix}\cos\alpha\\ -\sin\alpha\end{pmatrix}e^{+ik_on_2z}, \tag{10.30}$$

where v_1 and v_2 are electric field amplitudes of eigen modes, respectively. The relations between u, r, v_1, v_2, and t can be found by using the boundary conditions at the two surfaces of the Ch film: the tangential component of the electric field is continuous. The boundary conditions at $z = h$ are

$$t = v_1\cos\alpha e^{-ik_on_2h} + v_2\cos\alpha e^{ik_on_2h} \tag{10.31}$$

$$it = v_1\sin\alpha e^{-ik_on_2h} - v_2\sin\alpha e^{ik_on_2h} \tag{10.32}$$

Note that the local frame is the same as the lab frame at the entrance and exit planes of the Ch film, because the Ch film has an integer number of pitches. From (10.31) and (10.32), we can get

$$v_1 = \frac{t}{\sin(2\alpha)}(\sin\alpha + i\cos\alpha)e^{ik_on_2h} = \frac{t}{\sin(2\alpha)}ie^{ik_on_2h-i\alpha} \tag{10.33}$$

$$v_2 = \frac{t}{\sin(2\alpha)}(\sin\alpha - i\cos\alpha)e^{-ik_on_2h} = \frac{t}{\sin(2\alpha)}ie^{-ik_on_2h+i\alpha} \tag{10.34}$$

The boundary conditions at $z = 0$ are

$$u + r = v_1\cos\alpha + v_2\cos\alpha \tag{10.35}$$

$$u - r = -iv_1\sin\alpha + iv_2\sin\alpha \tag{10.36}$$

From (10.35) and (10.36), we get

$$u = \frac{1}{2}(\cos\alpha - i\sin\alpha)v_1 + \frac{1}{2}(\cos\alpha + i\sin\alpha)v_2 = \frac{t}{2\sin(2\alpha)}i\left(e^{ik_on_2h-i2\alpha} - e^{-ik_on_2h+i2\alpha}\right) \tag{10.37}$$

$$r = \frac{1}{2}(\cos\alpha + i\sin\alpha)v_1 + \frac{1}{2}(\cos\alpha - i\sin\alpha)v_2 = \frac{t}{2\sin(2\alpha)}i\left(e^{ik_on_2h} - e^{-ik_on_2h}\right) \tag{10.38}$$

The reflectance is given by

$$R = \left|\frac{r}{u}\right|^2 = \left|\frac{e^{ik_o n_2 h} - e^{-ik_o n_2 h}}{e^{ik_o n_2 h + i2\alpha} - e^{-ik_o n_2 h - i2\alpha}}\right|^2 = \left|\frac{\sin(k_o n_2 h)}{\sin(k_o n_2 h + 2\alpha)}\right|^2$$

$$= \left|\frac{\sin(2\pi n_2 h/\lambda)}{\sin(2\pi k_o n_2 h/\lambda + 2\alpha)}\right|^2 \tag{10.39}$$

n_2 is the effective refractive index given by (10.19) and is imaginary within the reflection band. Actually (10.39) is valid for wavelength near the reflection band (in and out of the reflection band). $\gamma = (\lambda/P)^2 - \bar{\varepsilon}$, $n_2 = \sqrt{\gamma^2 - \delta^2}/2\sqrt{\bar{\varepsilon}}$. Using (10.21) and (10.22), we can find:

$$\cos(2\alpha) = \frac{\gamma}{\delta}, \quad \sin(2\alpha) = \frac{\sqrt{\delta^2 - \gamma^2}}{\delta} \tag{10.40}$$

Equation (10.39) becomes

$$R = \left|\frac{\sin\left(\frac{2\pi n_2 h}{\lambda}\right)}{\sin\left(\frac{2\pi n_2 h}{\lambda}\right)\cos(2\alpha) + \cos\left(\frac{2\pi n_2 h}{\lambda}\right)\sin(2\alpha)}\right|^2$$

$$= \left|\frac{\sin\left(\frac{2\pi n_2 h}{\lambda}\right)}{\frac{\gamma}{\delta}\sin\left(\frac{2\pi n_2 h}{\lambda}\right) + \frac{\sqrt{\delta^2 - \gamma^2}}{\delta}\cos\left(\frac{2\pi n_2 h}{\lambda}\right)}\right|^2 \tag{10.41}$$

The reflection spectrum (reflection as a function of the wavelength of light) can be analytically calculated by using (10.41). The reflection spectrum with various cell thicknesses is shown in Figure 10.3(a).

When $\lambda = \sqrt{\bar{\varepsilon}}P \approx \bar{n}P$ (near the central wavelength of the reflection band), $\gamma = 0$ and $n_2 = i(1/2)\sqrt{\left[(\varepsilon_{//} - \varepsilon_{\perp})/2\right]^2/\bar{\varepsilon}} \approx i\Delta n/2$ and $2\alpha = \pi/2$

$$R = \left|\frac{\sin(ik_o h\Delta n/2)}{\sin(ik_o h\Delta n/2 + \pi/2)}\right|^2 = \left|\frac{(e^{k_o h\Delta n} - 1)}{(e^{k_o h\Delta n} + 1)}\right|^2 = \left|\frac{(e^{2\pi h\Delta n/\bar{n}P} - 1)}{(e^{2\pi h\Delta n/\bar{n}P} + 1)}\right|^2 \tag{10.42}$$

The peak reflection of the Ch LC film as a function of the product of the birefringence and film thickness is shown in Figure 10.3(b). The peak reflection is saturated when $\Delta nh/\bar{n}P = 1.0$. For example, for a Ch LC with the birefringence of 0.2, average refractive index of 1.6 and pitch of 400 nm (reflecting red light), the film thickness should be 3.2 μ in order to obtain saturated reflection.

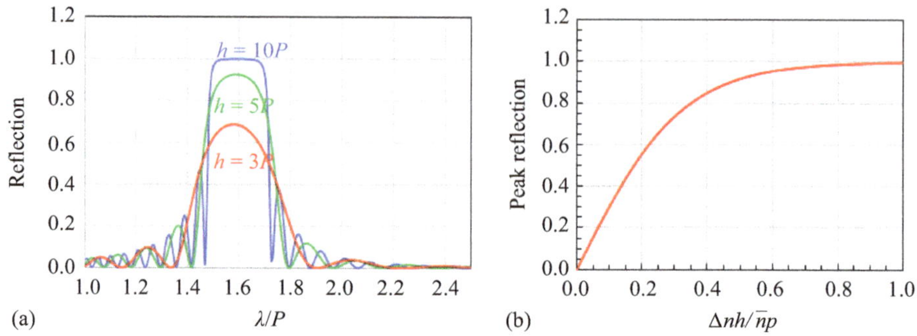

Figure 10.3 (a) Reflection spectrum of Ch LC with various cell thicknesses and (b) peak reflection as a function of birefringence and film thickness. The incident is circularly polarized with the same handedness as the helix of the Ch LC

If the incident light is unpolarized, it can be decomposed into two components: one component with right-handed circular polarization and the other component with left-handed circular polarization. One of the components is reflected by the Ch LC film and the other is transmitted. The maximum reflection of a single layer of Ch LC for unpolarized incident light is 50 percent. Two methods can be used to achieve higher reflection for unpolarized incident light. The first method is to stack two Ch LC films with left- and right-handed helixes. One of the films reflects the right-handed circular polarized component of the incident light and the other reflects the left-handed circular polarized component. The second method is to use two Ch LC films with the same helix and a half waveplate sandwiched between them. The top Ch LC film reflects the right-handed circular polarized component of the incident light. The unreflected left-handed circular polarized component goes through the half-waveplate and is converted into right-handed circular polarization. It is reflected by the bottom Ch LC film. When it goes through the half-waveplate again, it is converted back to left-handed circular polarization and then passes the top Ch LC film.

If the light is incident obliquely on the Ch LC cell, the reflection band is shifted to a short wavelength according to the Bragg law $\lambda = \bar{n}P\cos\theta$, where θ is the incident angle (defined with respect to the normal of the cell). Furthermore the peak reflection increases with the incident angle.

10.3 Ch LC reflective display

Ch LCs are used for reflective display because of their unique optical property of selective reflective reflection. The reflection is from the LCs themselves. They do not need external mirror. A more important property is that they do not need polarizers. They have higher reflectance than polarizer-based LC reflective displays. Furthermore, the feature of polarizer-free makes it possible to use plastic films as the display cell substrate. Plastic films have the advantages of light-weight, flexibility, ruggedness and roll-to-roll manufacture-capability. They, however, usually have nonuniform birefringence. If a display needs polarizers, its image quality will be degraded by the plastic films. Ch LC reflective displays do not need polarizers, and therefore are compatible with plastic substrates.

10.3.1 Display cell structure

In the reflective display, the Ch LC is sandwiched between two parallel substrates as shown in Figure 10.4. There are transparent conducting coating, such as indium-tin-oxide (ITO), on the inner surface of the substrates, which serves as the electrode through which voltages can be applied across the LC. On top of the electrode, there is an alignment layer which provides the proper orientation of the LC. The cell thickness is usually about 5 μ which is necessary to achieve high reflectance.

The optical properties of the Ch LC depend on the orientation of the helical axis [1,4]. When the LC is in the planar state (also called the planar texture or Grandjean texture), where the helical axis is perpendicular to the cell substrate as shown in Figure 10.4(a), the LC is reflecting. When the LC is in the focal conic state (also called focal conic texture), the direction of the helical axis is more or less random as shown in Figure 10.4(b). The focal conic state has a poly-domain structure and is scattering. If the LC has a positive dielectric anisotropy and a sufficiently high voltage is applied across the cell, the helical structure is unwound and the LC is switched to the homeotropic state (also called homeotropic texture), where the LC is perpendicular to the cell substrate as shown in Figure 10.4(c), and it becomes transparent.

10.3.2 Bistability

Ch LCs possess bistability: both the planar state and focal conic state could be stable at 0 V [4,10–12]. The bistability depends on the free energies of the two states and the energy barrier between the two states. The helical structure of the LC molecules is preserved in both states, and therefore these two states have similar elastic energies. The surface anchoring energies of the two states could be different

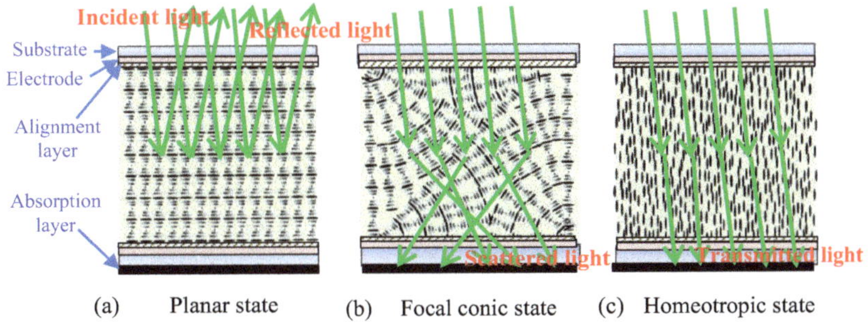

Figure 10.4 Schematic diagrams of the cell structure of Ch LC reflective display and the states of the Ch LC

because of the different orientations of the LC in the two states. In the planar state, the LC is parallel to the cell substrate on the cell surface, while in the focal conic state, the LC is parallel to the cell substrate in some regions and perpendicular to the cell substrate in other regions. Furthermore there is wall energy at the boundary between the domains in the focal conic state.

The bistability of the planar state and focal conic state can be achieved by using proper alignment layers, known as surface stabilization [13]. If a strong homogeneous alignment layer is used, then the planar state is stable at 0 V, but the focal conic state is not. If the LC is switched to the focal conic state by external stimuli such as electric field and pressing, it will slowly (on the order of minutes) relax back to the planar state. If a weak homogeneous alignment layer or a homeotropic alignment layer is used, the difference between the surface energies of the planar state and focal conic state is small. Both the planar state and focal conic state are stable at 0 V.

The bistability can also be achieved by using polymer networks, known as polymer stabilization [12]. For a cell having a strong homogeneous alignment layer, both planar and focal conic states can be stable at 0 V if a small amount of polymer is dispersed in the LC. The cell can be constructed in the following way. First the Ch LC is mixed with a small amount of photopolymerizable monomer. The mixture is filled into the cell. The monomer is then polymerized under UV light with the LC in the homeotropic state in the presence of an external electric field. The formed polymer network is phases separated from the LC, and is located at the boundaries between the LC domains. It creates a big energy barrier between the planar state and the focal conic state, and prevents the LC from relaxing from the focal conic state to the planar state.

Ch LC reflective display with bistability is called bistable Ch reflective display. The display makes use of the two bistable states which exhibit different optical properties. The planar state is reflecting while the focal conic state is non-reflecting. It is highly desirable that the planar state has a high reflectance, close to 50 percent and the focal conic state has a low reflectance, close to 0. As discussed in the previous section, the product of the cell thickness and the birefringence of the LC should be close to 1.0. As will be discussed later, the driving voltage is proportional to the cell thickness, and thus thin cell is desired. Therefore, large birefringence is highly desired, so that the cell can be thin. The focal conic state does not exhibit reflection, but exhibits light scattering. A small part of the incident light is scattered backward, which is undesirable. The backward scattering depends on the birefringence, cell thickness and domain size. The scattering is weaker for LCs with smaller birefringence and thinner cell thickness. The scattering is also weak for larger domain size. The scattering is strong when the domain size is comparable to or smaller than visible light wavelength. The birefringence, however, should be large, because of the required high reflectance of the planar state. The LC domain size in the focal conic state is the main parameter that can be adjusted to minimize the backward scattering. The factors affecting the domain size are the alignment layer, the dispersed polymer, the physical parameters of the LC and the applied voltage.

10.3.3 Viewing angle

Viewing angle is an important property of the Ch reflective display. As discussed in the previous section, the wavelength (color) of the reflection of Ch LC changes with the incident light angle. If the display has a strong homogeneous alignment layer and has no dispersed polymer, the LC will be in the perfect planar state (everywhere the helical axis is exactly perpendicular to the cell substrate), the color of the observed reflection will vary with the viewing angle. This is another reason that why Ch reflective display should not use strong homogeneous alignment layer. If the display has an imperfect planar state consisting of poly-domains as shown in Figure 10.5(a), the viewing would be better [7,10,14]. The helical axes of the domains are not exactly parallel to but distribute around the normal of the cell substrate. Under room light condition, the incident light is not collimated. Observed by the eye of the reader, the colors of the light reflected from different domains are different. There is an averaging effect on the observed color. When the viewing changes, the color-shift is much less severe. The poly-domain structure of the Ch LC can be achieved by using weak homogeneous or homeotropic alignment layers. It can also be achieved by dispersing a small amount of polymer in the LCs.

10.3.4 Gray-level reflectance

Although in the bistable Ch reflective display, the LC is bistable, it is possible to have gray-level reflectances, because of the poly-domain structure [10,11]. When all the domains are in the planar state, as shown in Figure 10.5(a), the display has a maximum reflectance. When a voltage pulse is applied across the cell, it tries to switch the LC from the planar state to the focal conic state. Because of the different domain sizes and different orientations of the helical axis, different domains are switched from the planar state to the focal conic state under different voltage pulse amplitudes and widths. When a low voltage pulse is applied, only some domains are switched to the focal conic state, as shown in Figure 10.5(b), and the reflectance

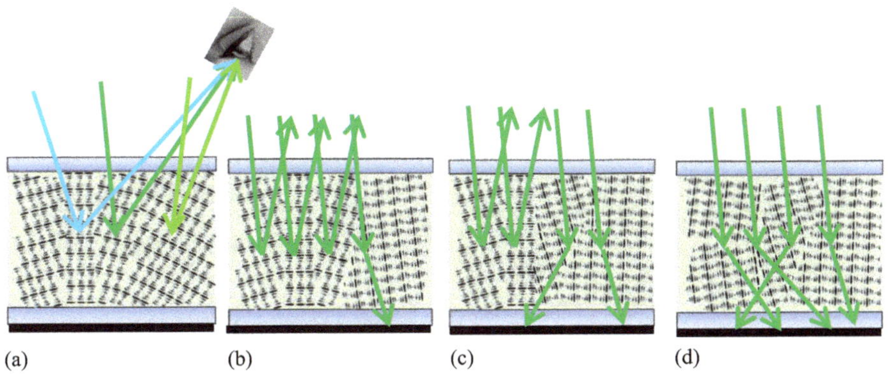

Figure 10.5 Schematic diagrams of the poly-domain structure of states with gray level reflectance of the Ch LC

decreases. When the amplitude of the voltage pulse is increased, more domains are switched to the focal conic state, as shown in Figure 10.5(c). When a sufficiently high voltage pulse is applied, all the domains are switched to the focal conic state, as shown in Figure 10.5(d), and the display has a minimum reflectance. The domain size is on the order of tens micron and cannot be seen by the naked eye.

10.3.5 Color

A single-layer Ch reflective display reflects mono-colored light. The color appearance of the display is determined by the LC and the absorption layer beneath the LC. If the absorption layer is black and the LC reflects green light, the display appears green when the LC is in the reflecting planar state, and black when the LC is in the non-reflecting focal conic state. A black–green display is shown in Figure 10.6(a). In some applications, a white background is preferred. This can be done by using a blue absorption layer (reflecting blue light) and a Ch LC reflects yellow light [15]. When the LC is in the reflecting planar state, yellow light is reflected by the LC and blue light is reflected by the absorption layer; the display has a white appearance. When the LC is in the non-reflecting focal conic state, only blue light is reflected by the absorption layer, the display has a blue appearance. A white-blue display is shown in Figure 10.6(b).

Full-color Ch reflective displays can be made by using color division. There are two schemes. In the first scheme, three Ch LC layers reflecting blue, green, and red light are stacked together [15–17]. Each layer has its own electrode such that its reflection can be controlled independently. When all the three Ch layers are in the focal conic state, the display has a black appearance. When the blue Ch layer is in the planar state and the other Ch layers are in the focal conic state, the display has a blue appearance, and so on. When the blue and green Ch layers are in the planar state and the red Ch layer is in the focal conic state, the display has a cyan appearance, and so on. When all three layers are in the planar state, the display has a white appearance. A full-color Ch reflective display made by Fujitsu Ltd. is

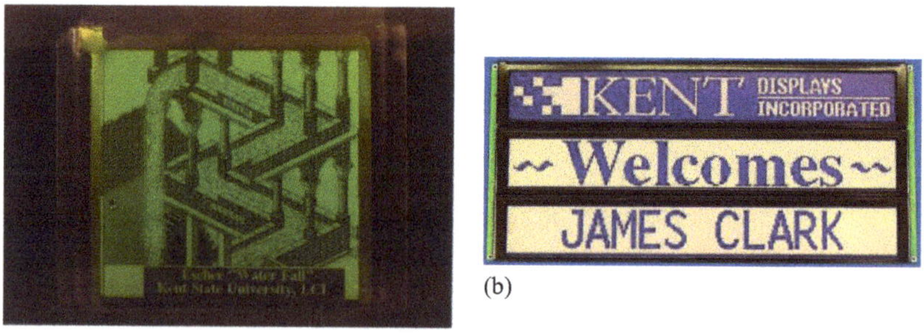

(a) (b)

Figure 10.6 Photographs of bistable Ch reflective displays: (a) black-green display and (b) white-blue display. Photo-courtesy of Kent Display Inc.

Figure 10.7 Photograph of full-color bistable Ch reflective displays. Photo-courtesy of Fujitsu Ltd.

shown in Figure 10.7 [18]. The advantage of this scheme is that the reflection is high. The disadvantage is that manufacturing is challenging. In the second scheme, alternating three Ch LC stripes reflecting blue, green, and red light are deposited in a single layer [17,19,20]. Partitions are placed between the stripes to prevent inter-diffusing. Any color appearance is obtained in a similar way as in the first scheme. The advantage of this scheme is that the manufacturing cost is lower. The disadvantage is that the reflection is low.

10.4 Electrical switching of bistable Ch reflective display

In the bistable Ch reflective display, the used Ch LC should have a positive dielectric anisotropy ($\Delta\varepsilon > 0$). It can be switched among the three states, as shown in Figure 10.8, by voltages applied across the display cell. The state of the LC depends on the amplitude and history of the applied voltage. The free energy of the system is mainly contributed by the elastic energy and electric energy and is given by

$$f = \frac{1}{2}K_{11}\left(\nabla \cdot \vec{n}\right)^2 + \frac{1}{2}K_{22}\left(\vec{n} \cdot \nabla \times \vec{n}\right)^2 + \frac{1}{2}K_{22}\left(\vec{n} \times \nabla \times \vec{n}\right)^2 - \frac{1}{2}\varepsilon_o \Delta\varepsilon\left(\vec{E} \cdot \vec{n}\right)^2. \tag{10.43}$$

Under a given voltage condition, the LC is in the state with the minimum free energy.

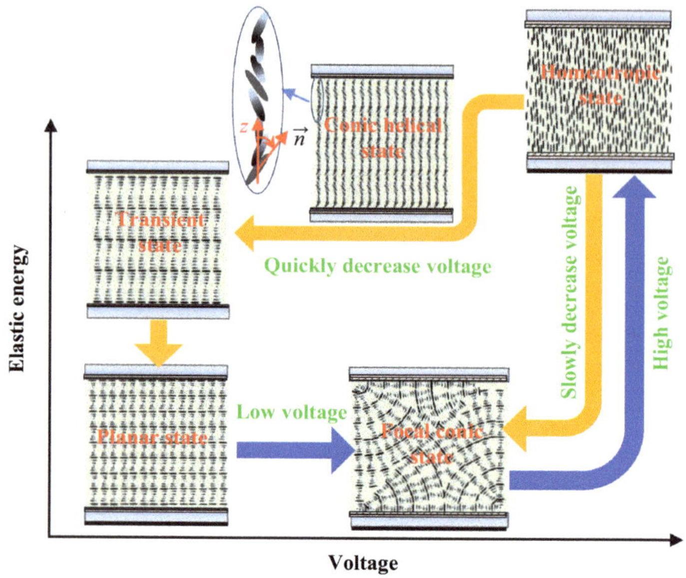

Figure 10.8 Schematic diagram of the transitions between states in bistable Ch reflective display

10.4.1 Transition from planar state to focal conic state

Initially if the applied voltage is 0, the Ch LC is in the planar state with the intrinsic pitch P, and its elastic and electric energies are 0. When a voltage is applied across the cell, the generated electric field is in the z direction (perpendicular to the cell substrate). On the one hand, the LC tends to rotate to align parallel to the electric field to reduce its electric energy, and thus the planar state is not stable under the electric field. On the other hand, the LC tries to retain the helical structure to keep its elastic energy low. If the applied voltage is not high, the LC is switched to the focal conic state where the helical structure is retained and thus the elastic energy is low, and the LC becomes parallel to the electric field in some regions and thus the electric energy is reduced (becomes negative). As mentioned earlier, there is wall energy between the domains and surface energy on the cell surface in the focal conic state. The electric field must be high enough so that the decrease of the electric energy can compensate for the wall and surface energies.

There are two mechanisms through which the LC transform from the planar state to the focal conic state. The first mechanism is the Helfrich deformation where the Ch layers are undulated [21–25]. In the undulation, two energies are involved. First, the undulation costs elastic energy which is against the transformation. Secondly, in some regions the LC is tilted toward the electric field direction and thus the electric energy is decreased. The electric energy is the factor driving the transformation. There is a threshold electric field above which the total free energy is decreased and the undulation takes place. The threshold electric field depends on the pitch P, elastic constants K_{22} and K_{33}, and the dielectric anisotropy $\Delta\varepsilon > 0$, and has a typical value around $2V/\mu m$. The Helfrich deformation is a homogeneous process and takes place simultaneously everywhere. Once the applied electric field is above the threshold value, the amplitude of the undulation increases with the electric field. When the electric field is high enough, the undulation collapses and the LC is transformed to the focal conic state. The transition is fast.

The second mechanism is the buckling of the oily streak where the Ch layers are bent [26]. Here, the involved energies are the electric energy, surface energy and wall energy. The electric energy drives the transition, and the other two are against the transition. There is also a threshold electric field in this mechanism, which is usually slightly lower than that of the Helfrich deformation. The oily streak buckling is a nucleation process that needs nucleation seeds and is low.

In summary, in order to switch the LC from the planar state to the focal conic state, a voltage pulse with the amplitude about 10 V and time interval about 10 ms (for 5 μ thick cell) should be applied. Once the LC is switched to the focal conic state, it will remain there even the applied voltage is turned off, because of the bistability of the focal conic state.

10.4.2 Transition from the focal conic state to the homeotropic state

In the bistable Ch reflective display, both the planar state and focal conic state are stable at 0 V. The LC can be directly switched from the planar state to the focal conic state, but not vice versa. In order to switch the LC from the focal conic state to the planar state, it must be switched first to the homeotropic state by a high voltage; it then relaxes to the planar state after the removal of the voltage, as shown in Figure 10.8. In the homeotropic state, the helical structure is unwound and the elastic energy is high. The voltage needed to switch the LC to the homeotropic state is given by [1,4,27]:

$$V_{FH} = V_C = \frac{\pi^2 h}{P}\sqrt{\frac{K_{22}}{\varepsilon_o \Delta\varepsilon}} \tag{10.44}$$

For example, $K_{22} = 5 \times 10^{-12}$ N, $\Delta\varepsilon = 10$, $h = 5$ μm and $P = 400$ nm, the threshold voltage is 30 V. The transition from the focal conic state to the homeotropic state is relatively slow, and the transition time is usually on the order of a few tens ms.

10.4.3 Transition from the homeotropic state to the focal conic state

The homeotropic state is unstable once the applied high voltage is removed. The LC can transform to either the focal conic state or the planar state, depending on how the applied voltage is removed. We first consider the transition from the homeotropic state to the focal conic state. When the applied voltage is decreased below V_C, the LC tends to restore its helical structure. However, it is impossible to form a helical structure with the one-handed twisting sense and the helical axis in the horizontal direction (parallel the cell substrate). Restoring the helical structure can be achieved only with the help of nucleation seeds [14]. There is also a hysteresis in the transition between the focal conic state and the homeotropic state: it transforms from the focal conic state to the homeotropic state at the threshold voltage V_C; it transforms from the homeotropic state to the focal conic state at a voltage V_{HF} which is lower than V_C. Experiments show that typically $V_{HF} = 0.9V_C$. Because it is a nucleation process, the transition from the homeotropic state to the focal conic state is slow, on the order of a few tens ms.

10.4.4 Transition from the homeotropic state to the planar state

The LC can also transform from the homeotropic state to the planar state [4,28–30]. In the transition, the LC first goes through a conic helical structure as shown in Figure 10.8. In the conic helical structure, there are both twist and bend deformations. In order to minimize the elastic energy, the conic helical structure has a pitch given by $(K_{33}/K_{22})P$. In order for the conic helical structure to start to form, the applied voltage must be below the threshold V_{HP} which is given by

$$V_{HP} = \frac{2}{\pi}\sqrt{\frac{K_{22}}{K_{33}}}V_C. \tag{10.45}$$

For most Ch LCs, $K_{33}/K_{22} \sim 2$, then $V_{HP} = 0.45V_C$, which is lower than V_{HF}. Once the applied voltage is decreased below V_{HP}, the conic helical structure simultaneously forms everywhere. If $K_{33} > K_{22}$, the conic helical structure is not stable. The polar angle θ (the angle between the z-axis and the LC director as shown in the inset in Figure 10.8) increases from 0 to 90° with time as the transition proceeds. When θ becomes 90°, a new planar state, called the transient planar state, is formed. The transition from the homeotropic state to the transient planar state is a homogeneous process and is fast. The transition time is given by

$$\tau_{HTP} = \frac{\gamma P^2}{K_{22}}, \tag{10.46}$$

where γ is the rotational viscosity coefficient of the LC. For example, $\gamma = 0.1\ \text{N}\cdot\text{s/m}^2$, $K_{22} = 5\times10^{-12}$ N and $P = 400$ nm, $\tau_{HTP} = 3$ ms.

We see that there are two possible relaxation modes from the homeotropic state: homeotropic-focal conic mode and homeotropic-transient planar mode. In the first mode, the LC transforms from the homeotropic state to the focal conic state; in the second mode, the LC transforms from the homeotropic state to the transient planar state. The realization of the modes depends how the applied voltage is removed. If the applied voltage is decreased to a value between V_{HP} and V_{HF} and kept there for a time interval longer than the homeotropic-focal conic transition time, the LC transforms to the focal conic state, while the homeotropic-transient planar mode is impossible. The LC remains there even if the applied voltage is then removed completely. If the applied voltage is decreased to a value below V_{HP}, the LC transforms to the transient planar state, because it is much faster than the homeotropic-focal conic mode. Note that the helical axis of the transient planar state is parallel to the LC direction in the homeotropic state, independent of the alignment layer.

The transient planar state has a pitch given by $(K_{33}/K_{22})P$, longer than that of the planar state. Its elastic energy is not zero. Therefore it is not stable. The LC will eventually relax to the planar state with the intrinsic pitch P [31]. The relaxation is a nucleation process and is slow, on the order of a few hundred ms.

10.5 Light switching of the bistable Ch reflective display

The color of the reflected light by Ch LCs can be tuned if their pitch is changed. For a Ch LC made from a nematic host and a chiral dopant, the pitch of the Ch LC depends on the concentration x and *HTP* of the chiral dopant, as shown by (10.1). Once the mixture is made and filled into a display cell, it is difficult to change the concentration. The *HTP* can be changed through molecular isomerization of the chiral molecule under light irradiation. This is known as light-driven cholesteric reflective display. This type of display has the advantages of remote controllability, high spatial and temporal resolution and full-color capability in single-layer cells.

Photoresponsive chiral dopants can be synthesized and subsequently added to the nematic hosts to yield Ch LCs. Light-driven chiral molecular switches and motors with very high helical twisting power (HTP) and exhibiting large variation in HTP upon photoisomerization have been synthesized and used as chiral dopants in the fabrication of Ch LCs. It should be noted that the molecular switches must have at least two different isomeric states which can be interconverted reversibly or irreversibly by light irradiation. Menthone, azobenzene, and dithienylcyclopentene, cyanostilbene-based molecular switches and overcrowded alkene-based molecular motors have been employed as light-driven chiral dopants in the fabrication of photoresponsive Ch LCs. Upon light irradiation, the molecular switches or motors undergo photoisomerization which leads to pitch modulation of the photoresponsive Ch LCs. This pitch modulation enables reflective cholesteric displays with different color [32–36].

Brehmer *et al.* have used a diblock copolymer containing menthone-based chiral photoresponsive moiety as a chiral dopant to fabricate Ch LCs, and they have demonstrated reflective cholesteric displays with different colors (Figure 10.9)

Figure 10.9 Photographs of three locally UV irradiated Ch films fabricated by adding a menthone-based diblock copolymer dopant. Reproduced with permission from Reference [37]. Copyright 1998 WILEY-VCH Verlag Gmbh

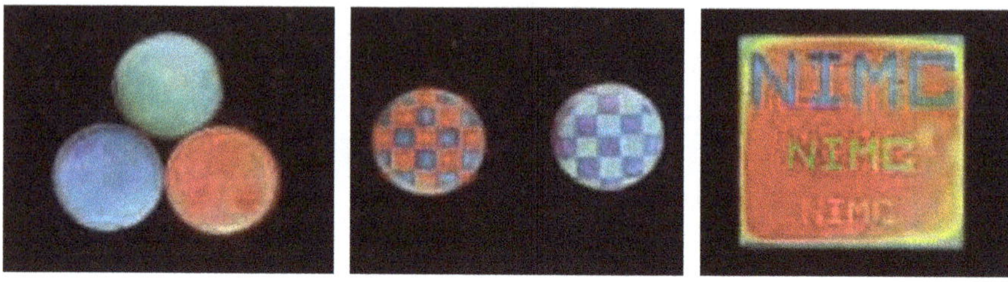

Figure 10.10 Photographs of super-cooled reflective cholesteric films. Left: Samples fixed in red, green and blue colors; middle: Samples with mesh images; and right: A sample with the letters of the alphabet. Reproduced with permission from Reference [38]. Copyright 2000 WILEY-VCH Verlag, GmbH

[37]. Menthone-based chiral dopants exhibit a large change in HTP upon E-Z photoisomerization. Taking advantage of this photoisomerization they were able to demonstrate local reflection color tuning in the Ch film by exposing to different doses of UV light.

Tamaoki *et al.* have fabricated reflective color displays that can be reversibly driven by light irradiation [38]. The photoresponsive Ch LC was obtained by doping an azobenzene derivative into the matrix of a cholesterol-based dimer LC. Full-color recording (Figure 10.10) was achieved by irradiating the Ch films with UV light and rapidly quenching the mesophase. The reflective colors were determined by the duration of UV light irradiation.

Kurihara *et al.* have demonstrated color displays reflecting red, green and blue colors [39]. For this purpose, a cholesteric mixture was prepared from nematic host E44 and two chiral dopants. One of the chiral dopants is photoresponsive due to the presence of azobenzene moieties in its chemical structure. Upon light irradiation, the cholesteric changes its pitch in response to the *trans–cis* photoisomerization of the light-driven chiral dopant. By exposing the cholesteric film through a gray mask, they have obtained patterned red, blue and green reflections (Figure 10.11).

Full-color reflection (Figure 10.12) was achieved in a Ch LC by van Feringa *et al.* using a light-driven molecular motor that exhibits large change in its HTP upon photoirradiation [40]. In this work the cholesteric film was made from nematic host E7 doped with the molecular motor.

Optical tuning of the reflection color of Ch LC containing azobenzene LCs has been demonstrated [41]. Stable and completely reversible photoinduced reflective states have been obtained with the help of both continuous wave and nanosecond laser pulses. Reflective displays have been demonstrated by Bunning *et al.* [42], where the information has been written by laser irradiation to the photoinduced isotropic state of the Ch LC mixture (Figure 10.13). In these displays, the information written can be erased by photoinduced phase transition to the isotropic phase and new information can be rewritten by focused laser scanning of the isotropic phase.

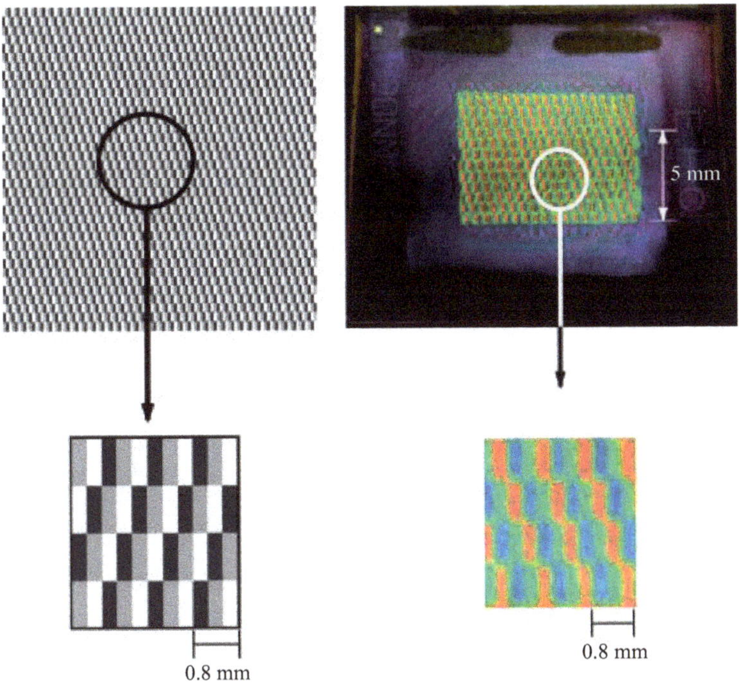

Figure 10.11 *Gray mask (left) and red, green and blue patterned (right) CLC film by UV irradiation through the gray mask. Reproduced with permission from Reference [39]. Copyright 2005 WILEY-VCH Verlag GmbH & Co.*

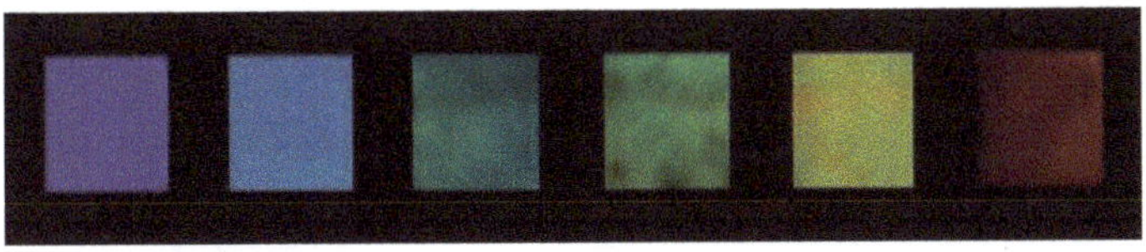

Figure 10.12 *Reflective Ch films obtained from Ch mixture containing molecular motor and irradiated with UV light for different durations of time. Reproduced with permission from Reference [40]. Copyright 2002 National Academy of Sciences*

Li *et al.* demonstrated reflective displays using axially chiral azobenzene-based chiral dopants. Binaphthyl-based chiral azodopants were designed and synthesized that possess large HTP and exhibit a large variation in its HTP upon *trans–cis* photoisomerization [43]. Color-reflective displays have been obtained from the molecular switch doped cholesteric mixtures using photomask techniques. Photo-chiral displays (Figure 10.14) that do not require either expensive drive electronics or complex addressing schemes have also been obtained using these light-driven chiral

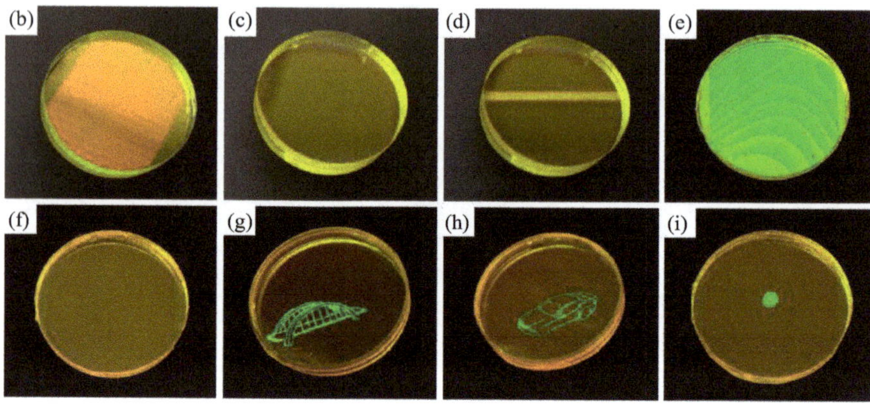

Figure 10.13 Ch reflective display obtained from photo-induced isotropic phase by focused laser scanning. Reproduced with permission from Reference [42]. Copyright 2007 WILEY-VCH Verlag GmbH & Co. KGaA, Weinheim

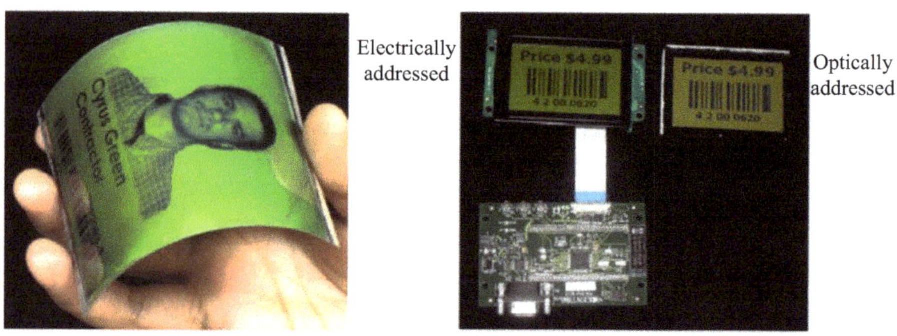

Figure 10.14 Optically addressable photochiral displays from Ch films containing the light-driven chiral molecular switch. Reproduced with permission from Reference [44]. Copyright 2008 Society for Information Display

molecular switches [44,45]. Interestingly, such cholesteric displays can be fabricated on flexible substrates. Such high resolution, low cost, and lightweight cholesteric displays are promising for security badges, advertisement cards, shelf-labels, and other applications where frequent updating of information is not necessary.

Li *et al.* have also reported a chiral molecular switch that enables reversible full-color tuning of the photoresponsive CLC across the visible region by UV and visible light [46]. Using these phototunable cholesteric mixtures, optically addressed displays have been fabricated using photomask technique and appropriate light irradiation. Another structurally related light-driven chiral switch has also been used in the fabrication of reflective cholesteric displays (Figure 10.15) where

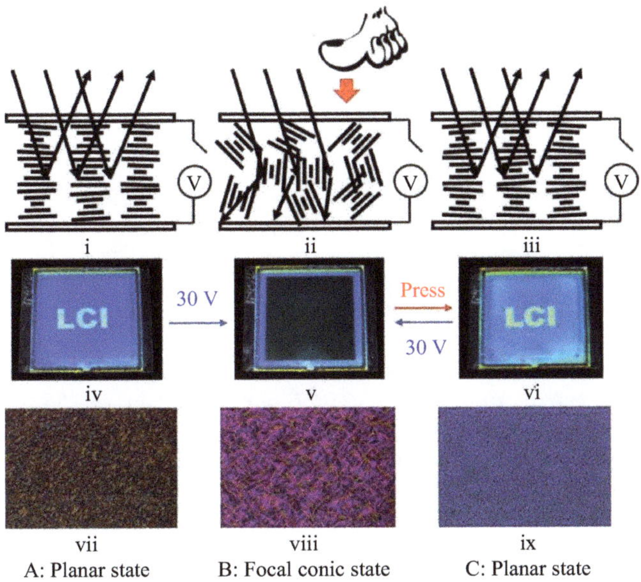

Figure 10.15 *An optically addressed photodisplay that was fabricated by photomask technique using a Ch liquid crystal containing chiral molecular switch that enable full-color tuning. The information written in the display can be hidden by a small voltage pulse and can be subsequentially restored by application of mechanical pressure. Reproduced with permission from Reference [47]. Copyright 2011 WILEY-VCH Verlag GmbH & Co. KGaA, Weinheim*

information can be hidden by applying a short pulse of voltage [47]. The written information can be retrieved when necessary by using simple mechanical shearing. Recently, reflective color displays have been demonstrated using a cholesteric mixture that undergoes handedness inversion upon light irradiation [48]. The cholesteric mixture has been formed by employing a photo insensitive chiral dopant and photoresponsive axially chiral dopant that undergoes dramatic HTP change upon photoisomerization.

The above-described reflective cholesteric displays involve UV light either to write the information or to display the written information. However, it would be more attractive to develop reflective displays that can alleviate the deleterious effects of unsafe UV light. Towards this end, development of visible light-driven reflective displays could be beneficial for practical applications. Accordingly, Li *et al.* have designed and synthesized a chiral molecular switch that exhibits reversible *trans–cis* photoisomerization in response to visible light irradiation and thus avoids the use of UV light (Figure 10.16) [49]. A cholesteric mixture fabricated using this chiral dopant was found to show selective reflection across the entire visible spectrum driven only by visible light. Such cholesteric mixtures are promising for the fabrication of visible light-driven reflective displays.

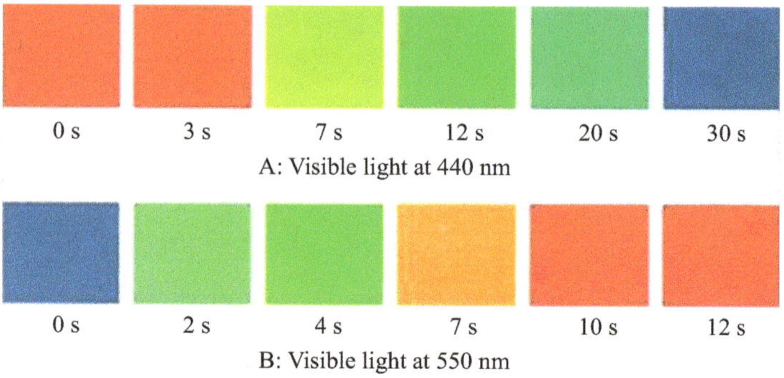

Figure 10.16 Reversible visible light-driven color tuning across the entire visible spectrum in a Ch film containing an azobenzene based axially chiral molecular switch as dopant. Reproduced with permission from Reference [48]. Copyright 2012 American Chemical Society

After achieving visible light-driven reflection color tuning, attempts have been made to take the benefits of near infrared light such as deep penetration, biological application and telecommunication, to tune the reflection color of cholesteric films. In this context, Li *et al.* have demonstrated near infrared light-driven reflection color tuning of chiral molecular switch doped cholesteric films across the visible spectrum using upconversion nanoparticles [50]. These upconversion nanoparticles absorb infrared radiation and emit UV or visible light in situ which brings about the *trans–cis* isomerization of the chiral dopant that leads to the modulation of the pitch of the cholesteric film. The infrared light induced pitch modulation enables reflection color tuning in the Ch films.

Compared to azobenzene-based chiral dopants, diarylcyclopentene-based light-driven molecular switches possess the unique characteristic of thermal stability in both photoisomeric states. Therefore, their use as chiral dopants in the fabrication of photoresponsive cholesteric mixtures could provide optically addressable displays with thermal stability. Feasibility of such an approach has been disclosed by Li *et al.* [51]. They have designed and synthesized a diarylcyclopentene-based chiral molecular switch and used it as the chiral dopant to produce cholesteric mixtures. Owing to its high HTP and the large variation in HTP upon photoisomerization, this molecular switch has enabled reflection color tuning across the visible spectrum (Figure 10.17). Reflective cholesteric displays have been developed using this light-driven chiral dopant.

Recently, cyanostilbene-based light-driven molecular switches have been developed and used in the production of cholesteric mixtures [52]. Such visible light reflecting cholesteric mixtures have been used in the fabrication of reflective displays that can be driven by light irradiation (Figure 10.18). Very recently, a halogen bonded supramolecular chiral molecular switch has been used in the fabrication of photoresponsive Ch LCs and light-driven reflective displays have been demonstrated using the developed cholesteric mixture (Figure 10.19) [53].

Figure 10.17 Reflective photodisplays without thermal relaxation obtained from cholesteric films containing a dithienylcyclopentene chiral dopant. Reproduced with permission from Reference [51]. Copyright 2012 American Chemical Society

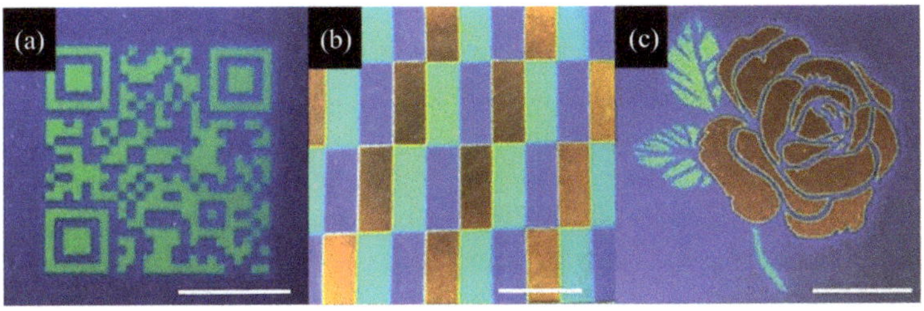

Figure 10.18 Reflective color displays developed from a cholesteric mixture containing a light-driven cyanostilbene derived axially chiral dopant. Reproduced with permission from Reference [52]. Copyright 2017 WILEY-VCH Verlag GmbH & Co. KGaA, Weinheim

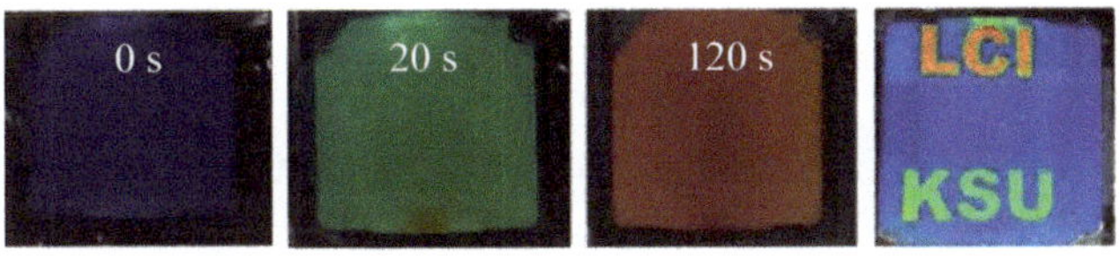

Figure 10.19 Reflective color displays fabricated from photoresponsive cholesteric thin films containing a halogen bonded chiral dopant. Reproduced with permission from Reference [53]. Copyright 2018 Wiley-VCH Verlag GmbH & Co. KGaA, Weinheim

Cholesteric reflective displays that can be driven by UV, visible and infrared light have been demonstrated by judicious design and synthesis of light-driven chiral molecular switches and motors based on different photochromic moieties. Cholesteric displays reflecting colors spanning the entire visible spectrum have been achieved due to the high HTP of the light-driven chiral molecular switches and motors and a significant change in their HTP values upon photoisomerization. Cholesteric reflective displays are cost-efficient, light weight and can operate in static and dynamic modes. The information displayed can be erased and rewritten in such color displays and when necessary the reflective displays can be driven by electric field or other stimuli to achieve special effects. Though appreciable development has been made on the realization of light-driven cholesteric reflective displays, there remains many issues that need to be addressed before they can be commercially viable for practical applications. Facile large area fabrication with uniform alignment throughout and long-term stability of the displays are issues that require focused attention. From a materials point of view, it is necessary to design and synthesize new light-driven chiral dopants with very high HTP, chemical and photochemical stability to keep the momentum of the further development of reflective cholesteric displays.

10.6 Conclusion

Cholesteric LCs possess periodic helical structure with periodicity in visible light region and exhibit the unique selective reflection. They can show various colors by themselves. Because they reflect ambient light and do not need power hungry backlights, they are very energy-saving. They do not need polarizers and exhibit high reflectance of 50 percent, close to those of regular pictures. They are compatible with flexible plastic substrates, and thus can be manufactured in roll-to-roll process with low cost. Furthermore, since they exhibit bistable states in the absence of applied voltage, they can be used to make multiplexed displays on a passive matrix. When they are used to display static images, they do not need to be constantly addressed, another attribute for low power consumption. They are suitable for many non-conventional display applications, such as E-book, E-wall paper, E-signage and E-cloth.

References

[1] de Gennes P. G. *The Physics of Liquid Crystals*, Clarendon, Oxford, 1974.
[2] Chandrasekhar S. *Liquid Crystals*, 2nd edn. Cambridge University Press, New York, 1997.
[3] Blinov L. M. and Chigrinov V. G. *Electrooptical Effects in Liquid Crystal Materials*, Springer-Verlag, New York, 1994.
[4] Yang D.-K. and Wu S. T. *Fundamental of Liquid Crystal Devices*, 2nd edn. Wiley, West Sussex, UK, 2015.
[5] Berreman D. W. 'Optics in stratified and anisotropic media: 4×4-matrix formulation'. *J. Opt. Soc. Am.*, **62** (1972), 502.
[6] Berreman D. W. and Scheffer T. J. 'Bragg reflection of light from single-domain cholesteric liquid crystal films'. *Phys. Rev. Lett.*, **25** (1970), 577.

[7] Wu S.-T. and Yang D.-K. *Reflective Liquid Crystal Displays*, John Wiley & Sons, Ltd., West Sussex, England, 2001.

[8] Xu M., Xu F.D. and Yang D.-K. 'Effects of cell structure on the reflection of cholesteric liquid crystal display'. *J. Appl. Phys.*, **83** (1998), 1938.

[9] Yariv A. and Yeh P. *Optical Waves in Crystals*, John Wiley and Sons, New York, 1984.

[10] Yang D.-K., Doane J. W., Yaniv Z. and Glasser J. 'Cholesteric reflective display: drive scheme and contrast'. *Appl. Phys. Lett.*, **65** (1994), pp. 1905–07.

[11] Yang D.-K., West J. L., Chien L. C. and Doane J. W. 'Control of the reflectivity and bistability in displays based on cholesteric liquid crystals'. *J. Appl. Phys.*, **76** (1994), pp. 1331–33.

[12] Yang D.-K. and Doane J. W. 'Cholesteric liquid crystal/polymer gel dispersions: Reflective displays'. *SID Intl. Symp. Digest Tech. Papers*, **23**, 759 (1992).

[13] Lu Z.-J., St. John W. D., Huang X.-Y., Yang D.-K. and Doane J. W., 'Surface modified reflective cholesteric displays'. *SID Intl. Symp. Digest Tech. Papers*, **26** (1995), 172.

[14] Yang D.-K. 'Chapter 16 Reflective cholesteric liquid crystal displays'. in Bhowmik A., Li Z. and Bos P. J. (eds.) *Mobile Displays*, ed., John Wiley & Sons, Ltd., West Sussex, England, pp. 443–66, 2008.

[15] Davis D., Kahn A., Huang X.-Y. and Doane J. W. 'Eight-color high-resolution reflective cholesteric LCDs'. *SID Intl. Symp. Digest Tech. Papers*, **29** (1998), 901.

[16] Davis D., Hoke K., Khan A., Jones C., Huang X. Y. and Doane J. W. 'Multiple color high resolution reflective cholesteric liquid crystal displays'. *Proc. Intnl. Display Research Conf.*, **242** (1997).

[17] Hashimoto K., Okada M., Nishguchi K., Masazumi N., Yamakawa E. and Taniguchi T. 'Reflective color display using cholesteric liquid crystals'. *SID Intl. Symp. Digest Tech. Papers*, **29** (1998), 897.

[18] http://vivaebooks.blogspot.com/2009/10/ereaders-to-come-fujitsus-flepia.html?m=0 (Accessed on 3 August, 2018)

[19] Chien L.-C., Muller U., Nabor M.-F. and Doane J. W. 'Multicolor reflective cholesteric displays'. *SID Intl. Symp. Digest Tech. Papers*, **26** (1995), 169.

[20] Braganza C., Bowser M., Krinock J., *et al.* 'Single layer full color cholesteric display'. *SID Intl. Symp. Digest Tech. Papers*, **42** (2011), 396.

[21] Helfrich W. 'Deformation of cholesteric liquid crystals with low threshold voltage'. *Appl. Phys. Lett.*, **1970** (1970), pp. 531–32.

[22] Helfrich W. 'Electrohydrodynamic and dielectric instabilities of cholesteric liquid crystals'. *J. Chem. Phys.*, **55** (1971), pp. 839–42.

[23] Gerritsma C. J. and Van Zanten P. 'Periodic perturbations in the cholesteric plane texture'. *Phys. Lett. A*, **37** (1971), pp. 47–48.

[24] Hurault J. P. 'Static distortions of a cholesteric planar structure induced by magnetic or AC electric fields'. *J. Chem. Phys.*, **59** (1973), pp. 2068–75.

[25] Yu M., Yang H. and Yang D.-K. 'Stabilized electrically induced Helfrich deformation and enhanced color tuning in cholesteric liquid crystals'. *Soft Matter*, **13** (2017), 8728.

[26] Lavrentovich O. D. and Yang D.-K. 'Cholesteric cellular patterns with electric-field-controlled line tension'. *Phys. Rev. E*, **57** (1998), *Rapid Communications*, pp. R6269–72.
[27] Meyer R. B. 'Distortion of a cholesteric structure by a magnetic field'. *Appl. Phys. Lett.*, **14** (1969), 208.
[28] Yang D.-K. and Lu Z.-J. 'Switching mechanism of bistable cholesteric reflective displays,' *SID Intl. Symp. Digest Tech. Papers*, **26** (1995), 351.
[29] Kawachi M., Kogure O., Yosji S. and Kato Y. 'Field-induced nematic-cholesteric relaxation in a small angle wedge'. *Jpn. J. Appl. Phys.*, **14** (1975), 1063.
[30] Greubel W., Wolf U. and Kruger H., 'Electric field induced texture changes in certain nematic/cholesteric liquid crystal mixtures'. *Mol. Cryst. Liq. Cryst.* **24** (1973), 103.
[31] Watson P., Anderson J. E., Sergan V. and Bos P. J. 'The transition mechanism of the transient planar to planar director configuration change in cholesteric liquid crystal displays'. *Liq. Cryst.*, **26** (1999), 1307.
[32] Wang Y. and Li Q. 'Light-driven chiral molecular switches or motors in liquid crystals'. *Adv. Mater.*, **24** (2012), pp. 1926–45.
[33] Li Y. and Li Q. 'Photoresponsive cholesteric liquid crystals', in Q. Li (ed.) Chapter 5 in *Intelligent Stimuli Responsive Materials: From Well-defined Nanostructures to Applications*, John Wiley & Sons, Hoboken, NJ, 2013.
[34] Tamaoki N. 'Cholesteric liquid crystals for color information technology'. *Adv. Mater.*, **13** (2001), pp. 1135–47.
[35] Bisoyi H. K. and Li Q. 'Light-driven liquid crystalline materials: From photo-induced phase transitions and property modulations to applications'. *Chem. Rev.*, **116** (2016), pp. 15089–15166.
[36] Bisoyi H. K. and Li Q. 'Light-directing chiral liquid crystal nanostructures: From 1D to 3D'. *Acc. Chem. Res.*, **47** (2014), pp. 3184–3195.
[37] Brehmer M., Lub J. and van de Witte P. 'Light-induced color change of cholesteric copolymers'. *Adv. Mater.*, **10** (1998), pp. 1438–1441.
[38] Tamaoki N., Song S., Moriyama M. and Matsuda H. 'Rewritable full-color recording in a photon mode'. *Adv. Mater.*, **12** (2000), pp. 94–97.
[39] Yoshioka T., Ogata T., Nonaka T., Moritsugu M., Kim S.-N. and Kurihara S. 'Reversible-photon-mode full-color display by means of photochemical modulation of a helically cholesteric structure'. *Adv. Mater.*, **17** (2005), pp. 1226–29.
[40] van Delden R. A., Koumura N., Harada N. and Feringa B. L. 'Unidirectional rotary motion in a liquid crystalline environment: Color tuning by a molecular motor'. *Proc. Nat. Acad. Sci.*, **99** (2002), pp. 4945–49.
[41] Hrozhyk U. A., Serak S. V., Tabiryan N. V. and Bunning T. J., 'Optical tuning of the reflection of cholesterics doped with azobenzene liquid crystals'. *Adv. Mater.*, **17** (2007), pp. 1735–42.
[42] Hrozhyk U. A., Serak S. V., Tabiryan N. V. and Bunning T. J. 'Photoinduced isotropic state of cholesteric liquid crystals: Novel dynamic photonic materials'. *Adv. Mater.*, **19** (2007), pp. 3244–47.

[43] Li Q., Green L., Venkataraman N., *et al.* 'Reversible photoswitchable axially chiral dopants with high helical twisting power'. *J. Am. Chem. Soc.*, **129** (2007), pp. 12908–09.

[44] Montbach E., Venkataraman N., Khan A., *et al.* 'Novel optically addressable photochiral displays'. *SID Digest*, **39** (2008), pp. 919–22.

[45] Venkataraman N., Magyar G., Lightfoot M., *et al.* 'Thin flexible photosensitive cholesteric displays'. *J. SID*, **17** (2009), pp. 869–73.

[46] Ma J., Li Y., White T., Urbas A. and Li Q. 'Light-driven nanoscale chiral molecular switch: Reversible dynamic full range color phototuning'. *Chem. Commun.*, **46** (2010), pp. 3463–65.

[47] Li Q., Li Y., Ma J., Yang D.-K., While T. J. and Bunning T. J., 'Directing dynamic control of red, green, and blue reflection enabled by a light-driven self-organized helical superstructure'. *Adv. Mater.*, **23** (2011), pp. 5069–73.

[48] Wang L., Chen D., Gutierrez-Cuevas K. G., *et al.* 'Optically reconfigurable chiral microspheres of self-organized helical superstructures with handedness inversion'. *Mater. Horiz.*, **4** (2017), 1190–95.

[49] Wang Y., Urbas A. and Li Q. 'Reversible visible-light tuning of self-organized helical superstructures enabled by unprecedented light-driven axially chiral molecular switches'. *J. Am. Chem. Soc.*, **134** (2012), pp. 3342–45.

[50] Wang L., Dong H., Li Y., *et al.* 'Reversible near-infrared light directed reflection in a self-organized helical superstructure loaded with upconversion nanoparticles'. *J. Am. Chem. Soc.*, **136** (2014), pp. 4480–83.

[51] Li Y., Urbas A. and Li Q. 'Reversible light-directed red, green, and blue reflection with thermal stability enabled by a self-organized helical superstructure'. *J. Am. Chem. Soc.*, **134** (2012), pp. 9573–76.

[52] Li J., Zhang Z., Tian J., Li G., Wei J. and Guo J. 'Dicyanodistyrylbenzene-based chiral fluorescence photoswitches: An emerging class of multifunctional switches for dual-mode phototunable liquid crystals'. *Adv. Opt. Mater.*, **5** (2017), 1700014.

[53] Wang H., Bisoyi H. K., Wang L., Urbas A. M., Bunning T. J. and Li Q. 'Photochemically and thermally driven full-color reflection in a self-organized helical superstructure enabled by a halogen-bonded chiral molecular switch'. *Angew. Chem. Int. Ed.*, **57** (2018), pp. 1627–31.

Chapter 11

Enhancement of optical output and low-power consumption in nanoparticle-embedded liquid crystal devices

Yukihide Shiraishi[1], Shunsuke Kobayashi[2], Tomohiro Miyama[3], Naoki Toshima[2], Hirokazu Furue[4], Kenzo Takeishi[5], Haruyoshi Takatsu[5], Kai-Han Chang[6], and Liang-Chy Chien[6]

We present unique optical effects in nanoparticle (NP)-embedded liquid crystal devices. For conducting this research, we synthesized Pγ CyD-ZrO_2 and we prepared NP system comprised of seven NPs of Aerosil R-812 (EVONIK) through a sonication, where the original Aerosil has the size of 400 nm. LCD device cells embedded with these NPs exhibit the optical enhancement effects such as reduction of laser speckle patterns, increase of output intensity, viewing angle, and color gamut and further reduction of power consumption evaluated through luminance efficiency.

This chapter is written by updating the previous publications [1–8] through describing the details of syntheses of NP and giving physical modeling of light scattering based on Rayleigh–Gans scattering theory accompanying an updated figure and also the effect of raising effective temperature called Kobayashi temperature and pretilt increasing effect with simulation.

11.1 Introduction

The motive and purpose of the doping of NPs into LCD is to modify physical properties of liquid crystal (LC), actually more or less all the physical properties and parameters change by doping NPs into LCD depending on the combination of NP and LC, thus it is expected to change electro-optical characteristics of LCD [1–3].

[1]Department of Applied Chemistry, Sanyo-Onoda City University, Japan
[2]Tokyo University of Science, Yamaguchi, Japan
[3]Center for Liberal Arts and Science, Sanyo-Onoda City University, Japan
[4]Department of Materials Science and Technology, Tokyo University of Science, Japan
[5]Technology of Liquid Crystal Materials, DIC Corporation, Japan
[6]Liquid Crystal Institute, Kent State University, USA

In a previous paper, we reported that the doping of NPs into LC electro-optical devices brings enhancements of their characteristics such as high-speed response, wide viewing angle, large color gamut reaching 140 percent of NTSC specification, implementation of field-sequential color LCDs, and optical homogeneous effect including laser speckle patter reduction [1–8]. This chapter is updated though detailed description of the NP of PγCyD-ZrO_2 and preparation of NP system of Aerosil R-812 (EVONIK) whose size is 7 nm and contains 7 silica NP covered with organic ligand molecules through a special sonication and explanation of obtained unique optical effects based on Rayleigh–Gans optical scattering effect accompanying updated relevant figure. And further obtained enhancements of optical effects are explained thorough effective raising of temperature called Kobayashi temperature due to the existence of NP and based on the new index: luminance efficiency (LE).

11.2 Experimental

1. Nematic LC: NTN-02(DIC), $n_e = 1.625$, $n_o = 1.499$, $\Delta n = 0.126$
2. Electrically controllable birefringence (ECB) cells: cell gap of 13 μm, with and without doping NPs (Shiraishi Lab) [4]
3. Narrow gap field-sequential (color-filter-less) twisted nematic (NTN) modules (10 cm × 10 cm, 100 × 100 pixels) (OKAYA) [5] with and without doping NPs of Aerosil R-812 (EVONIK) and ECB-cells with PγCyD-ZrO_2.
4. NPs:
   (A) Properties and synthesis of PγCyD-ZrO_2 (Shiraishi Lab) NP size is 7.2 nm and the refractive index of $n = 2.2$ doped quantity was 0.075 weight percent and the mean distance between NPs is 40 nm. The synthesis of PγCyD-ZrO_2 was done as shown in Figure 11.1.

   The PγCyD-ZrO_2 protected NPs were prepared with a microwave reactor equipped with ultrasonic nozzle (Microwave 2,450 MHz and 150 W, ultrasonic 20 kHz, Shikoku Keisoku, Japan). PγCyD (0.014 mmol) in monomeric units, 0.1 times the total amount of zirconium ethoxide (0.14 molo) were mixed in tetra ethylene glycol to form 185 cm^3 solution. the mixed solution filled with pure nitrogen and then expose to the microwave (2,450 Hz) and ultrasonic wave (150 W, 29 kHz). Contaminations like ions in the dispersion were removed by washing the dispersions three times with ethanol by using an ultrafiltration. Finally, complete removal of the solvent and volatile byproducts by vacuum evaporation leads to produce PγCyD-ZrO_2 protected NPs with the size of 7.2 nm (7–8 nm). These total operations need an expertise accomplishment.

   (B) Aerosil R-182 (EVONIK) of SiO_2 that is usually available product has the size of 400 nm and contains 7,000 NPs; we prepared Aerosil R-182 called NP system of having the size of 40 nm containing seven NPs with $n = 1.46$ and protected with ligand organic molecules. Reason why we prepared 40 nm-sized NPs was that we are interested in the size effect on

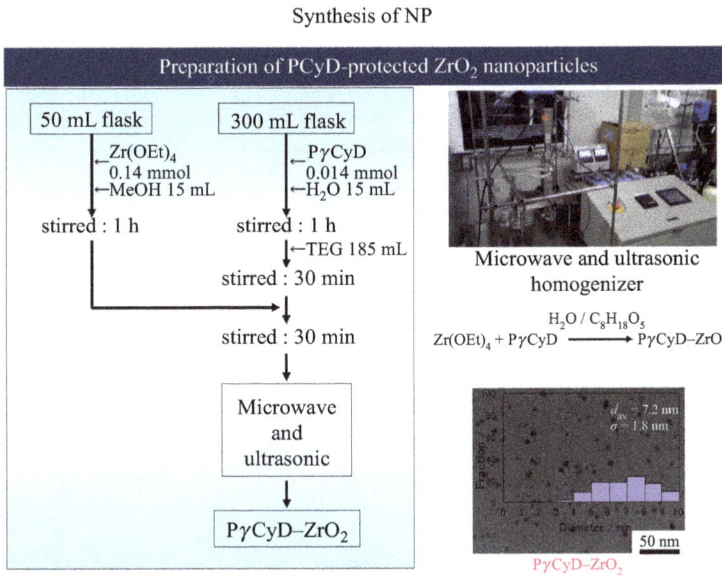

Figure 11.1 Synthesis and TEM of PγCyD-ZrO_2 NPs

the optical effect of NP-embedded LC, even though the doped concentration was very low like 0.1 weight percent. Method for producing 40 nm NPs from the original NPs with 400 nm-size was as the following way:

We doped original NPs into NLC (NTN-02) to gather with a chiral dopant having the pitch of 12 μm and then, the bottle containing this mixture was subjected to a sonication for 2 h with an ordinary laboratory ultrasonic syringe.

TEM micro-photographs of the mixture that was obtained after the procedure of a freeze break technique are shown in Figure 11.1(a) before the sonication and (b); after the sonication demonstrating successful getting of NP system with 40 nm size in which there exist seven silica NPs. Those TEM treatments were done by Tokuyama Corp.

11.3 Results and considerations

11.3.1 Laser speckle reduction

For the observation of optical transmitted pattern through ECB cells with and without doping NPs of PγCyD-ZrO_2, we used a laser with wavelength 488 nm (LGK7872M LASOS). A white screen was placed at a distance of 30 m from the cell. Figure 11.3 demonstrates the scattering patterns projected on a screen of an Argon laser beam passing through the ECB cells without NPs (Figure 11.2(a)) and with NPs (Figure 11.2(b)). The laser speckles and some bright crosses are visible on the screen as seen in Figure 11.3(a). The bright crosses are the diffraction from the

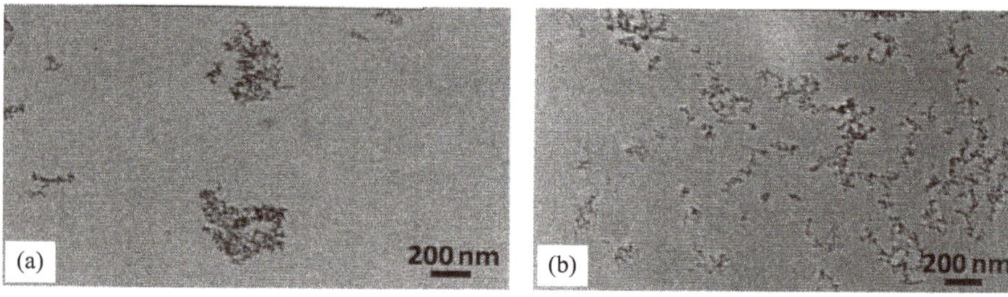

Figure 11.2 TEM photographs: (a) before sonication and (b) after sonication (©SPIE)

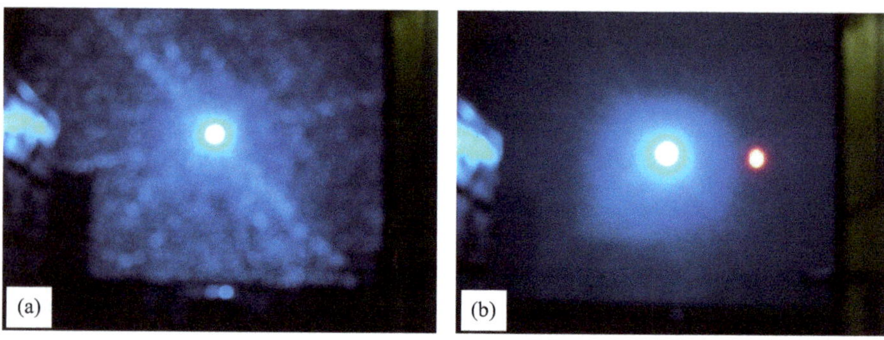

Figure 11.3 Optical patterns projected on a screen of an argon laser beam passing through ECB cells: (a) without NPs and (b) with NP doped (©SPIE)

periodic structure on the substrates of the ECB cell. By contrast, the speckles and diffraction patterns are smooth out after passing through the NP-embedded LC cell as seen in Figure 11.3(b). There is no external voltage applied on the LC cell during observation, which indicates that it is zero-power consumption laser speckle reduction device. The benefit of using PγCyD-ZrO_2 NP rather than that of Aerosil silica is that there is large refractive index difference between NPs and LC host. The scattering intensity increases as the refractive index difference between NP and LC host is large, where the necessary quantity of doped NPs was only 0.075 weight percent.

The reduction of speckle pattern and crosses may be attributed to the optical scattering by the NPs and thus de-coherence occurs for passing light though the NP-embedded ECB cell, where the size of NP (7 nm) is much smaller than optical wave length. This kind of optical scattering will be analyzed with Rayleigh–Gans criteria [10] such that:

$$2ka|m-1| \ll 1 \tag{11.1}$$

where k is the wave number for light with wavelength λ propagates in a medium with the refractive index ratio m between n-NP (2.2) and n-LC (1.4699); $m = 1.47$

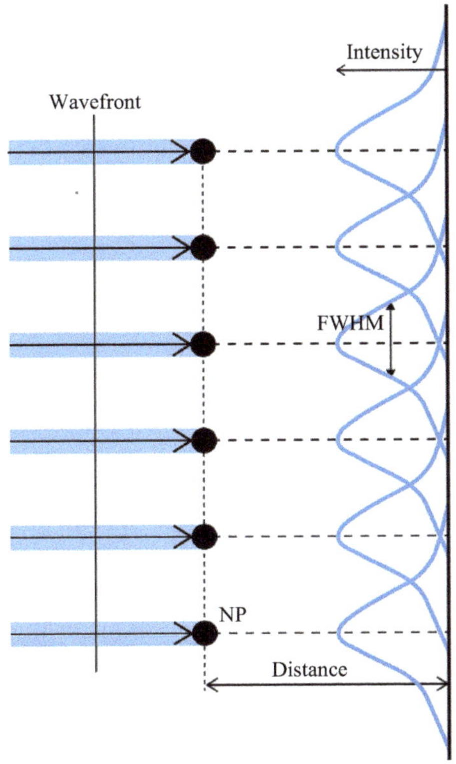

Figure 11.4 Illustration of laser speckle reduction by Rayleigh–Gan's scattering of NPs. Reprinted with permission from reference [7]. Copyright (2018), Elsevier B.V

and $k = 2\pi/\lambda = 2\pi n_{\mathrm{medium}}/\lambda_{\mathrm{vac}}$, and a is the radius of the NP. In this system, the diameter of NP is ~7 nm and the wavelength of light is 488 nm, which satisfy Rayleigh–Gans criteria. The mechanism of laser speckle reduction with NPs is illustrated in Figure 11.4 that was drawn by updating original figure in Ref. [7]. As an idealized situation, we take use one-dimensional NP array with the distance of between NPs is as example to explain the laser speckle reduction contributed by the scattering effect of NPs. Light (six right arrows in Figure 11.4) shines on the NPs from the left-hand side, which creates a forward scattering pattern on the right-hand side. The projection of scattering pattern from each NP has Gaussian-like intensity distribution. The width between the 50 percent intensity location is defined as full-width half-maximum (FWHM). The distribution of NPs and scattering pattern are critical for laser speckle reduction. For smoothing the speckles, the FWHM has to be larger than the speckle periodicity σ,

$$\sigma \approx 1.2\frac{\lambda D}{D_{\mathrm{laser}}}, \tag{11.2}$$

where D is the observation distance as denoted in Figure 11.4, λ is the wavelength of incident light, and D_{laser} is the diameter of the laser spot [8,9,11]. The mean NP distance has to be smaller than the FWHM so that the scattering pattern of NPs does not create another speckle-like intensity distribution. In order to have minimized intensity loss, the scattering has to be occurred from multiple NPs.

In the experiment, the FWHM of the scattering pattern from single NP on the screen calculated by Rayleigh–Gan's scattering, which is ~46.5 m, is larger than the average distance between NPs that is about 400 nm. From calculation, the laser speckle periodicity from the laser, which is ~17.6 mm, is smaller than the FWHM of the scattering pattern of the NP in the system [10,11]. The number of NPs covered within the laser spot is $\sim 10^{10}$. Therefore, the scattering can reduce the laser speckle efficiently without losing the light intensity.

We compare scattering effect between NP of PγCyD-ZrO_2 and that of Aerosil R-812. With the Aerosil R-812, the concentration of NP was 30 weight percent and need for applying electric field [10]. Thus, the result was that PγCyD-ZrO_2 shows efficient optical scattering effect even though amount of doping of NPs is 0.074 weight percent and without applying electric field . Benefit of using PγCyD-ZrO_2 is due to the high refractive index of the PγCyD-ZrO_2 with $n = 2.2$; this causes a large refractive index discrepancy between NPs and LC host with $n = 1.499$, whereas the refractive index of silica is $n = 1.46$ bringing a weak optical scattering. A detailed research for this topic particularly physical simulations will be published elsewhere. A precise and quantitative measurement of the laser speckle patterns is now underway by Furue-Shiraishi group with a measuring instrument Adobe Photoshop Element 10; results will be published elsewhere.

11.3.2 Enhancement of the viewing angle property and output optical intensity

We measured the angular-dependence of optical output of the TN Modules without and with NPs of PγCyD-ZrO_2 with ELDIM EZ Contrast MS Series at 25 °C and applied 2.5 V square voltage at 100 Hz and the results are shown in Figure 11.5(a) and 11.5(b). The experimental results show that the optical output has widened uniformity with NPs of PγCyD-ZrO_2. In other words, the viewing angle of the TN module is widened with the existence of NPs. We further extracted the angular dependence of optical output at the horizontal line of polar plot in Figure 11.6 (i.e., the observation is at azimuthal angle of 0° and scan through polar angle from −90° to 90°). The results show that not only the viewing angle is widened, but also the optical intensity increases with the existence of NPs.

The results of an independent experiment demonstrate that the threshold voltage of a TN module deceases by −7 percent due to the existence of NPs and this decreasing is reproduced by raising temperature by 10° on the pure TN module; they call this temperature Kobayashi temperature [1]. Almost the same results shown in Figures 11.4 and 11.5 were obtained on the ECB cell.

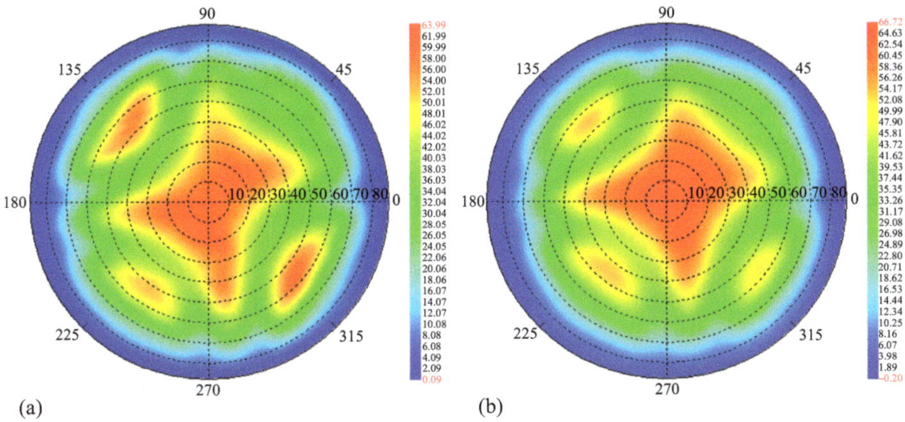

Figure 11.5 Optical output viewing-angle-dependent profiles of TN-module: (a) without NPs and (b) with NPs of PγCyD-ZrO_2 (©SPIE)

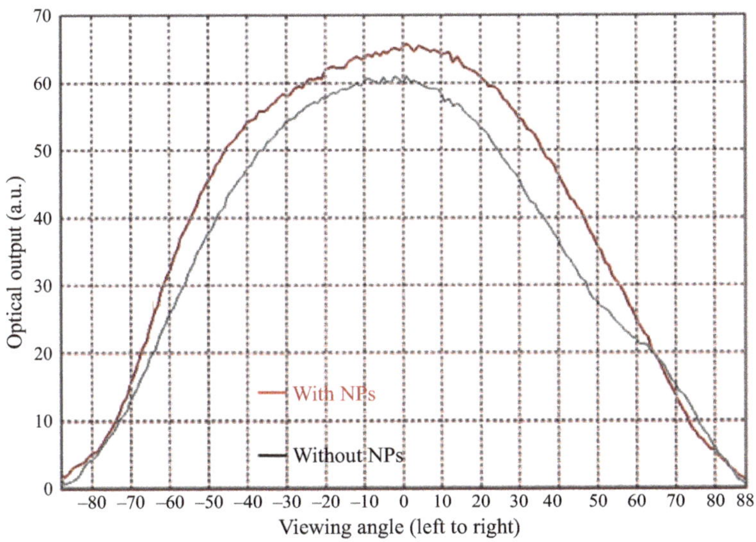

Figure 11.6 Viewing angle property at the azimuthal direction 0° of TN-module without NPs (lower line) and with NP of PγCyD-ZrO_2 (upper line) (©SPIE)

An approximate analytical relation between the relative decrease of threshold voltage and that of order parameter is given such that:

$$\delta V_{\text{th}}/V_{\text{th}} = \frac{1}{2}\delta S/S. \tag{11.3}$$

Since, $V_{th} = \pi(K_{eff}/\varepsilon_0\Delta\varepsilon)^{1/2}$ and here, K_{eff} (effective elastic constants) $\propto S^2$ and $\Delta\varepsilon \propto S$. By the experiment, $\delta V_{th}/V_{th}$ was -7 percent so that $\delta S/S$ becomes $= -14$ percent. This decrease of $\delta S/S$ was realized by the raising of temperature by 10 °C. This is the definition and the determination of the Kobayashi temperature.

The reason why output intensity of the TN module is enlarged in comparison with that of the pure TN modules may be attributed to the following two origins: one is increasing the forward scattering by NPs and the other is the effect of the following situations: according to our independent experiments, (1) the *V–T* curve in the normally black mode of the NT module shift toward low voltage by 7 percent owing to the doing of NPs; (2) both *V–T* curves saturated (reduced transmittance $T = 100$ percent) at the applied voltage of 4 V; and (3) at $T = 90$ percent for 2.8 V, there occurred increase of output intensity by 5 percent due to the doping of NPs; this means that our TN module will be operated at a low voltage by 7 percent compared to that for the TN module without NPs. This shift in the *V–T* curve is reproduced by the raising of temperature by 10° temperature. Besides this effect the increase of pretilt angle from 1° to 3° contribute the increase of viewing angle; a detailed simulation is under process and the results will be published elsewhere.

11.3.3 Analysis of surface topography of ECB cells with and without NPs

The OPTIPRO-MICRO (SHINTECH) is an instrument for analyzing surface topography of LC cells whose substrates are treated by rubbing or photoalignment for aligning LC molecules. Figure 11.7(a) demonstrates the surface topography in an LCD cell fabricated by rubbing process that reflects the existence of

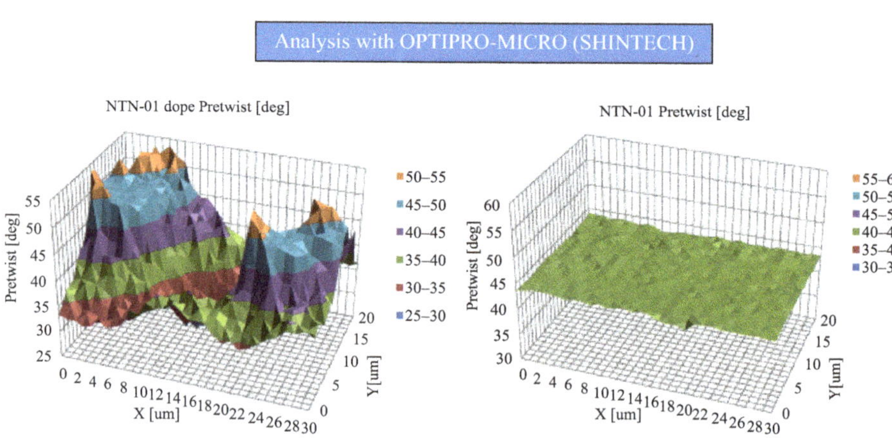

(a) NTN-01 (Pure) (b) NTN-01 + Nanoparticles

Figure 11.7 Surface topography of an LC cell with substrates fabricated by rubbing with SHINTECH OPTOPRO-MICRO. The LC cell (a) without NPs and (b) with NP. Reprinted with permission from reference [7]. Copyright (2018), Elsevier B.V micro-meter order optically anisotropic LC segments; whereas, in (b) we see almost completely smooth appearance, even though on the similar sample. These phenomena may be understood that this is attribute by the Rayleigh–Gan's optical scattering may completely diminish optical interference (diffraction) caused by the surface structure, since the OPTIPRO-MICRO works as an optical instrument. This means that phenomena shown in Figure 11.7 may be attributed to optical phenomena, but there is an possibility of the smoothing effect due to the existence of NPs at the substrates of LCD cell.

11.3.4 Increase of color gamut by doping NPs in NTN FSC-LCD

We measured the color gamut of both the two field-sequential color NTN LCD modules without and with NPs of Aerosil 812 silica with the above mentioned luminance colorimeter (Topcon BM-7A) and placing LCD modules in the bench-top Type Temperature Chamber SU-641 (serial no. 92008614 (ESPEC)) and connecting between the LCD-module and the luminance colorimeter through a peripheral optical fiber.

Figure 11.8 shows that the color gamut of NTN-field-sequential color (FSC) LCD module is increased by doping NPs of Aerosil R-812. At room temperature, the color gamut of our NTN-FSC LCD without NPs has the color gamut of 140 percent of NTSC standard due to color filter less and a 5 percent increase from that of the LCD with NPs. Even at −10 °C, the LCD with NPs has the color gamut of 100 percent of NTSC standard. This kind of increase of color gamut will be primarily due to color filter-less and the situation that the LCD is kept in a high temperature by 10 °C (called Kobayashi temperature) due to a reduction of order parameter by 16 percent [1]. Reason why color gamut increases with the raising of temperature is recognized that it is inherent to FSC-LCD and this may be attributed

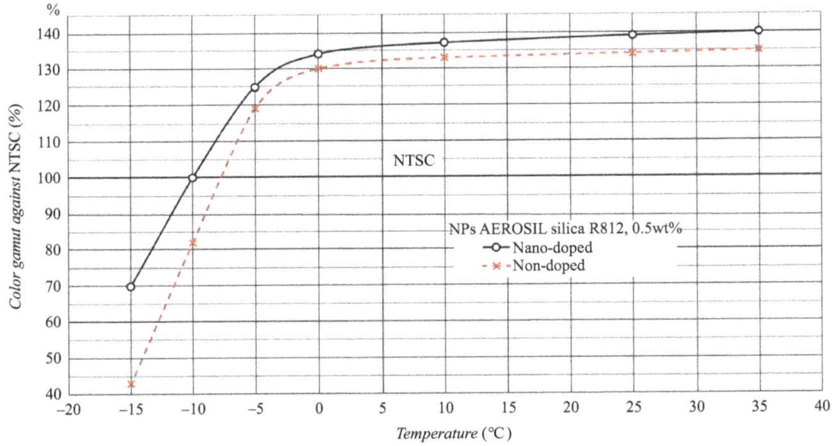

Figure 11.8 Temperature dependence of color gamut of NTN-FSC-LCDs with and without the doping of NPs of PγCyD-ZrO_2. Reprinted with permission from Reference [7]. Copyright 2018, Elsevier B.V

Table 11.1 Evaluation of LE of relevant LCDs

	LCD-TV (40″)	FSC-LCD (with NPs)
[A] Luminance (cd/m^2)	350	1,124
[B] Power consumption	80/0.44 = 182	2.7/0.01 = 270
[A]/[B] LE (cd/W)	1.92	4.16

to the reduction of response time of both the rising and falling process as temperature rises that brings a god formation of RGB color and reduction of color mixing of two adjacent colors. Contrary to this, at extremely low temperature, due to the increase of response time there occurs a shrinkage of color gamut as shown in Figure 11.8.

11.3.5 LE of NTN-FSC-LCD with NPs

We suggest a new index for expressing energy efficiency called LE such that: LE = [Luminance (cd/m^2)]/[W/m^2] [3,6] and we evaluated the values LE of our NTN-FSC-LCD with NPS and existing LCD-TV: SHARP Model LC-40H40 and further compared these values based on The Energy Program Version 5-3(2011). Table 11.1 shows a comparison of the LE of the LCD-TV and our FSC-LCD module. The Energy Star Program (ESP) Version 5-3 (2011) declare that: power consumption of an LCD-TV with the diagonal of 50″ must be 108 W or less. According to this regulation the LE value of LCD will be 3.9 cd/W by taking the maximum value of luminance is to be 1,000 cd/m^2, even though the ESP does not specify the value of luminance. The catalogue of the SHARP LCD-TV with 60 fps declares that its power consumption is 80 W with that of the tuner.

If we take its luminance as an appropriate value to be 350 cd/m^2, then its LE value will be 1.92 as shown in Table 11.1, while our FSC-LCD module with sequential RGB colors each of whose frame rate is 60 fps. The values of luminance were measured with a LUMINANCE COLRIMETER BM-7A (Topcon) and electrical power consumption was evaluated by measuring applied voltage and electrical current. Thus, we get its LE-value is 4.16 as demonstrated in Table 11.1.

Our FSC-LCD can achieve the ESP standard. In SID 2016, Sadashivan *et al.* (QD Vision Inc.) reported that they got LE of 3.38 by adopting QD back light for LCD-TV [12].

11.4 Considerations and concluding summary

1. Laser speckle pattern was almost completely reduced for the coherent laser light passing through an ECB cell doped with NP of PγCyD-ZrO$_2$ at the concentration of 0.075 weight percent without applying electric field to the LC cell. This effect was analyzed based on Rayleigh–Gan's scattering theory.
2. Optical output characteristics ECB-LCD module doped with PγCyD-ZrO$_2$ was shown to be largely enhanced in terms of the increase of total output intensity

accompanying with the broadening of viewing angles. This effect was explained as follows: (1) The existence of NPs corresponds to raising of temperature by 8.3 °C (called Kobayashi temperature), which is attributed to the decrease of order parameter by 6.2 percent. (2) Local pretilt angle increases from 1° to 3°.

3. In a field-sequential color NTN-LCD module doped with NPs of Aerosil R-812, color gamut reaches 140 percent that of NTSC standard at room temperature and 100 percent at −10°. This FSC-LCD exhibited LE = 4.1 that exceeds that of the Energy Star Program Version 5-3 (2011) LE = 3.8. These effects may be attributed to the situation that doping NPs brings a situation of as if LCD is kept at elevated temperature by 10 °C (called Kobayashi temperature).

Acknowledgments

Part of this research was carried out as MEXT Japan Yamaguchi Green Materials Cluster project. We also thank Mr. Yukiya Yamashita of NIPPON AEROSIL CO. LTD for providing us with Aerosil NPs. And K.-H. Chang and L.-C. Chien thank Vol. 10125 the financial support from Ohio Third Frontier (OTH) Venture Support Fund under grant number TECH2015-0128.

References

[1] Kobayashi S., Akimoto M., Takatoh K., *et al.* 'Electro-optical Properties of LCD Doped with Nanoparticles and With Optical Compensators: Ways for Fast Response'. *Mol. Crys. Liq. Crys.*, 2014, **594**, pp. 21–30.

[2] Kobayashi S., Shiraishi Y., Sawai H., Toshima N., Takatsu H. and Kotani K. 'Optical Properties of NTN-FSC-LCD and ECB cells with the Doping of Nanoparticles'. *Proc. of SPIE 8642*, 2013, p86420M.

[3] Kobayashi S., Chapter 1, *Progress In Liquid Crystal Science and Technology*, In Honor of Shunsuke Kobayashi's 80th Birthday, (Editors) H.-S. Kwok, S. Naemura, and H.L Ong. Singapore, World Scientific (2013).

[4] Sawai H., Shiraishi Y., Miyama Y., Kobayashi S. and Toshima N. 'Zirconia Nanoparticles Having a Nanospace of Poly(cyclodextrin: Preparation and Application to Liquid Crystal Devices'. *J. Nanoscience and Nanotechnology* 2013, pp. 11–13.

[5] http://www.okayaelec.co.jp (Accessed on 14 August, 2018).

[6] Kobayashi S., Shiraishi Y., Furue H., Chang K.-H. and Chien L.-C. 'Optical Homogeniging Effects in Nanoparticle-Embedded Liquid-Crystal Devices'. (invited) *Proceedings of SPIE-The International Society for Optical Engineering*, 1012511-1012511-6 (2017).

[7] Kobayashi S., Shiraishi Y., Toshima N., *et al.* 'Further Study of Optical Homogeneous Effects in Nanoparticle-Embedded Liquid-Crystal Devices' *J. Mol. Liq.*, 2018, **267**, pp. 303–307.

[8] Chang K.-H., Shiaishi Y., Toshima N., Kobayashi S., Furue H. and Chien L.-C. 'Laser Speckle Reduction Using Nanoparticle-Embedded Liquid Crystals' SID 2017 Digest of Technical Papers, 38, 804–07 (2017).

[9] Van De Hulst H. C., Light Scattering by Small Particles, Dover Publications, Inc., New York (1957).

[10] Ishikawa H., Weng W., Ono M., Furue H. 'Reduction of Laser Speckle Noise by Using Fine-Particle-Dispersed Liquid Crystals', 26th International Liquid Crystal Conference 2016 (Kent State University), ILCC-2016-97.

[11] Dainty C. (Ed), *Laser Speckle and Related Phenomena*, 1984, Berlin, Heidelberg, Springer-Verlag, ISBN 0-387-13169-8.

[12] Sadasivan S., Bausemer K., Coliss S. and Prat R. 'Performance Bench Marking of Wide Color Gamut Televisions and Monitors'. SID 2016, Digest of Technical Papers, 37, pp. 333–35.

Chapter 12
Ferroelectric micro LC display

Masaki Oishi[1]

Ferroelectric micro LC display is one of the transmissive liquid crystal on silicon (LCOS) displays, and it has some advantages for small and high resolution with low power consumption. This technology using ferroelectric LCs is useful for wearable device, such as head mounted display.

12.1 Introduction

Till date, many micro displays have been investigated [1–3]. In this chapter, features of ferroelectric micro LC display and differences from other micro LC display, trend of ferroelectric micro LC display technology development and introduction of market are discussed in detail. Also particularly due to the widespread use of the Internet, as called Internet of Things (IoT), the demand for accessing necessary information and bringing in information regardless of where it is being expanded has expanded, and at the same time the display element importance is also increasing. Under such circumstances, there is an increasing demand for a wearable device that can display necessary information when necessary and is equipped with. In this chapter, ferroelectric micro LC display, which is one of the display elements for head-mounted display (HMD) among wearable devices, is also discussed.

12.2 What micro LC display and LCOS are

There are mainly two kinds of micro liquid crystal display (LCD), transmission type and reflection type. The difference between them is whether to transmit light from the light source to display (transmissive type LCD) or reflective to display (reflective type LCD).

An example of a transmissive micro LCD element will be introduced. In the transmission type micro LCD, it is possible to display an image by selecting transmission/non-transmission of light from the light source by liquid crystal.

[1]Open Innovation Department, Citizen Watch Co., Ltd., Japan

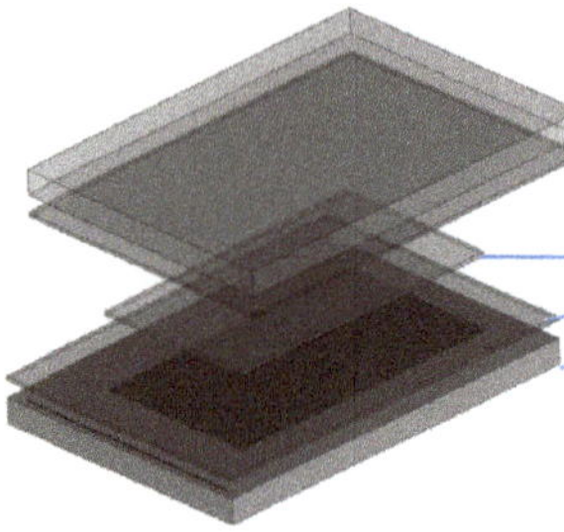

Figure 12.1 Each layer of reflective micro LCD showing LCOS construction (liquid crystal on silicon)

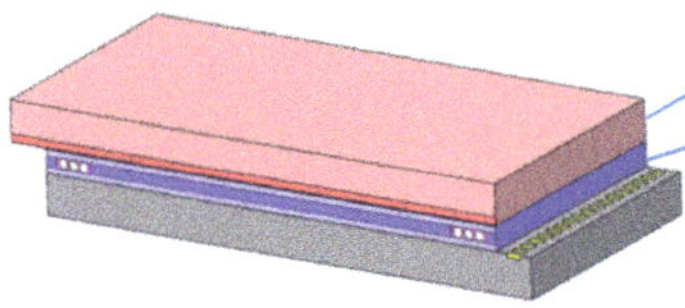

Figure 12.2 Cross section of LCOS (liquid crystal on silicon)

For example, transmissive micro LCD has a liquid crystal between two glass substrates. The configuration and cross section of transmissive LCOS (liquid crystal on silicon) are explained in Sections 12.1 and 12.2, respectively (Figures 12.1 and 12.2). At that time, pixels are formed on the substrate on one side, and various patterns can be expressed by controlling the light from the light source in transmission/non-transmission in each pixel.

Regarding colour display, colour filter method and field-sequential method can be cited, and these will be explained in Section 12.3.

It is possible to arrange the light source and the display linearly as a feature of the transmissive micro LCD element. Therefore transmissive micro display is possible to construct a simple optical system, which can be advantageous in that an optical system can be constructed in a space-saving manner for a wearable device which is required to be miniaturised.

On the other hand, there are disadvantages such as reduction in light efficiency. Because, the transmissive micro LCD element generally controls each pixel, therefore has a TFT (Thin Film Transistor) circuit in each pixel and the TFT circuit part does not transmit light, so that transmissive micro LCD has low light efficiency.

Next is an example of a reflection type of micro LCD, especially LCOS (liquid crystal on silicon) type (Figure 12.1). Image can be displayed by selecting reflection/non-reflection of light from a light source by a liquid crystal. A structure in which a liquid crystal is provided between a glass substrate and a silicon substrate can be cited as an LCD device structure. At that time, pixels are formed on the silicon substrate, and various patterns can be expressed by controlling the light from the light source to reflect/non-reflection at each pixel. The colour display is the same as the transmission type micro liquid crystal display element. Since the reflection type micro liquid crystal display element is a reflection type, the structure with the light source is complicated as compared with the transmission type micro LCD element. However, it is possible to use a pixel on the surface of the silicon substrate as a reflective electrode, and there is an advantage that the effective efficiency of light can be increased by forming a circuit for driving the pixel in the silicon. This means it is possible to increase the brightness, which is effective when simultaneously checking both the real world and the micro LCD image using the HMD. And particularly in the see-through type HMD, like an AR (augmented reality) and MR (mixed reality) function is remarkably effective.

12.3 Advantage point of ferroelectric micro LC display

A ferroelectric micro LCD is a micro LCD having a ferroelectric liquid crystal. It is characterised by one of a reflection type LCD having a ferroelectric liquid crystal.

Ferroelectric liquid crystal (FLC) has ferroelectricity, that is, spontaneous polarisation. This is biggest difference compared with other liquid crystal (like a nematic liquid crystal). Because FLC has spontaneous polarisation, unlike the induced polarisation caused by the external electric field of other liquid crystals, high-speed electric field response is possible. As the superiority of ferroelectric liquid crystal, the following items can be mentioned.

1. High response time can bring field-sequential mode, also can bring high frame rate.
2. Suppression of contrast reduction due to the lateral electric field of the pixel electrode edge.

Time division colour display (field-sequential method) can be cited as a method to exploit the superiority of ferroelectric liquid crystal. In a colour twisted nematic (TN) type of LCD using a conventional backlight, the white light source is constantly lit. That is, as a space division display method (colour filter method), one pixel is divided into three (red, green and blue), each red, green and blue micro colour filters are provided for each pixel, and colour display is realised by performing spatial colour synthesis. That is, colour reproduction is performed in the spatial domain.

In the ferroelectric micro LCD, the same pixel is temporally divided into three by using red, green and blue colour light sources such as LEDs. As a time division display method (field-sequential method), and the liquid crystal light shutter is opened in synchronism with monochromatic light emission of monochromatic of each red, green and blue. That is, a method of performing colour reproduction in the time domain. Therefore, RGB micro colour filter is unnecessary for a ferroelectric micro LCD, and it is unnecessary to divide one pixel into three spatially.

Figure 12.3 shows field-sequential method and colour filter method. As a trend of the latest micro-LCD, small size and high resolution can be cited, but each pixel is miniaturised in a compact and high definition micro-LCD element, and a micro

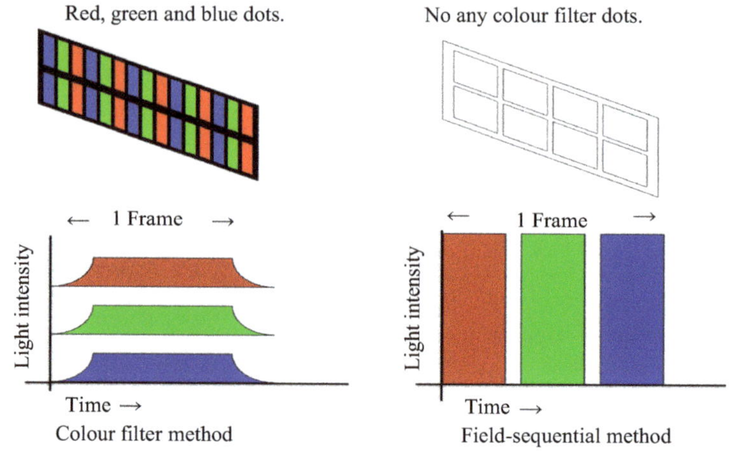

Figure 12.3 Frame image of colour filter method and field-sequential method

colour filter is provided for each miniaturised pixel. The colour filter method in the spatial domain is extremely difficult.

For example, it is extremely difficult to form an RGB micro colour filter corresponding to a high resolution with a pixel pitch like a 5 μm or less, and at the same time, a colour filter method for absorbing light by a colour filter for colouring is preferred from the viewpoint of light utilisation efficiency is also disadvantageous.

That is, the field-sequential method is indispensable in the micro-LCD which is required to have small size and high resolution. Furthermore, in the field-sequential method, high-speed screen rewriting which does not cause flicker in the time domain is also indispensable element. For these reasons, ferroelectric liquid crystal having a high response speed is the most powerful method of providing realistic devices for micro-LCD elements of small and high definition time division type.

Without using a micro colour filter, it is possible to display a high resolution by performing colour display by time division with the same pixel, and realise a high aperture ratio. Then, a micro-liquid crystal characterised by small size with high resolution.

The narrow gap of the ferroelectric liquid crystal (gap: the gap between the glass substrate and the silicon substrate and this thickness is for the liquid crystal layer) and in-plane motion suppresses the contrast reduction of the pixel electrode edge against the lateral electric field.

Figure 12.4 shows the operation by the electric field of ferroelectric liquid crystal and nematic liquid crystal. The long ellipse represents liquid crystal molecules, and the major axis direction of the ellipse represents the alignment direction of the liquid crystal. Arrows indicate the direction of polarisation of the liquid crystal. The electric field (E) applied to the liquid crystal portion is represented by a vertical line written in the vertical direction, which means that the lateral electric field is generated as the pixel electrode edge becomes.

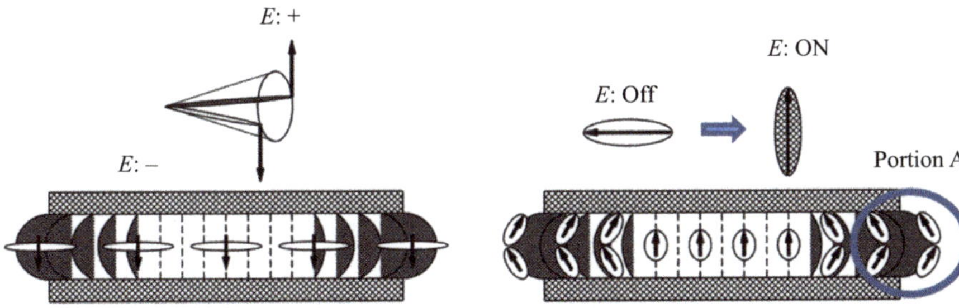

(a) Electric field of FLC (b) Electric field of nematic

Figure 12.4 Electric field of FLC and nematic LC

While the nematic liquid crystal has electrostatic polarisation in the major axis direction of the liquid crystal molecule, the ferroelectric liquid crystal has spontaneous polarisation parallel to the substrate normal direction. Therefore, due to the electrical coupling with the electric field parallel to the substrate normal direction, the direction of the liquid crystal molecules changes according to the direction of the driving electric field, the light is phase-modulated, light and dark are generated through the polarising plate, image display is performed.

In the case of a ferroelectric liquid crystal, the liquid crystal is oriented in the horizontal direction and the polarisation is upward in the case of the ferroelectric liquid crystal, and when it is −E, the liquid crystal moves on the cone shown in Figure 12.4(a). Therefore, when liquid crystal molecules are projected to the substrate surface, they are oriented in the horizontal direction while rotating toward the front side and the polarisation direction is downward. In this way, it is driven by in-plane motion that moves in the same plane with respect to the substrate. The nematic liquid crystal is driven in the vertical direction with E-On from the horizontal direction of the initial orientation by E-Off.

With respect to the electric field, in the nematic liquid crystal, the liquid crystal driving direction takes a movement from the direction orthogonal to the optical axis to the parallel direction, whereas in the ferroelectric liquid crystal, it is driven in a direction in which the relationship with the optical axis always changes. Therefore, in the nematic liquid crystal, the orientation direction of the liquid crystal is not completely aligned with the optical axis due to the transverse electric field at the edge of the pixel electrode (Portion A in Figure 12.4), causing poor alignment and lowering the light transmittance. However, such a phenomenon is not seen in the ferroelectric liquid crystal, and the contrast loss is much smaller. When the resolution becomes higher, the influence of the transverse electric field generated at the edge of the pixel electrode with respect to the pixel size increases, so that the ferroelectric liquid crystal, has very advantageous in the small and high resolution micro display.

Ferroelectric liquid crystals usually have a helical structure, but narrow liquid crystal layer (gap) is needed for the in-plane motion. And then it will be used as a display. Further, due to the narrow gap, the effective voltage applied to the liquid crystal layer is increased, and as a result, liquid crystal driving under low voltage can be realised, and high response speed can be obtained in low-voltage driving.

In order to make use of the ferroelectric liquid crystal, it is essential to form a liquid crystal element with a narrow gap, and by using a liquid crystal element with a narrow gap, ferroelectric liquid crystal can be effectively used as a display.

In general, it is difficult to achieve narrow gaps because stable production is very difficult, and mass production is difficult. This is because it becomes difficult to maintain the gap uniformity in the micro display element with narrow gap. Especially gap non-uniformity appears as display unevenness in micro display, which can be a factor of deteriorating quality. By using precision automation technology for these tasks, stable mass production and supply can be realised by controlling in-plane uniformity for micro-display elements having a gap of 1 μm or less.

12.4 Silicon backplane

Due to the space margin of the backplane and the high carrier mobility, the micro-LCD using the DRAM as the back plane, as shown in Figure 12.5, not like other micro LC displays, ferroelectric micro-LCD has liquid crystal driver and the frame buffer memory, the video processing and other functions. It means peripheral control system can be incorporated in the same backplane. And this is the one of advantage point of ferroelectric micro-LCD. This is because the ferroelectric liquid crystal can be driven at low voltage it is possible to form a driving section with a CMOS circuit normally used in a silicon backplane.

Also, in the field-sequential method, 'colour break', which is called colour break, in which the observer perceives switching of temporal colour, becomes a problem, but this backplane used in the ferroelectric micro-LCD which has high response speed. The backplane can show three same images at one frame (It will be 180 fps), the so-called triple speed display within one frame (60 fps) taking advantage of the high response speed of the above ferroelectric liquid crystal, and this triple speed display is effective against colour break.

Furthermore, in order to correspond to the temperature characteristics of the liquid crystal, by providing the temperature compensation function in the backplane, it is possible to adopt different liquid crystal driving methods at a plurality of temperature points, realise optimum liquid crystal driving at each temperatures, it brings high image quality in the wide temperature ranges. The biggest feature is 'single chip,' it has all the above functions in the same backplane.

Therefore, a peripheral control circuit such as a liquid crystal drive circuit or the like becomes unnecessary externally, and it is possible to respond to market demands which are required to downsize the entire device.

By providing the necessary driving circuits in the same backplane in single chip, what is necessary as external input (communication) when using a ferroelectric micro-liquid crystal element is video interface/power supply/serial interface.

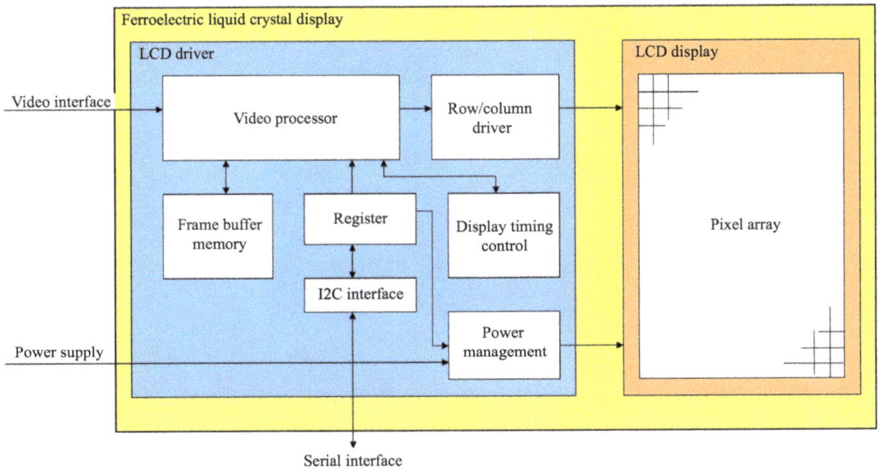

Figure 12.5 Silicon backplane of ferroelectric micro-LCD

12.5 Application, market

Ferroelectric micro LC display has two models as product group.

1. Ferroelectric micro LC display with optics
2. Ferroelectric micro LC display without optics.

With optics model has light source and illuminator, without optics model does not have any optics (light source or illuminator). These are to cope with both the market requiring optical systems (electrical view finder) and the markets that need to deal with the optical systems of each customer (projector or need high brightness market.)

A simplified cross-sectional view of with optics model is shown in Figure 12.6. Light from a light source (e.g., LED) becomes a surface light source from the point light source based on the optical path of its own optical design, obtains the polariser, and outputs it as deflected light. Thereafter, only the light polarised by the ferroelectric liquid crystal of the ferroelectric micro-LCD is reflected by the analyser to the ferroelectric micro-LCD and can be confirmed as transmitted light by transmitting through the analyser. These optical systems are designed based on the cultivated technologies and experiences, and stable mass production supply is possible by precision automation technology. Further, with the silicon backplane described above, it is possible to control luminance, contrast, etc. of each device, and mass production supply with stable quality can be realised.

For without optics model, the same optical system is prepared according to each application. However, the light emission timing of the light source corresponding to field-sequential mode is output as a digital signal from the silicon backplane.

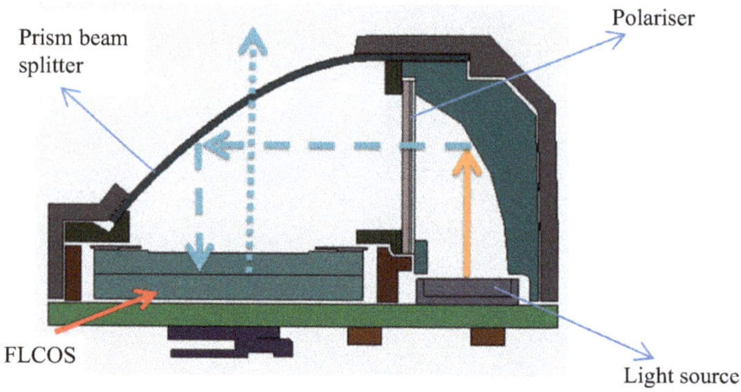

Figure 12.6 Cross section of ferroelectric micro LC display with optics

With optics model uses the LED driver built in the silicon backplane, so it can not send a large current to the light source. Therefore, without optics model is preferred for applications requiring high luminance (large current).

With optics model is mainly used for electronic viewfinder, monocular, binocular market, and without optics model is mainly used for projector and HMD (head-mounted display) market.

In the HMD market, it can be roughly divided into four categories.

1. Monocular small type (transmissive type/non-transmissive type).
2. Monocular large size (transmission type/non-transmission type).
3. Binocular small type (transmissive type/non-transmissive type).
4. Binocular large type (transmission type/non-transmission type).

Transmissive and non-transmissive type exists in each of monocular and binocular. Differentiation is made according to the application. Transmissive HMD is called see-through, and many are characterised by AR (augmented reality) function.

For recent B2B (business-to-business) applications, used cases are given which provide smooth work instructions by combining work instruction images with the real world, such as work support and remote area support. In that case, transparent type is very effective. Further, in the transmission type of both eyes, it is possible to display a large amount of information and also a high definition display element is required. MR (mixed reality) is based on such transmission type of both eyes.

A particularly large non-transmissive HMD is called an immersive type, and there are ones characterised by a VR (virtual reality) function. Most recently, it is expected to be useful in amusement applications and game applications, and many applications are being developed. For the HMD market, in the ferroelectric micro LC display has product group shown in Table 12.1.

qHD (960 × 540) and 720p (1,280 × 720) of 0.24 active area inch are effective for the monocular and binocular small type of HMD. Both of them can deal with the miniaturisation required for wearable, and miniaturisation of the entire device

Table 12.1 Ferroelectric micro LC display for HMD market

Model	**0.24 inch**	**0.38 inch**
Resolution	960 × 540 (qHD)	1,280 × 720 (720p)
Aspect ratio	16:9	16:9
Pixel pitch	5.5 μm	6.6 μm
Frame rate	47–63 Hz	47–63 Hz
Clock	25–75 MHz	24–75 MHz
Colour field rate	360 Hz (Typ.)	
Video interface	(a) RGB or YCbCr(4:4:4)-Parallel (24 data, Hd, Vd, Clock) (b) YCbCr(4:2:2)-Parallel (16 data, Hd, Vd, Clock) (c) YCbCr(4:2:2)-Serial (8 data, Hd, Vd, Clock)	
Input video resolution	480 × 270–1,368 × 768	640 × 360–1,368 × 768
Control interface	I2C Serial	
Colour image method	Field-sequential mode	
Input signal level	CMOS 1.8–3.3 V	
Input power	1.8 V, 3.3 V, I/O	1.8 V, 3.3 V, 5.0 V, I/O
Model	**0.24 inch**	**0.40 inch**
Resolution	1,280 × 720 (720p)	1,280 × 960 (QuadVGA)
Aspect ratio	16:9	4:3
Pixel pitch	4.1 μm	6.3 μm
Frame rate	47–63 Hz	47–120 Hz
Clock	25–75 MHz	24–103 MHz
Colour field rate	540 Hz (Max)	540 Hz (Max)
Video interface	(a) RGB or YCbCr(4:4:4)-Parallel (24 data, Hd, Vd, Clock) (b) YCbCr(4:2:2)-Parallel (16 data, Hd, Vd, Clock) (c) YCbCr(4:2:2)-Serial (8 data, Hd, Vd, Clock) (d) MIPI DSI Subset	(a) RGB888-Parallel 24-bit DATA (b) YCbCr(4:4:4)-Parallel (24 data, Hd, Vd, Clock) (c) YCbCr(4:2:2)-Parallel (16 data, Hd, Vd, Clock)
Input video resolution	640 × 360–1,368 × 768	VGA–QuadVGA
Control interface	I2C Serial	
Colour image method	Field-sequential mode	
Input signal level	CMOS 1.8 V	
Input power	1.8 V, 3.3 V, 5.0 V, I/O	

can be realised by taking advantage of the single chip. In addition, it is also possible to cope with a transmission type HMD (see-through) by making use of the high light use efficiency of the reflection type micro liquid crystal. Especially 0.24″ 720p has newly added functions effective for HMD market, and it has panel self-refresh (PSR) function. The PSR function can repeatedly display the final image data on the ferroelectric micro LC display side when the data from the image output side is interrupted.

This function makes it possible to display on the ferroelectric micro LC display side even if the image output side corresponding to the parent machine of the HMD is in the sleep (standby) state, and as a result it is possible to reduce the power consumption of the whole system.

For monocular and binocular large types, it is possible to correspond to 0.38″ 720p (1,280 × 720), 0.40″ QuadVGA (1,280 × 960), 0.50″ UXGA (1,600 × 1,200). Although these product groups are small size, they have realised HMD compatible with high angle of view and high definition. It is possible to realise miniaturisation of the whole device by taking advantage of single chip.

In these product groups, high luminance correspondence is realised by taking advantage of the high light use efficiency of the reflection type micro LC display. Technologies of small, high resolution, single chip, ferroelectric micro-LC display and reflective micro display device make these product groups feasible.

12.6 Wearable (HMD) market

There are three main categories when looking at the HMD market.

1. Virtual reality (VR)
2. Augmented reality (AR)
3. Mixed reality (MR).

Virtual reality is full immersion in a virtual world, and real world not visible.

Augmented reality is additional information available when viewing the real world. It is like a personal head-up display. MR is virtual world mixed with real world, and it can also be an augmented reality system.

At this point (March 2018), VR segment is approaching tail end of early market stages, and AR and MR segments are within early market stage. AR and MR segments will be slower to move through market stages than VR.

When considering AR, MR system, combiner is the optical component that allows virtual content to mix/augment real world.

1. Wave guide type. *Shown in Figure 12.7.
2. Mirrored reflector. *Shown in Figure 12.8.

Wave guide type is the most compact system, but this system can use limited light angles, so self-emission display (like OLED) emits light at very broad range of angles is very optically inefficient.

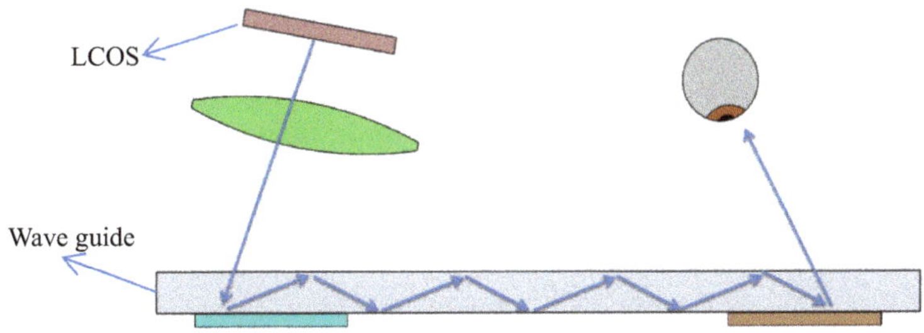

Figure 12.7 Head-mounted display wave guide type

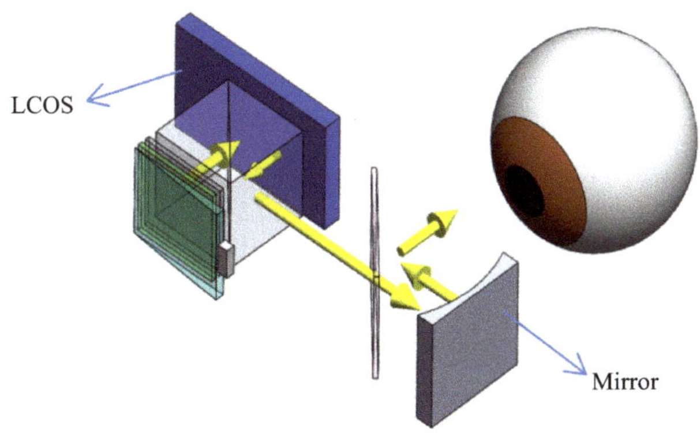

Figure 12.8 Head-mounted display mirrored reflector type

Based on these technical back ground, reflect LCOS has an advantage. Mirrored reflector type, polarised reflectors needed, so that 2× more efficient is required. Self-emission display cannot reach required brightness.

Based on these technical back ground, reflect LCOS also has an advantage. For all optics systems, important thing is brightness and light efficiency. Especially for wave guide system, light efficiency of limited angle is required. When consider other characteristics which are needed for HMD.

1. Motion artefact must be minimised.
2. Accommodation, vergence conflict must be addressed.
3. Compact system is required.
4. Low power consumption is required.

To minimise the motion artefacts, high colour field rate and low persistence are needed. Ferroelectric micro LC displays are capable of much higher field rates than nematic LCOS. This is based on high response time of ferroelectric liquid crystals. Scanning MEMS are limited in speed of scan, and modulation of light source. Micro-mirror device must trade-off bit-depth for field rate. So that Ferroelectric micro LC display has advantage of high colour rate. Also ferroelectric liquid crystal switching speed allows very short colour fields with minimal time between each colour.

Some accommodation/vergence conflict correction methods rely on light wavelength control, so that reflect LCOS (including ferroelectric micro LC display) has advantage. To make compact system with wide field of view with minimal size requires small pixels. Colour filter method needs each colour filter, so always have larger pixels than field-sequential method. Nematic LCOS has fringe field, and that will make limitation for minimum pixel size. So, field-sequential method without fringe field issue will have advantage, and ferroelectric micro LC display has all elements.

Also single chip (integrated circuit) of ferroelectric micro LC display can minimise system size from mechanical dimensions point of view. And single chip also works for low power consumption. Power consumption of ferroelectric micro LC display is lower than other micro display, when compared at system level. (Not only display but also the driving circuit.)

12.7 Technical improvement

High light efficiency is very important for all markets; to achieve high light efficiency, these two items are need to improve.

1. Throughput
2. Diffraction ratio.

Throughput is critical for light efficiency, and many factors affect to throughput. Light from light source to LCOS and to LCOS to eye, light takes following path light source to glass to liquid crystal to refractive pixel to liquid crystal to glass transmittance and reflectance of reflective pixel, and fill factor are important. Usually, glass transmittance is about 98 per cent, and reflectance of reflective pixel is based on reflectance of aluminium (about 88 per cent), and fill factor is from pixel size and pixel gap width.

If fill factor is 95 per cent, and assuming that there is no absorption in the liquid crystal layer then calculated throughput will be 98 (glass) $\times$ 95 (fill factor) $\times$ 88 (reflectance) $\times$ 98 (glass) $=$ 80.3 per cent.

This numerical value is the theoretical value, and the goal is to make actual throughput closer to this numerical value. And important factor to affects throughput is the fill factor, fill factor is calculated by the ratio of pixel and pixel gap. If pixel size is big, such as 10 μm, then it is easy to get big fill factor, but as explained small pixel size with high resolution is needed for micro display. For example, 0.24″ 720p has 4.13 μm pixel pitch (pixel pitch is distance between each pixel centre) and has about 92 per cent fill factor, it means pixel itself is 3.83 μm and 0.30 μm pixel gap. To get higher fill factor, pixel gap should be smaller, and smaller pixel gap will bring short problem to each pixel. In addition, the liquid crystal is strongly influenced by the transverse electric field.

As explained ferroelectric micro LC displays has advantage for transverse electric field, from this reason, ferroelectric micro LC display is be able to realise small pixel with high fill factor. To improve fill factor, improvement of both design side and process side are needed. And of course in addition LC material is important. Wave guide optics system allows to use limited light angles, so it is important that diffraction ratio needs to be improved too.

To get high diffraction ratio, small pixel gap is efficient, because wide pixel gap will generate interference of light. Also flatness of each pixel is important to minimise interference of light. To make small pixel gap with good flatness uniformity is effective for both throughput and diffraction ratio.

12.8 Summary

Micro display has strong potential for future devices. There are various kinds of micro display technologies, and each micro display has each advantages. This chapter is focused on ferroelectric micro LC display, and ferroelectric micro LC display has these advantages.

1. High response time
2. Single chip
3. Small and high resolution.

Ferroelectric liquid crystal has high speed with low power, then it can allow to use high colour rate, and this advantage is very strong for high frame rate image such as 120 frames per seconds (120 fps). For AR and MR, real image and displayed image are shown at the same time, so the display image needs to switch with high speed to remove discomfort. Therefor high speed liquid crystal is important factor for both augmented reality and MR.

Since ferroelectric liquid crystal can drive with low power, silicon background itself can have all circuits to drive ferroelectric liquid crystal and for other functions such as integrated frame buffer, video processing and LED driver.

Single chip can cover such integrated circuits, at the result, single chip can decrease total system parts, and it will bring big benefit to make system with low power consumption and make small system with few parts. Small and high resolution is needed for every situation. And small display needs small pixel, but needs to care for the transverse electric field at the edge of the pixel electrode.

As explained in Section 12.3, ferroelectric liquid crystal can be realised small pixel without any effect of transverse electric field. From these reasons, ferroelectric micro LC display is suitable for small and high resolution market, especially for small system such as wearable devices like a HMD.

Important development items of ferroelectric micro LC display are low power consumption and light efficiency to meet demand of future devices.

References

[1] http://www.holoeyesystems.com/uploads/media/BinocularHolographic_Waveguide_Visor_Display_01.pdf (Accessed on 26 July 2018).

[2] https://stanford.edu/class/ee267/lectures/lecture8.pdf (Accessed on 26 July 2018).

[3] Kasai I., Morimoto T., Noda T. and Tanijiri Y. 'The Development of a Head-Mounted Display (HMD) Incorporating a Holographic Optical Element (HOE) and Wom like Eyeglasses', KONICA MINOLTA TECHNOLOGY REPORT, Vol. 1, pp. 39–44, 2004.

Chapter 13
LC GRIN lens technology

Shinichi Uehara[1]

Liquid crystals (LCs) are attractive to optical devices which are expected to make the optical function or performance variable. Many technologies for various applications have been investigated, and LC GRIN (GRadient-INdex) lens is a typical of those technologies. Based on the purpose of variable optical device, it is important to clarify the target of function and performance, and in order to achieve the target, the device structure and materials need to be developed. Also, optical performance evaluation is significant because the optical devise using LC generally presents specific characteristics such as polarisation dependence. In this chapter, display applications of LC GRIN lens technology is focused, and future applications, breakthrough issues, or performance limitation are discussed.

13.1 Introduction

LCs are suitable for active optical devices due to some benefits such as a low voltage operation and, low power consumption. The typical examples are switchable (or variable/active) lenses or prisms, and they are sometimes used in displays, imaging optics, and optical instruments. A GRIN LC optics are generally defined as the optics covering optical effects produced by a gradual variation of the refractive index of LC materials, and several types have been developed [1–4].

Figure 13.1 shows the typical types of LC GRIN lens. In a narrow sense, as shown in the type (a), LC GRIN lens consists of only LC, electrode and substrate, and when the voltage is applied, the electric field induces gradual variation of refractive index, which means electronic field controllable. In this case, when the voltage is off, the device shows completely isotropy in optical. In the narrowest sense, the LC GRIN lens is limited to be used in this variable type of refractive index. Instead of the electric field, the LC GRIN lens can be switched by the polarisation of incident light (polarisation control) [5]. In this case, the outer polarisation control device (i.e., TN LC cell) is often used. Sometimes, this type is classified to one of the LC GRIN lens.

In a broad sense, as shown in Figure 13.1(b), the LC GRIN lens can include a concave polymer structure, which are optically isotropic in typical, and then the inside

[1]AGC Inc., Japan

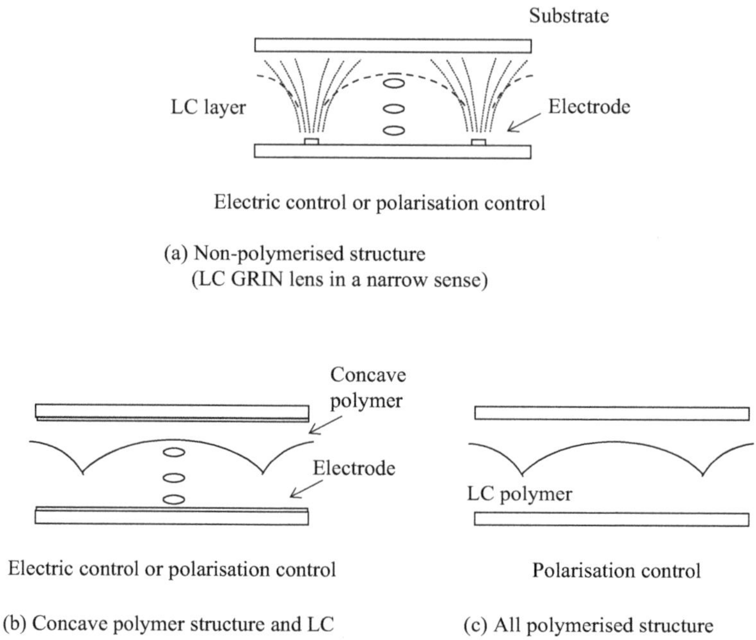

Figure 13.1 Classification of LC lens: there are two kinds of definitions of LC GRIN lens, that is, in a broad sense and in a narrow sense

of concave structure, an LC material is filled. This type can be controlled by not only the electric field, but also the polarisation state of incident light. If the concave structure is easy to made, it will be effective. However, sometimes there are difficulties in the LC alignment, and it is difficult to achieve optical isotropy due to the concave structure, which means that the viewing angle dependence exists. As shown in Figure 13.1(c), instead of the LC material, LC polymer can be applied. This is completely a passive structure like a GRIN fibre. In this case, the LC polymer cannot be controlled by the electric field, and polarisation control is necessary. Sometimes, these types of (b) and (c) are included in the LC GRIN lens in a broad sense.

The type (a), that is, the narrowest sense of LC GRIN lens seems an attractive candidate for variable optical device, because lens characteristics (i.e., focal length, lens aperture size, array pitch and array direction) can be changed. The structure of this type is simple, because it consists of only electrodes, LC materials and substrates. Compared with an LCD, the pattern of electrode is special, and thicker LC layer is necessary. However, generally the fabrication equipment of LCD can be applied, and this means that if an LCD is made, the same size of LC GRIN lens will be also made. For display applications, this is very important point, because additional investment can be minimised.

13.2 Application and usage

The design of LC GRIN lens depends on its application and usage, and it is important to clarify the target performance. There are several applications for LC

GRIN lens, and in this chapter, autostereoscopic 3D displays [6], which need special eye-glasses for depth perception, is focused first, because in these several years the LC GRIN lens has been developed and evolved for this application [7–10]. Recently, an 8K × 4K resolution display is becoming common, and its high resolution is suitable for 3D display applications [11].

As shown in Figure 13.2, in front of the display (i.e., LCD), the array of LC GRIN lens is attached, and it distributes the light rays from display pixels to many directions. The light rays towards the same direction are used for showing each parallax images, and then the depth perception is induced. In the simplest, the right and left images are used, and this is called as 'two-view autostereoscopic display'. Of course, several images can be applied as 'multi-view autostereoscopic display' or 'integral imaging', which are recently called as 'light-field display'. There are slight differences to show parallax images or to reproduce light rays of objects, but the basic structure is almost the same. Two-dimensional lens array is ideal for showing not only horizontal, but also vertical parallaxes, but due to the resolution limitation, one-dimensional lens array (semi-cylindrical lens array) is often used.

Of course, if there is no necessity for switching 2D/3D modes, plastic lens arrays can be applied. And, even in this case, if the pixel density is optimised, 3D and 2D images can be shown at the same time [12]. However, in this case, the plastic lens effect cannot be off, and it tends to cause some undesired diffused reflection. Still, strong demands on switching 2D/3D modes exist.

The slit array or pinhole array called as 'parallax barrier' is often used also, and its structure is similar. However, compared to the parallax barrier, the LC GRIN lens has many advantages, i.e., brighter screen, lower power consumption, suitable for higher number of views and so on.

13.3 Fundamental structure and design

13.3.1 Optical design

As explained, the main component of LC GRIN lens is LC material layer and electrode, and these two affects the optical performance. As shown in Figure 13.3, the significant LC parameter is the retardation consisting of the thickness of LC layer d, and the optical anisotropy (the difference between the extraordinary and the

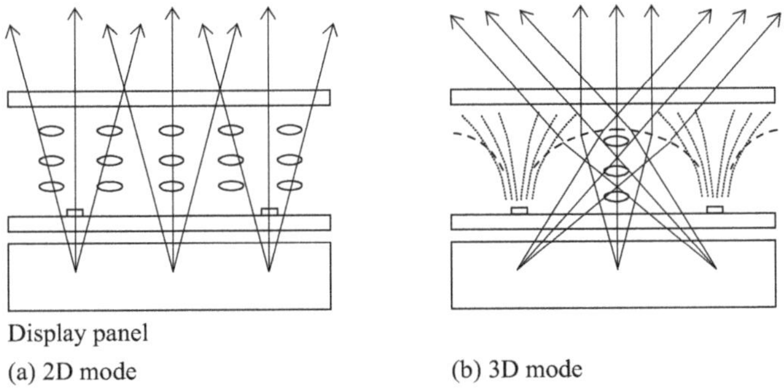

Display panel

(a) 2D mode (b) 3D mode

Figure 13.2 LC GRIN lens for 3D display application

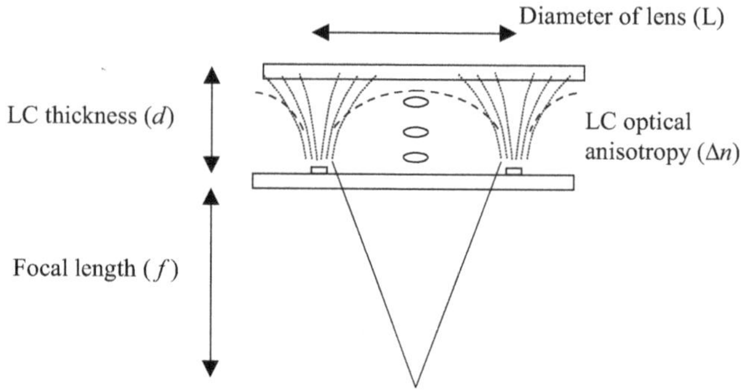

Figure 13.3 LC design parameter for LC GRIN lens

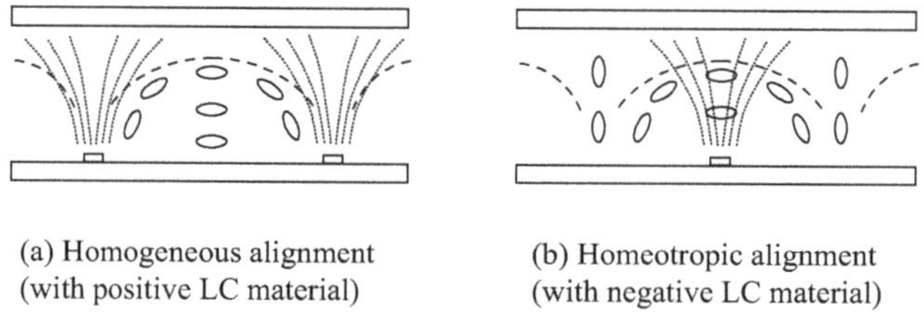

Figure 13.4 Typical two types of LC molecule direction distribution

ordinary refractive indices) Δn ($= n_e - n_o$). For the paraboloid lens, the focal length f is calculated as follows:

$$f = \frac{(L/2)^2}{2\Delta nd} \tag{13.1}$$

where L is the diameter of the lens [13–16]. This means that larger Δn can reduce the thickness of LC layer, or can shorten the focal length. For larger diameter, larger Δnd or longer focal length is necessary. Generally, in the optical design of 3D displays, the diameter (pitch) of the lens and the focal length is decided first, and then the LC material (Δn) and the LC thickness is decided. The focal length is about the distance between the lens and the pixel layer of display, and the LC thickness will be around several ten microns, that is, about ten times larger than that of LCD. Therefore, the fabrication of this larger thickness of LC layer is the key of LC GRIN lens fabrication.

The LC material is classified into the positive and negative types, and there are two major LC alignments: the homogeneous and homeotropic alignments. Any combination can be applied, but mainly the positive LC material with homogeneous alignment is used. Depending on the initial direction of LC alignment, Figure 13.4 shows typical two types of LC molecule directions. It should be noted that in any cases, the polarisation dependence exists.

The response time, T_{on} and T_{off}, is shown as follows:

$$T_{on} = \frac{\gamma_1 d^2}{\varepsilon_0 \Delta\varepsilon (V_{on} - V_{th})^2} = \frac{\gamma_1 d^2}{K_{eff}\pi^2 \left(\frac{V_{on}}{V_{th}} - 1\right)^2}$$
$$T_{off} = \frac{\gamma_1 d^2}{K_{eff}\pi^2} \tag{13.2}$$

This means that the response time of LC GRIN lens is about 100 times of LCD, and is about a second, if the same LC material is applied. Of course, larger dielectric constant, larger elastic constant, and small viscosity will be effective for faster switching. However, for 3D display applications, the user is sensitive to change of image contents, but not to the lens effect, and therefore in most cases it doesn't cause issues. For other applications which need higher switching speed, there are some development of applying the blue phase LC mode [17,18].

In order to achieve the parabolic distribution of refractive index, the electrode design is also important. Figure 13.5 shows several types of electrode design for LC GRIN lens with the homogeneous alignment. In the simplest design (a), the electrode is arranged only on a part of lens edge, and common electrode is used on the other substrate. For this electrode, transparent conducting materials like ITO can be

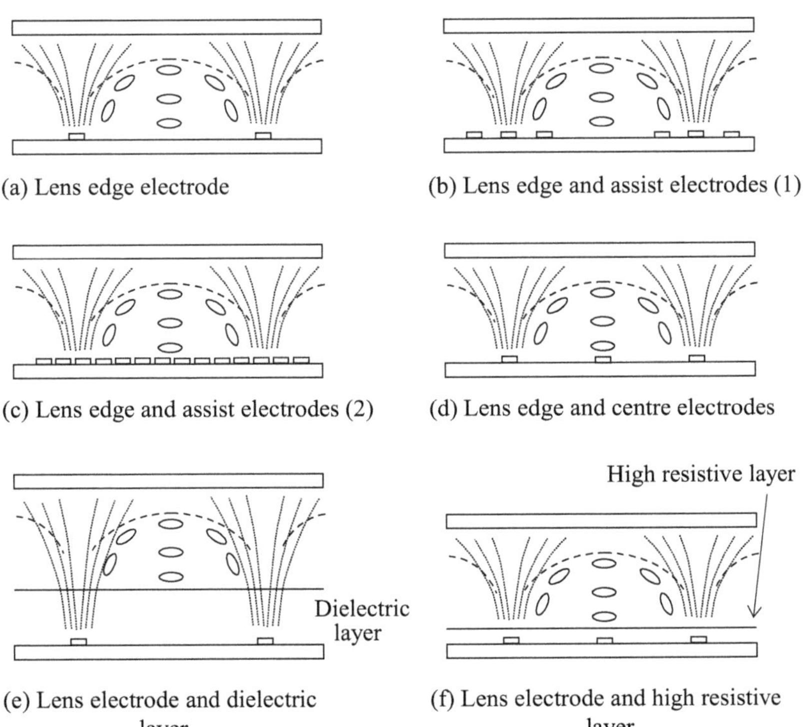

Figure 13.5 Several types of electrode design for LC GRIN lens

used appropriately. In this design, several parameters, such as the dielectric constant and elastic constant of LC materials, the LC retardation, the electrode width and pitch needs to be optimised. One or several assist electrodes can be arranged around the lens edge electrode in order to adjust the parabolic shape (see (b) and (c)) [19]. Sometimes, the additional electrode is applied around the lens centre (see (d)), but this is not so important if the voltage is the same as that of the common electrode. For these electrode designs, the overlap of electrical potential should be taken into account, and therefore LC director simulator can be applied [20]. In addition, there are several other structures, using thick dielectric materials (see (e)) [21] or high resistive layer (see (f)) [18]. However, these methods tend to cause higher driving voltage or power consumption.

13.3.2 Fabrication design

The fabrication process of LC GRIN lens is similar to that of LCD, and typically the following process is applied: (1) electrode forming, (2) LC alignment layer forming, (3) gap spacer dispensing/forming, (4) seal printing and (5) LC injection/ODF. However, it is noted that the LC layer thickness is much larger compared to LCD. When a column spacer is applied as gap spacer, high aspect ratio will be necessary, and it will affect the alignment later treatment. For example, the ordinary rubbing process might be difficult to apply. Sometimes, ball spacer dispensing brings good results. The larger LC thickness (cell gap) means that in the LC injection or ODF process, moving speed of LC material becomes high, and air babbles defect tends to occur. In addition, we should avoid 'gravity mura', which is the LC thickness variation caused by the gravity when the LC cell stands for a long time, but this is not specific to the LC GRIN lens.

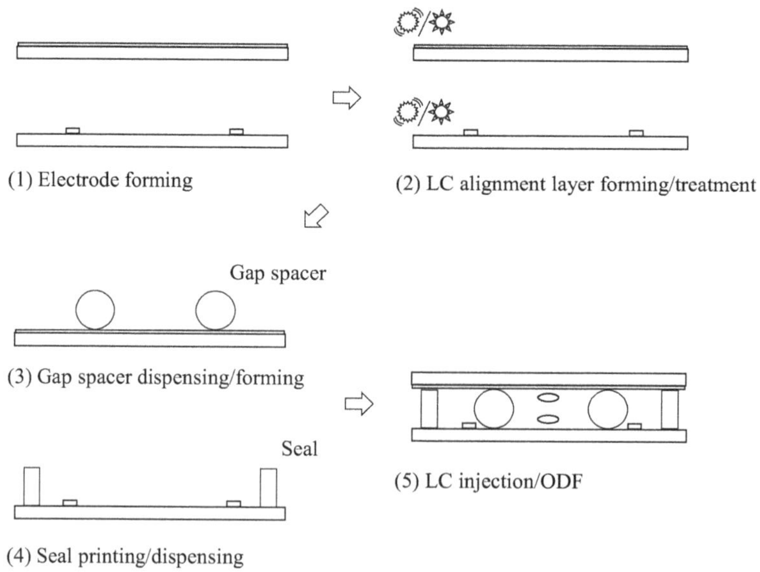

Figure 13.6 Typical fabrication process of LC GRIN lens

13.3.3 Application design

In addition to the basic design methods mentioned above, it is important to consider the application design, which clarifies the final target of functions to be implemented and specifications to be fulfilled. For the application design, considering the similarity of LCD, the following three keys can be focused: LC mode of GRIN lens, electrode design and driving method. Some examples of application and related LC GRIN lens design will be explained.

13.3.3.1 Array direction switchable lens

When smart phones or tablets are used, the user easily rotate them based on the contents (i.e., from landscape to portrait, and vice versa), and this rotation function will be necessary even if 3D display is equipped. Of course, if two-dimensional lens array, such as a fly-eye lens is applied, both horizontal and vertical parallax can be shown, but in this case 3D image resolution will be lower. In order to achieve high 3D display performance in each landscape and portrait orientations, the 2D/3D rotatable display needs to combine two lenses with different array directions, and to switch the lens direction according to the screen orientation, as shown in Figure 13.7(a). The relation among two array directions of the lens electrode and LC alignment direction will be significant factor, because LC molecule directions will be changed when driving voltages are applied to each lens electrode [22,23].

As widely known, LC molecules has anisotropy, and therefore causes viewing angular dependence. Until now, for an LC display application, in order to reduce the viewing angle dependence, many LC modes, such as IPS, VA, film compensated TN, and so on, are developed. In the same manner, it is expected to develop wider angular performance in LC mode for GRIN lens.

The LC director simulator is available to clarify the lens performance, that is, the viewing direction dependence of lens retardation profile, with changing the angular difference between the LC alignment direction and electrode array direction. It is obvious that when the lens is neither tilted nor panned, the lens retardation

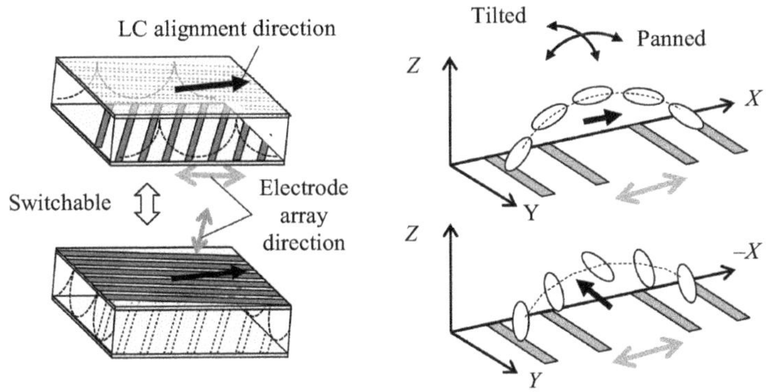

(a) Electrode and LC alignment direction of array direction switchable lens

(b) Viewing angle dependence on display tilted/panned

Figure 13.7 Array direction switchable lens for landscape/portrait rotatable 3D display profiles represent small difference among all angles, because the LC molecule directions are aligned almost symmetrically.

However, when tilted at ±30°, if the angular difference is over 70°, the retardation profile is degraded like a prism shape or the retardation profiles have sag around the middle. This is due to the influence of LC molecule aligned directions around the centre of lens. Near the centre of the lens, when the angular difference is 0°, the direction of most molecules is moved toward the *z*-direction in the *x*–*z* plane by the electric field which is generated from the electrode arrayed in the *x*-direction. When this optical model is tilted at ±30°, most molecules are tilted only in the *x*–*z* plane, and therefore the apparent refractive index is not so changed. On the other hand, the optical model with 90° of angular difference contains the molecules tilted in *y*-*z* plane, the apparent refractive index is significantly changed depending on the display tilting.

Based on this result, the angular difference among both lens electrode array directions for landscape and portrait is made small, and the viewing angle performance of 3D display is improved compared to the conventional two orthogonal electrodes type.

13.3.3.2 Array pitch switchable lens: Fresnel lens and over driving

The 3D display with adapting the 3D viewing angle has been developed. When several persons use the 3D display at the same time, the 3D viewing angle is made wider as a multi-user mode. In a personal mode, the narrower viewing angle can be accepted. Instead, the resolution becomes higher, and the personal user can obtain higher 3D image quality. This function can be realised by switching the lens array pitch, and in particular, switching between non-Fresnel and Fresnel lenses should be effective, as shown in Figure 13.8(a) [24].

Figure 13.8(b) shows the development examples of electrode design method to switch between the non-Fresnel lens for narrow pitch lens and Fresnel lens for wide pitch lens. In this design, a minimum electrode group consisting of an essential electrodes of non-Fresnel lens and Fresnel lens are arrayed symmetrically. This design can reduce the driving voltage source needed, and realises good 3D performance, and the 3D viewing angle can be switched between 15° and 23°.

For making the 3D viewing angle wider, the Fresnel lens design is attractive because the wider lens pitch can be realised without increasing the LC thickness. However, it should be noted that the effect of electric field generated by electrodes of lens is completely different from that of LCD. In a usual LCD, as shown in Figure 13.8(c), the pixel electrode width is much larger than LC thickness. Each pixel electrode drives the corresponding LC molecules independently, and generally the influence of neighbouring pixel electrodes can be ignored. However, in LC GRIN lens, each lens generally has several lens electrodes with different driving voltages in order to make the retardation profile parabolic. The influence of this synthesised electric field generated from these several lens electrodes should be taken into account.

For example, in the Fresnel type of LC GRIN lens, each lens consists of lens edge electrode and Fresnel electrodes, and unbalance of these electric fields cause

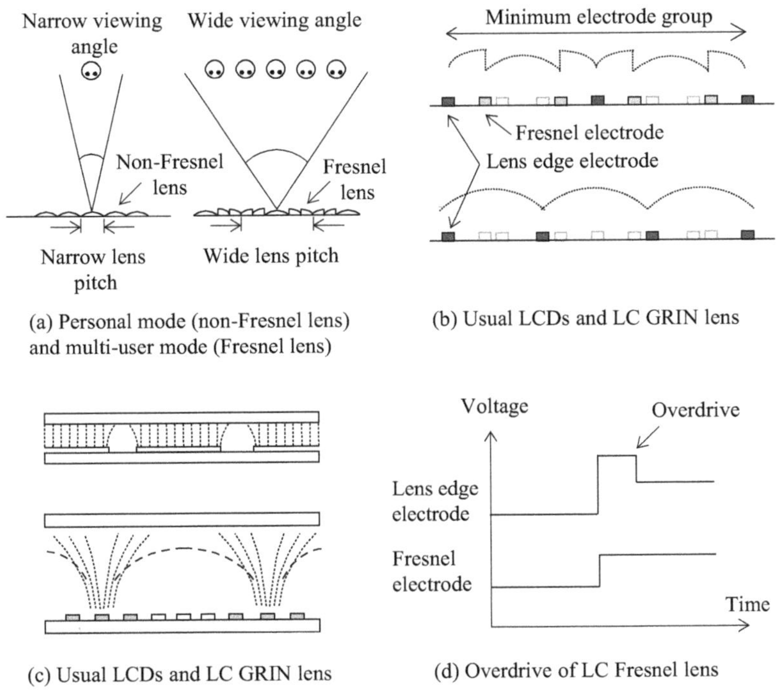

Figure 13.8 Array pitch switchable lens for 3D viewing angle switchable display

the disclination of LC alignment on the Fresnel electrodes. Generally, the disclination is induced by discontinuous LC molecule orientation, such as a reverse tilt phenomenon, and the disclination results in degrading the lens performance and making it unstable. It is found that the disclination can be suppressed when the electric field generated from the edge electrode is stronger on the position of Fresnel electrode, because the LC molecule direction becomes continuous. No reverse tilt is induced.

In order to obtain the stronger effect of the edge electrode, the over driving of edge electrode is effective, as shown in Figure 13.8(d) [25]. In usual LCDs, over driving is applied to improve the motion image quality, but in here the over driving is to improve the electric field of the edge electrode. It is reported that the over driving is effective for achieving and stabilising good lens effects. Of course, by increasing the voltage of edge electrode can also realise the similar effect. However, the increasing voltage sometimes degrades the lens performance. The edge electrode voltage cannot be determined independently. On this point, the effect of over driving is transient, and it does not affect the final voltage balance. Once the disclination is suppressed, the stable LC alignment is achieved. A 17-inch 3D display with adapting 3D viewing angle using the above method has been developed.

13.3.3.3 Partially switchable lens

When 3D images and 2D images such as characters are shown in the same screen, the partial switching is necessary. For this function, a rectangular coupled electrode structure has been developed [26].

As shown in Figure 13.9(a), each rectangular area includes slanted lens edge electrodes with vertically connected and common electrodes with horizontally coupled. Then, for 3D window area, the voltage V_a is applied to the lens edge electrode, while the voltage V_b for the other area. In the same manner, for the 3D window area, the voltage $-V_{\text{com}}$ is applied to the common electrode, while the voltage V_{com} for other area. As a result, there are four kinds of rectangular areas, those are, A_{3D}, A_{2D1}, A_{2D2} and A_{2D3}, as shown in Figure 13.9(a), and the voltage difference of each area will be the following:

$$
\begin{aligned}
&\text{The 3D window area:} && V_{A3D} = V_a + V_{\text{com}} \\
&\text{Other area for 2D:} && V_{A2D1} = V_a - V_{\text{com}} \\
& && V_{A2D2} = V_b + V_{\text{com}} \\
& && V_{A2D3} = V_b - V_{\text{com}}
\end{aligned}
\tag{13.3}
$$

In order to activate the lens inside the 3D window, V_{A3D} should be more than $V_{\text{th\_lens}}$, that is, the threshold voltage of the LC GRIN lens. When V_{A2D1}, V_{A2D2} and V_{A2D3} are under $V_{\text{th\_lens}}$, the lens effect does not appear, and 2D image can be

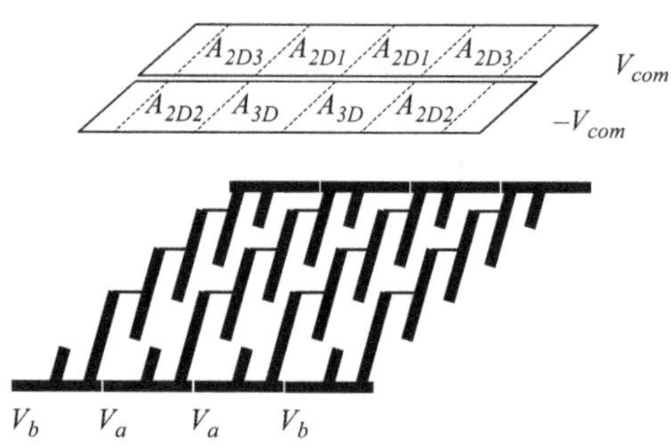

(a) Electrode structure, applied voltage and 2D/3D areas

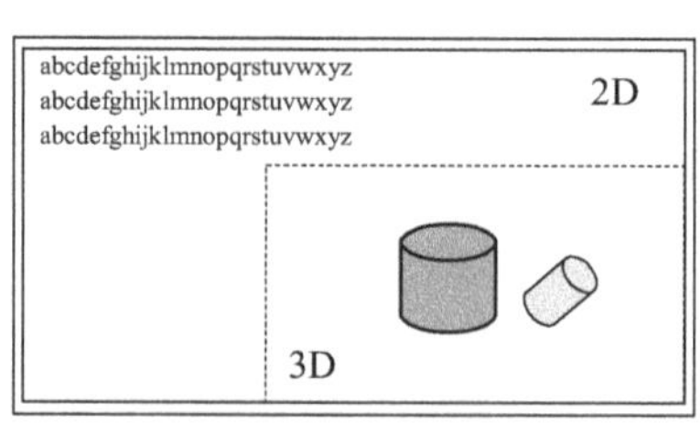

(b) Partially switchable 2D/3D display

Figure 13.9 Partially switchable lens for 2D/3D display

observed, as shown in Figure 13.9(b). A 17″ partially switchable 2D/3D display using the above method has been developed.

13.3.3.4 Focal length switchable lens

Sometimes, 3D display is used not only in the stand-up posture, but also flatbed posture, and therefore flatbed/stand-up switchable 3D display is necessary. In this type, the focal length of LC GRIN lens needs to be changed, because in the flatbed mode longer focal distance is necessary due to the influence of oblique viewing direction. The flatbed type 3D display is also known as 'phantogram' or 'tabletop type', and it is effective for improving the depth perception.

In the flatbed mode, the driving voltage of LC GRIN lens is reduced, so that the focal length of lens is made to be longer [27]. It is also reported that by reducing the driving voltage from 5 to 3.8 V, the focal length is about 1.3 times larger. This will be also appropriately used in smart phones or tablets.

13.4 Performance evaluation

For developing the new technologies, appropriate evaluation methods should be developed at the same time, so as not to mislead the technology development. The key evaluation methods of LC GRIN lens are (1) retardation distribution, (2) lens performance and (3) application performance (3D performance in this case).

In the evaluation of retardation distribution, first we need to consider that the retardation value is large, and its order number is also high. There exists several methods to evaluate the retardation value, such as a polarimeter, but this influence of high order should be noticed. In addition, this evaluation needs to measure the in-plane distribution of retardation, two-dimensional measuring device is better rather than spot measuring type.

Regarding the lens performance, the lens spot analysis can be applied, but this is common to the ordinary lens. The 3D performance evaluation, such as crosstalk measurement is not specific to LC GRIN lens also, but these are essential and significant.

13.5 Future challenge

The LC GRIN lens can be applied to not only 3D displays, but alto other type displays, screen, camera, optical instruments and so on. The following are some typical examples of applications.

13.5.1 Active screen

The LC GRIN lens can also be used for projector screens. For example, in order to increase the light ray and parallax image density, time multiplex projector array is equipped with an angular steering screen module including LC prism arrays [28,29]. For this kind of screen application, electro wetting technology is often used. The polarisation dependence of LC GRIN lens can improve the optical performance by using several array screens at the same time.

13.5.2 ARVR head-mounted display

The LC GRIN lens is also one of attractive components for ARVR head-mounted display, and there are several purposes to apply as follows:

- doubling the pixel density or reducing the screen door effect [30],
- vision correction for AR images or actual objects, or reducing accommodation issue [31,32],
- expanding the depth of light-field mode or depth-fused mode [33,34], and so on.

13.5.3 Camera

The LC GRIN lens is also used for a camera system in order to change the performance of light-field camera, or to increase the image resolution [35–36]. Sometimes, a microscope uses this as the variable optical elements [37].

13.5.4 Ophthalmic equipment, glasses and contact lenses

Ophthalmic applications are important for the LC GRIN lens, and a wavefront sensor or refractometer is tried to be applied [38]. The correcting glasses with adjustable focus for hypermetropia have already been developed and available in the market [39]. For eyeglasses applications, there are several keys such as the transparency, polarisation dependence, haze, focus range, response time, field-of-view size, and so on. Recently, the contact lens type or intraocular lens type has been developed [40], and these should be one of big future challenges.

13.6 Summary

For active/variable optical devices, LCs are one of attractive candidates, and many applications including 3D displays have been developed. Similar to LCDs, there are three significant keys: the LC mode, electrode design and driving method. In actual designs, it is important to clarify the target of device functions and performance, because some characteristics of LC materials (i.e., birefringence, LC thickness, . . .) are limited. The application of LC GRIN lens is not limited to displays, and in future more applications will be developed.

References

[1] Uehara S., Kashiwagi M., Takagi A., Kizu Y. and Baba M. '2D/3D Displays with LC GRIN Lens for Medical Systems', SID Symposium Digest, 18.1, 2016.

[2] Harrold J., Wilkes D. J. and Woodgate G. J. 'Switchable 2D/3D Display–Solid Phase Liquid Crystal Microlens Array', IDW '04, 3Dp-6L, 2004.

[3] Montgomery D. J., Woodgate G. J., Jacobs A., Harrold J. and Ezra D. 'Analysis of the Performance of a Flat Panel Display System Convertible Between 2-D and Autostereoscopic 3-D Modes', SPIE Proc. 4297, 148–59, 2001.

[4] de Zwart S. T., Ijzerman W. L., Dekker T. and Wolter W. A. M. 'A 20-in. Switchable Auto-Stereoscopic 2-D/3-D Display', Proc. IDW '04, 3D2-5L, 2004.

[5] Takagi A., Saishu T., Kashiwagi M., Taira K. and Hirayama Y. 'Autostereoscopic Partial 2D/3D Switchable Display Using Liquid Crystal Gradient Index Lens', SID Symposium Digest, 30.3, 2010.

[6] Itoh G. 'Hardware and Software Technologies for Glasses-Free 3D TVs and PCs', SID Symposium Digest, 17.1, 2012.

[7] Hong H., Jung S.-M., Lee B.-J., Im H.-J. and Shin H.-H. 'Autostereoscopic 2-D/3-D Switching Display Using Electric-Field-Driven LC lens (ELC lens)', SID Symposium Digest, 25.3, 2008.

[8] Uehara S. '3D Display Technologies for Integral Imaging Method', IMID15, 58-2, 2015.

[9] Oka S., Sugita T., Naganuma T., Saito T., Komura S. and Miyazawa T. 'Improvement of the Crosstalk in Autostereoscopic 3D Display using LC GRIN Lens with High-Resolution LCD', IDW '11, LCT6-3L, 2011.

[10] Matsushima J., Ibata Y., Masumura K., Asai T., Sato T. and Shigemura K. 'A Two-Way Multi-View 2D/3D Display Combining a LC Lens and HVxDP Panel Using a Novel Pixel Arrangement', SID Symposium Digest, 73.4, 2016.

[11] T. Koito, Saito K., Hayashi S. and Takizawa K. 'Newly Developed Light Field Display with Ultra-Wide Viewing Angle and High Resolution', SID symposium digest, 21.4, 2018.

[12] Uehara S., Ikeda N., Takanashi N., *et al.* 'A 470 × 235-ppi Poly-Si TFT-LCD for High-Resolution 2-D and 3-D Autostereoscopic Displays', *Journal of the SID* 13/3, pp. 209–14, 2005.

[13] Kowel S. T., Cleverly D. S. and Kornreich P. G. 'Focusing by Electrical Modulation of Refraction in a Liquid Cell', *Applied Optics*, Vol. 23. P. 278, 1984.

[14] Bao R., Mai H., Zhang G. and Ye M. 'Dependence of the Optical Power of an LC Lens on Cell Gap', SID Symposium Digest, 61.2, 2015.

[15] Li L., Shi L., Bryant D. and Bos P. J. 'Design and Modeling of a Refractive Liquid Crystal Lens for Tunable Optical Correction in 3-D Stereoscopic Displays', SID Symposium Digest, 3.3, 2011.

[16] Kasano M., Ichihashi K., Asai Y., Hayashi K., Tanaka Y. and Mimura K. 'Design for Reducing Autostereoscopic Display Crosstalk Using a Liquid Crystal Gradient-index Lens', SID Symposium Digest, 51.3, 2014.

[17] Li Y. and Wu S. -T. 'Polarization Independent Adaptive Microlens Using a Blue-Phase Liquid Crystal', SID Symposium Digest, 3.1, 2011.

[18] Liu Y., Lee C.-T., Lin H.-Y., Li Q. and Wu S.-T. 'Polarization-Independent and Fast-Response Blue Phase Liquid Crystal Lens with a PEDOT:PSS Film', SID Symposium Digest, 16.3, 2012.

[19] Hong H., Jung S.-M., Lee B.-J., Im H.-J. and Shin H.-H. 'Electric-Field-Driven LC Lens for 3-D/2-D Autostereoscopic Display', *Journal of the SID*, 17/5, pp. 399–406, 2009.

[20] http://www.shintech.jp (Accessed on 24 July 2018).

[21] Sato S. 'Liquid-Crystal Lens-Cells with Variable Focal Length', *Japanese Journal of Applied Physics*, Vol. 18, No. 9, pp. 1679–84, 1979.9.

[22] Kashiwagi M., Uehara S., Takagi A. and Baba M. 'LC GRIN Lens Mode with Wide Viewing Angle for Rotatable 2D/3D Tablet', SID Symposium Digest, 15.2, 2013.

[23] Kashiwagi M., Uehara S., Takagi A. and Baba M. 'LC GRIN lens mode with wide viewing angle for rotatable 2D/3D tablet', *Journal of the SID,* 22/01, pp. 68–75, 2014.

[24] Takagi A., Uehara S., Kashiwagi M., Kizu Y. and Baba M. 'Variable Lens-Pitch LC GRIN Lens for Adapting 3D Viewing Angle', SID Symposium Digest, 61-1, 2015.

[25] Uehara S., Kashiwagi M., Takagi A., Kizu Y. and Baba M. 'Overdriving LC GRIN Lens to Stabilize Lens Profile for 2D/3D Display', SID Symposium Digest, 60.4, 2014.

[26] Takagi A., Uehara S., Kashiwagi M. and Baba M. 'Function-Integrated LC GRIN Lens for Partially Switchable 2D/3D Display', SID Symposium Digest, 15.4, 2013.

[27] Uehara S. 'LC GRIN Lens Technology for Multi-Functional 3D Display', IDW'15, 3D2/LCT3-1, 2015.

[28] Cheng H.-H., Bhowmik A. and Bos P. J. 'Large Angle Image Steering Using a Liquid Crystal Device', SID Symposium Digest, 51.2, 2014.

[29] Xia X., Zhang X., Zhang L., Surman P. and Zheng Y. 'Projection-based Multi-view Three-dimensional Display with Angular Steering Screen', SID Symposium Digest, 70.2, 2018.

[30] Zhan T., Lee Y.-H. and Wu S.-T. 'Doubling the Pixel Density of Near-to-Eye Displays', SID Symposium Digest, 3.4, 2018.

[31] Lin Y.-H. and Wang Y.-J. 'Liquid Crystal Lenses in Augmented Reality', SID Symposium Digest, 18.2, 2017.

[32] Jamali A., Bryant D., Zhang Y., Grunnet-Jepsen A., Bhowmik A. and Bos P. 'Design Investigation of Tunable Liquid Crystal Lens for Virtual Reality Displays', SID Symposium Digest, 72.3, 2017.

[33] Huang H. and Hua H. 'An Integral-Imaging-Based Head Mounted Light Field Display Using a Tunable Lens and Aperture Array', SID Symposium Digest, 72.1, 2017.

[34] Liu M., Li H. and Liu X. 'A Deep Depth-of-Field Near-to-Eye Light-Field Display Utilizing LC Lens and Dual-Layer LCDs', SID Symposium Digest, 10.2, 2018.

[35] Kwon H., Ito M., Kizu Y., *et al.* 'Switchable Liquid Crystal Micro Lens Arrays for the Light-Field Camera Application', SID Symposium Digest, I1-2, 2015.

[36] Gao K., Cheng H.-H., Bhowmik A., McGinty C. and Bos P. 'Ultra-Compact Non-Mechanical Zoom Lens for Enhanced Machine Vision and Computer Input Applications', SID Symposium Digest, 61.3, 2015.

[37] Hsieh P.-Y., Lee Y.-H., Luo Z. and Wu S.-T. 'Depth Enhancement of a Light Field Microscope with Hexagonal Liquid Crystal Lens Array', SID Symposium Digest, 30.2, 2016.
[38] Lin Y.-H., Chen H.-S., Chen M.-S., Wang Y.-J. and Chang C.-M. 'Liquid Crystals for Ophthalmic Lenses and Biosensing Applications', SID Symposium Digest, 39.4, 2014.
[39] https://www.mitsuichem.com/en/release/2017/171002.htm (Accessed on 24 July 2018).
[40] Galstian T. 'Liquid Crystals for Dynamic Vision Correction', SID Symposium Digest, 71.5, 2018.

Chapter 14
Essence and practical use of field-sequential colour LCD

Masaya Okita[1] *and Shunsuke Kobayashi*[2]

Herein, we present and discuss essential virtue of field-sequential colour liquid-crystal display (FSC-LCD) and how we technically resolved fundamental problems in the FSC-LCD such as the occurrence of flickering, hue rotation at low temperature and colour breakup. And in addition, we present demonstration of the implementations of actual examples of indicators in airports and railway stations, a transparent LCD and an LCD exhibiting good legibility under the direct sunshine.

14.1 Introduction

Nowadays, field-sequential colour (FSC) LCD is a representative display method for representing colour images without using colour filters that brings very low power consumption, very wide colour gamut, and very high legibility even in the outdoor as will be presented in this chapter and other chapters of 11, 15 and 16.

Historically, FSC technology dates back to the 1940s and it was developed by CBS, USA, and reported by the *New York Times* reported on 4 September 1940 such that 'Colour Television Achieves Reality developed by CBS' [1]. Regarding LCD, Michael G Clark and Ian A Shanks [2] and Ralf Vatne, Philip A Johnson and Philip J Bos [3] reported FSC-LCD by combining a CRT-tube and an LCD screen in 1982 and 1983, respectively. In 1985, Hiroshi Hasebe and Shunsuke Kobayashi demonstrated a flat-panel FSC-LCD with a narrow gap TN-LCD and a ferroelectric LCD (FLCD) [4] and Toshinori Tanaka, Hiroshi Hasebe and Shunsuke Kobayashi reported an FSC-LCD with LED backlight in 1986 [5].

In this article, we shall present and discuss how difficult problems in the FSC-LCD such as low temperature colour performance, flickering, and colour breakup for implementing practical FSC-LCD; and we shall demonstrate some early work and updated results such as elimination scheme of colour breakup based on a patent by Masaya Okita [6].

[1]HDT Inc., Japan
[2]Tokyo University of Science, Yamaguchi, Japan

Before discussing updated research work, we shall present examples of our early work though which we made the succeeding progress in this field. A practical implementation of FSC-LCD installed in airports and railway stations; Figure 14.1 demonstrates an airport indicator installed at Tokyo International Airport Narita since 2006 and still well works; where, panels comprises modules with the size of 10cm × 10cm and 24 × 24 pixels fabricated by HDT [7] and OKAYA ELECTRIC [8]. This kind of module is operated by static driving but suitable to fabricate LCD without colour filter. It is also featured by excellent power efficiency denoted by luminance efficiency of 4.21 cd/W that was obtained by the Shunsuke Kobayashi *et al.* [9]. This type of FSC-LCD indicators are also installed in the five airports in China and in the railway station of KTX in Korea.

Another early work was a-Si TFT driven 1.5 inch FSC-LCD with QVGA (320 × 240 dots) 265 ppi (pixel per inch) fabricated by Masaya Okita in 1999 as shown in Figure 14.2.

撮影協力：成田国際空港株式会社

Figure 14.1 FSC-LCD at Narita Air Port since 2006

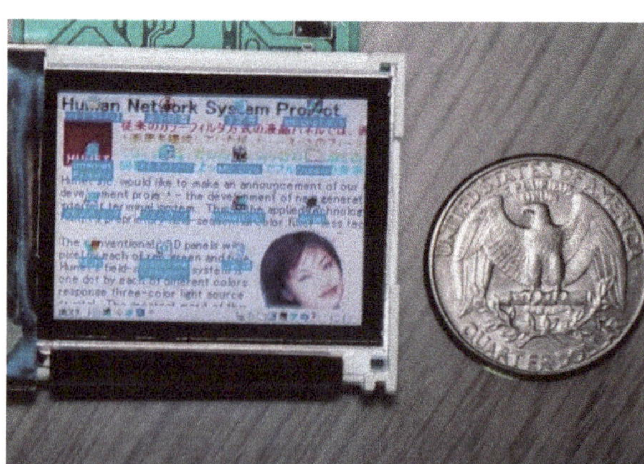

Figure 14.2 1.5″a-Si TFT-driven FSC-LCD

Through the evaluations of performances of this prototype, we detected two problems: one was the special flicker cause from FSC method and the other was hue rotation in the colour space at the lower temperature under 0 °C, which colour filter LCD does not have this hue rotation (these problem is a kind of Valley of death); in this chapter, we shall describe and discuss how we have resolved these problems based on a patent by Masaya Okita [6].

14.2 Basic method of FSC-LCD

In the FSC-LCD system, the following definitions are given. First of all, a minimum cycle in which three colour light sources of red, green and blue sequentially turn on and off to display any given colour is herein called a 'display frame', and its length of time is called 'one display frame period'. A minimum cycle for displaying any one of the three colour is herein called a 'colour field', and its length of time is called 'one colour field period'.

In the existing FSC methods, red, green and blue light sources are sequentially turned on and off once for each colour as shown in Figure 14.3 and the display frame comprising red => green => blue, for example, is repeated.

Figure 14.4 is a block diagram that illustrates a drive circuit for an existing FSC-LCD.

Figure 14.5 is a timing chart showing timings in the existing FSC-LCD. With reference to Figures 14.4 and 14.5, timings of image signals in the conventional FSC-LCD are explained. The drive circuit of Figure 14.4 for the existing field-sequential display device includes two frame memories that are a first frame memory and a second frame memory. These frame memories are controlled by a memory arbitration circuit to each write and read an image of one frame independently from each other.

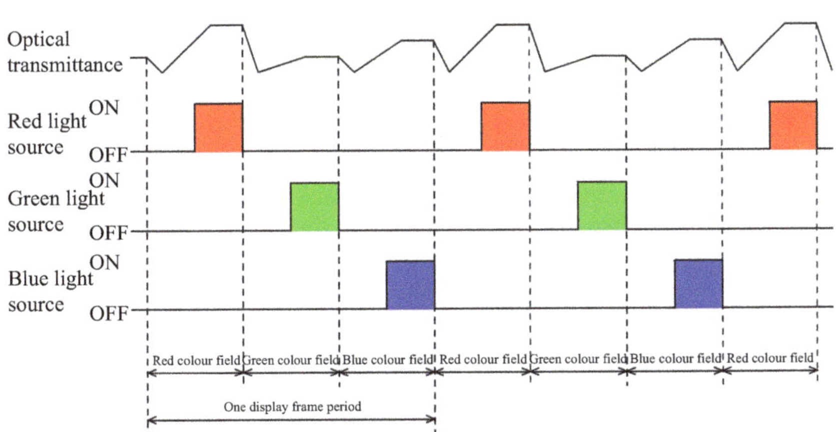

Figure 14.3 Timing chart of the existing FSC-LCD

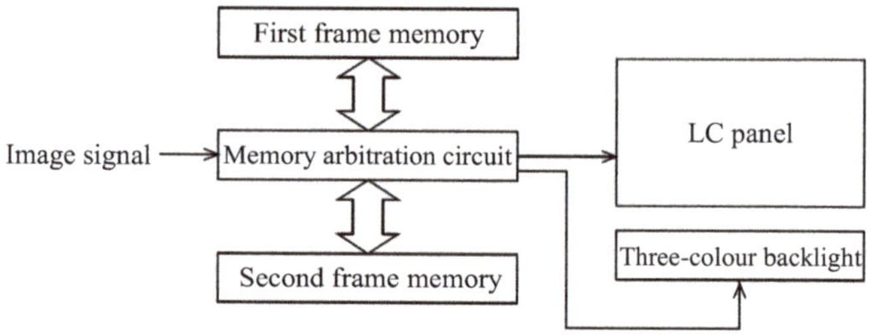

Figure 14.4 Block diagram of the existing FSC-LCD

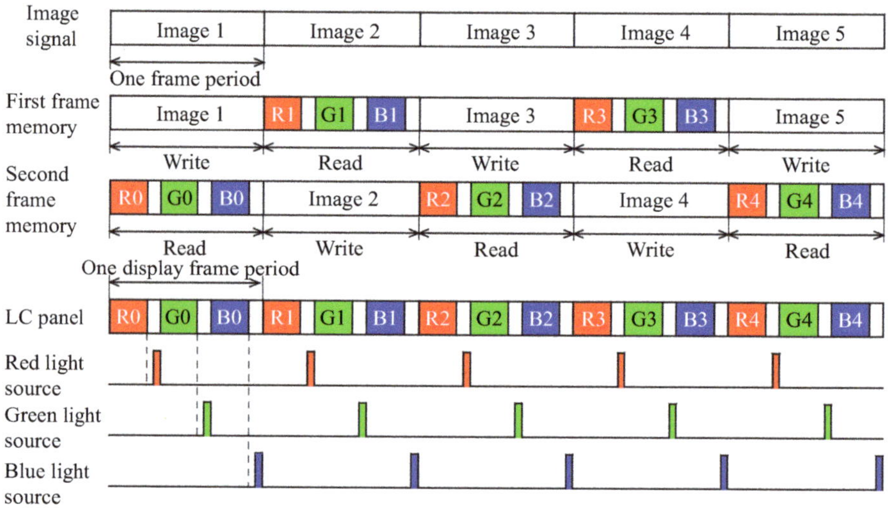

Figure 14.5 Timing chart of the existing FSC-LCD

With reference to Figure 14.5, during a frame period in which 'Image 1' is input as an image signal, data of 'Image 1' is written in the first frame memory. Simultaneously in the same frame period, data of 'Image 0', which was written in the second frame memory in the preceding frame period, is read out in the second frame memory in the order of the red colour field, green colour field and blue colour field. After the data of each colour image is written, the light source of each colour is turned on and off, thereby representing a colour image.

In the next frame period in which 'Image 2' is input, data of 'Image 2' is written in the second frame memory. Simultaneously in the same frame period, data of 'Image 1', which was written in the first frame memory in the preceding frame period, is read out in the first frame memory in the order of the red colour field, green colour field and blue colour field. After the data of each colour image is written, the light source of each colour is turned on and off, thereby representing a colour image.

14.3 Flickering in FSC-LCD caused from backlight blinking

One of the well-known and common problems in the flat-panel display is the issue of flickering. The cause of the flickering is due to the reversal of the polarity of driving voltage that accompanying a delicate asymmetry in the amplitude of the driving voltages. Besides this kind of origin, in the FSC-LCD another kind of flickering occurs by the colour sequential of backlight; this is a special phenomenon of the FSC-LCD.

At SID 2000 Symposium, Fumiaki Yamada, Hajime Nakamura, Yoshitami Sakaguchi and Yoichi Taira reported that the magnitude of observed flickering depends on colour, RGB, where green colour produces two hold strong flickering over red and blue colour and at the luminance of 85 nits, the threshold frequencies for producing flickering were: red 30 Hz, green 50 Hz, and blue 35 Hz, respectively [10].

With this knowledge, Masaya Okita devised a four colour fields (RGBG) FSC method for reducing flickering instead of conventional RGB colour system [11]. As shown in Figure 14.6 the colour sequential in a frame is to be: R => G => B => G => in this method and repeats this cycle, where the green sources are used twice in the one frame. As a result, this colour sequential system is recognised that experimentally effective.

Masaya Okita observed the occurrence of flickering in the red colour at the luminance of 300 nits with the display frame frequency of 60 Hz; however, with the display frame frequency of 75 Hz the flickering is effectively reduced. In an actual implementation like that of Narita airport, we adopt this 70 Hz.

Appearing of flickering depends on level luminance of display and that of ambient light. The display frame frequency should be decided in consideration of flickering caused from backlight blinking.

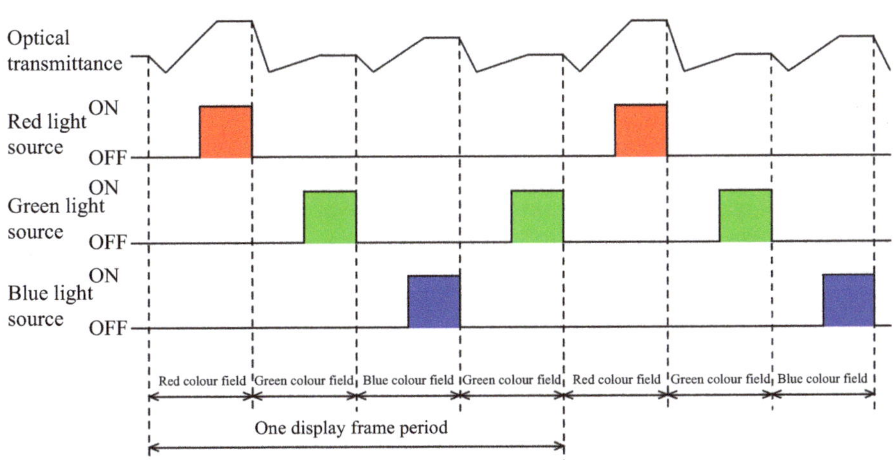

Figure 14.6 Timing chart of FSC-LCD having RGBG 4 colour fields

14.4 Dependence of flickering on driving: a way of its reduction

Appearance of flickering in an ordinary colour filter type and that in the FSC-LCD is essentially different: in the former, driving is done with 60 Hz and then the reversal of the polarity could accompany of flickering at 30 Hz. And in these cases the reduction of flickering is done by controlling voltages of common electrode in TFT driving.

Masaya Okita tried similar way to reduce the flickering on driving of the FSC-LCD. But the response time of LC in FSC-LCD is very short. Therefore, Masaya Okita found out that it is impossible to reduce the flickering in acceptable level with all area and all grey level.

Instead of this way, we devised this polarity change within each colour field and we completely eliminate flickering on driving; this way is needed to increase field frequency by twofold or more. But the utilisation of ordinary light guide type backlight also is needed to increase field frequency.

14.5 Reduction of colour shift at low temperature

An additional issue in the conventional FSC-LCD with RGB three colour fields FSC method is the colour shift at low temperature. In existing FSC methods in which one frame consists of three colour fields, if LC response time becomes slow at low temperature, it cannot complete its full response before the light source of the next colour field turns on, and the colour of the next colour field is undesirably mixed accordingly. As a result, the actually displayed colour of the red colour field appears yellowish with mixture of green. Similarly, the actually displayed colour of the green colour field becomes cyan-tinged, and the actually displayed colour of the blue colour field becomes magenta-tinged. Therefore, the colour purity will be deteriorated. These phenomena are observed by a rotation of RBG colour that is called a hue rotation in the x–y chromaticity diagram. The hue rotation of R => G => B scheme is rotated anti-clockwise and the hue rotation of R => B => G scheme is rotated clockwise [12].

In the method of Figure 14.6 for lighting the green colour twice in each display frame, each red colour field and each blue colour field intervene respectively between two green colour fields, and each green colour field results in being interposed between colour fields of magenta that is a mixture of red and blue in average. Because of mixture of magenta that is a complementary colour of green, the colour displayed in the green colour field approaches towards white colour on green and magenta lines on the chromaticity diagram. Thus, the hue rotation can be stopped using RGBG four colour fields FSC methods.

14.6 Colour break up in motion picture: a way of its reduction

There is the problem of colour breakup peculiar to FSC-LCD. It is generally said that colour breakup occurs when motion pictures are displayed on FSC-LCD. More specifically, with reference to Figure 14.7, when a white square image moves in a

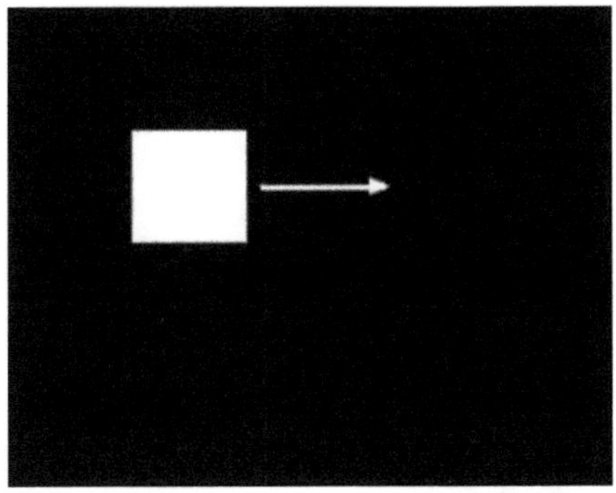

Figure 14.7 Chart for explaining colour breakup

black background, false colour are perceived along the boundaries intersecting the moving direction of the white square image.

'Colour breakup in motion pictures' derives from the timings of storing general image signals in frame memories and retrieving them for respective colour fields that are peculiar to FSC-LCD. Figure 14.8 illustrates how such 'colour breakup in motion pictures' appears on the conventional FSC-LCD.

Explained below are causes of 'colour breakup in motion pictures' of existing FSC-LCD with reference to Figures 14.5, 14.7 and 14.8. Taking an existing FSC-LCD with which the white square image is being displayed as moving at a uniform speed in the black background as shown in Figure 14.7. Figure 14.8(a) shows at the motion of the white square image along the horizontal axis and its time along the vertical axis. As shown in Figure 14.5, data of images of an identical position are outputted to the liquid crystal panel in one frame period and in one display frame period. Observer's eye tracking, however, moves at a constant speed following the motion of the image. This is shown by an inclined line in Figure 14.8(a). Figure 14.8(b) also shows relative positions of the image being displayed when assuming that the observer's eyes stay at a constant position. The image being displayed is recognised by the observer's eyes as an integrated, synthesised image. In the case where the white square image is moving at a uniform speed, positions displayed by respective colour are constant respectively as shown in Figure 14.8(b), and this causes the observer to perceive colour breakup shown by references III and IV in Figure 14.8(b).

As to 'colour breakup in motion pictures' of FSC-LCD, which has been considered difficult to solve heretofore, the following novel driving scheme can largely reduce colour breakup without the need for large modification of existing FSC driving circuit. Figure 14.9 is a timing chart showing timings of FSC-LCD with the novel driving scheme eliminating 'colour breakup in motion pictures'.

Figure 14.10 illustrates how such 'colour breakup in motion pictures' eliminates with the novel driving scheme of FSC-LCD. A drive circuit used in the novel driving scheme to realise the timing shown in Figure 14.9 is basically the same as

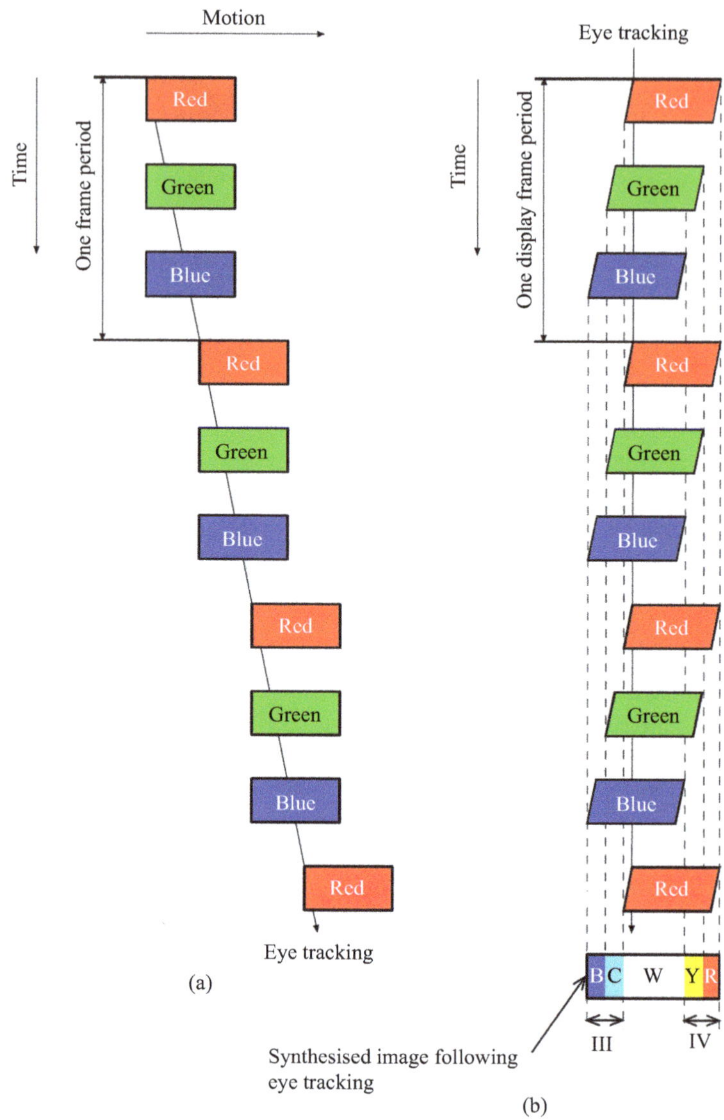

Figure 14.8 Diagram for explaining causes of colour breakup in motion pictures of existing FSC-LCD

the drive circuit of Figure 14.4 used in the existing FSC-LCD. However, the drive circuit used in novel driving scheme is different from the existing one in that the timings of the memory arbitration circuit for reading data from the first and second frame memories to supply the LC panel with display data are different from those of existing FSC-LCD as best understood in comparison with the timing of Figure 14.5 (existing FSC-LCD).

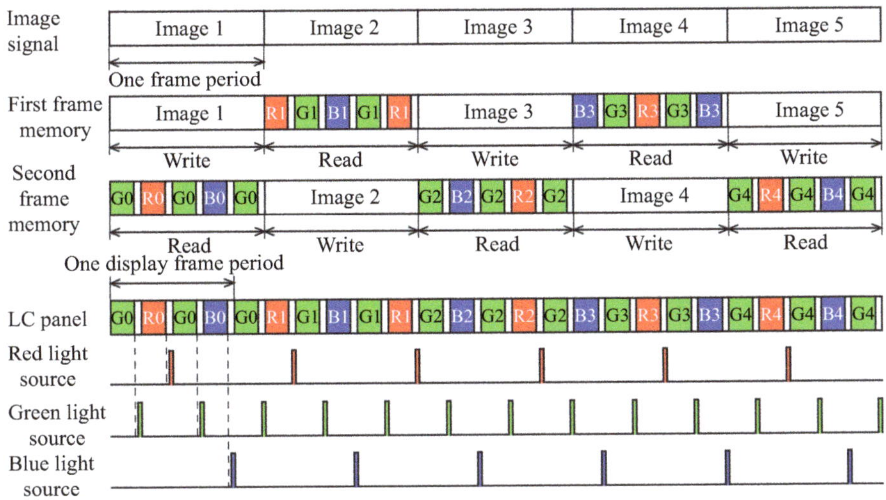

Figure 14.9 Timing chart of FSC-LCD with the novel driving scheme eliminating colour breakup in motion pictures

Masaya Okita already proposed an FSC method in which one frame is composed of four colour fields, i.e., red, green, blue and green colour fields to constitute one display frame. For preventing hue rotation and flickering, the embodiment of the novel driving scheme drives the system by composing each display frame of four colour fields in the order of red, green, blue and green.

Features of the novel driving scheme are explained below with reference to Figures 14.9 and 14.10. In this embodiment, one frame period of image signals to be inputted is composed of five colour fields. One display frame is composed of four colour fields, i.e. red, green, blue and green colour fields. Therefore, the composition of the colour field in one frame period of image signals to be inputted is not constant as show in Figures 14.9 and 14.10. When a certain frame period is composed of RGBGR, the next frame period composed of GBGRG. It continues with BGRGB and GRGBG. In the image following observer's eye tracking in Figure 14.10(b), colour of colour fields at the same display positions change for every frames in the order of red, green, blue and green. As a result, the image displayed in Figure 14.10(b) is perceived by the observer as an integrated, synthesised image. Thus, it is recognised as white as indicated by references I and II in Figure 14.10(b) and the observer does not perceive colour breakup in motion pictures.

The novel driving scheme does not need to change the write timings for the frame memories. Instead, it is sufficient to change the read-out cycle and colour order from the frame memories. Therefore, a device intended by the novel driving scheme can be realised by slightly changing an existing FSC driving circuit.

The most common frame frequency of image signals is 60 Hz. In the embodiment of the novel driving scheme, the display frame frequency is 5/4 times the frame frequency. Accordingly, the display frame frequency is 75 Hz. Thus, the

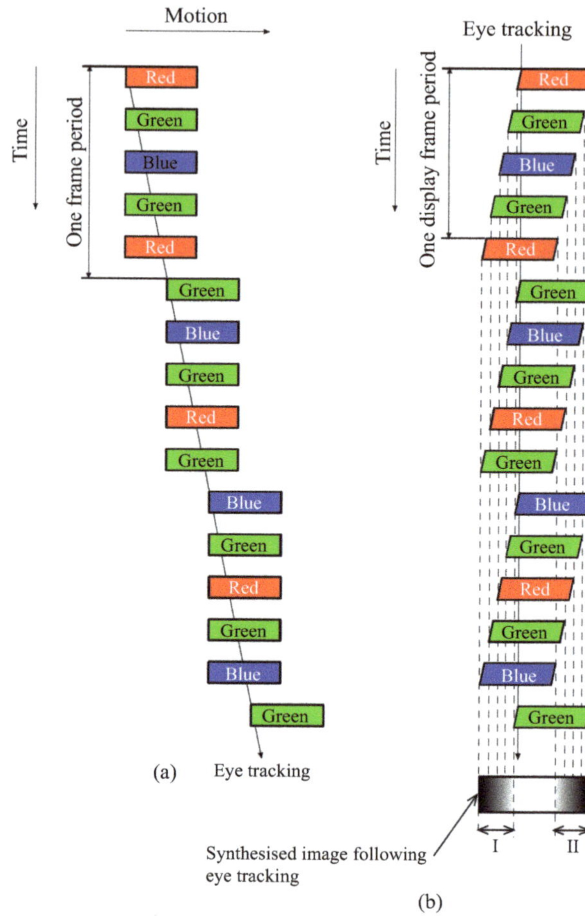

Figure 14.10 Diagram for explaining the novel driving scheme eliminating colour breakup in motion pictures

on–off frequency of the green light source whose flickering are most perceivable for human eyes is 150 Hz, and the on–off frequency of the red light source and the blue light source is 75 Hz, respectively. Considering that CRT displays for personal computers are implemented with means for users to set the frame frequency at 75 Hz to reduce flickering, the on–off frequency of 75 Hz is acceptable for users having usual levels of sensitivity to flickering. Thus, the field-sequential system according to the embodiment of the novel driving scheme can realise the display frame frequency that keeps observers unperceivable to flickering even when the frequency of the image signal inputted is maintained at 60 Hz.

In this embodiment of the novel driving scheme, in case the frame frequency of the image signal is 60 Hz, the display frame frequency is 75 Hz. However, in case the frame frequency of the image signal is 30 Hz or lower, the colour field frequency may be adjusted such that colour of first colour fields of respective frame periods vary from frame to frame. In this manner, it is still possible to reduce colour breakup in motion pictures.

In this embodiment of the novel driving scheme, the frame period is an integral multiple of the colour field period. Synchronisation, however, is not necessary between them. Only if it is ensured that colour of first colour fields of respective frames vary from frame to frame at almost all occurrences, colour breakup can be reduced as well.

The reduction of colour breakup is independently investigated by the group of Han-Ping D. Shieh [13] and that of Tatsuo Uchida [14].

14.7 Saccadic colour breakup

'Colour breakup in motion pictures' can be reduced by an appropriate control of timings of frame memories. Heretofore, however, colour breakup could not be removed completely even by such a novel driving scheme. This is because there is the problem of 'saccadic colour breakup' in addition to 'colour breakup in motion pictures'. 'Saccadic colour breakup' is a phenomenon in which a false colour of a high saturation, not contained in the image displayed, is perceived randomly, momentarily. This occurs when eyes are moved swiftly because it causes a large movement of the viewing point in one frame period, which invites a large displacement of the image of red, green and blue.

It is a characteristic nature of 'saccadic colour breakup' that it occurs not only with motion pictures but also with still pictures on the FSC-LCD. It prominently occurs at boundaries between white and black of images exhibiting a high contrast. Increasing the frame frequency is one idea to decrease this problem. Heretofore, however, this 'saccadic colour breakup' could not be removed even in projectors using the FSC method that were doubled in frequency (120 Hz) as compared with the normal frequency (60 Hz). This problem has been an obstacle against promulgation of colour display devices using FSC method. Furthermore, since the approach by increasing the frame frequency results in increasing the power consumption of the display device, it has been difficult to increase the frame frequency in display devices using certain kinds of monochromatic display elements because of limited speeds of behaviour of such display elements.

Figure 14.11 shows a good idea to reduce 'saccadic colour breakup' significantly. As referred to earlier, Figure 14.6 illustrates a relationship in time between optical transmittance of a display element and on–off switching of red, green and blue light sources in a formerly proposed FSC-LCD in which the red light sources and the blue light sources are switched once respectively and the green light source is switched twice in one display frame period. In this former proposal of Figure 14.6 in which the on–off switching frequency of the green light source is two times the display frame frequency, flickering is less perceivable. Even with this example of Figure 14.6, however, 'saccadic colour breakup' could not be reduced. This is because the length of time for displaying red, for example, is only

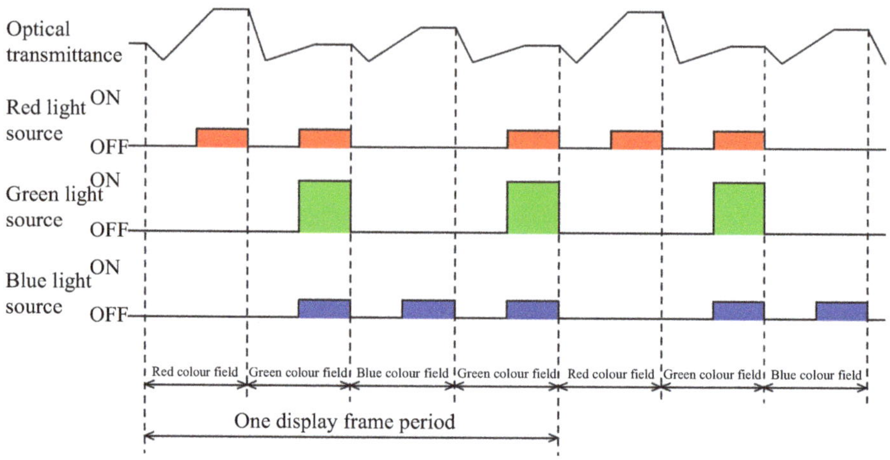

Figure 14.11 Timing chart of FSC-LCD reducing saccadic colour breakup

one quarter of colour fields in one display frame period, and this causes observers to perceive red instantaneously because of offsetting of respective colour images when their eyes move quickly. Especially, red is more perceivable than the other colour. Therefore, unless the ratio of light-up time of the red light source is increased, saccadic colour breakup is perceived.

Referring back to Figure 14.11, in a more preferable embodiment of the novel driving scheme, when the green light source is turned on and off in each green colour field, at least the red light source is additionally turned on and off in each green colour field. More preferably, it is desirable to turn on and off all of the three light sources of green, red and blue in each green colour field.

In Figures 14.6 (existing system) and 14.11, the intensity of red light in the red colour field in the system of Figure 14.11 is lowered as compared with that in the red colour field in the existing system of Figure 14.6, and the decreased quantity of red light in the red colour field is distributed to emission of the red light source in the green colour fields such that the total quantity of red light in one display frame in the system of Figure 14.11 be substantially equal to the quantity of red light in one display frame in the existing system of Figure 14.6. This is for the purpose of maintaining the white balance. Therefore, in case the display element displays a white image, a red image is displayed not only in the red colour field but also in the green colour field. Thus, the red image is displayed in three quarters of colour fields in one display frame.

The same is applicable to quantity of blue light in the blue colour field as well. To maintain the white balance, quantity of blue light in the blue colour field is lowered as compared with that in the blue colour field in the existing system of Figure 14.6, and the decreased quantity of blue light in the blue colour field is distributed to emission of the blue light source in the green colour fields such that the total quantity of blue light in one display frame be substantially equal to the quantity of blue light in one display frame in the existing system of Figure 14.6.

Therefore, a blue image, as well, is displayed in three-quarters of colour fields in one display frame. As a result, 'saccadic colour breakup' is removed to an unnoticeable level even without increasing the frequency of display frames. As to quantity of green light in the green colour field, the number of colour fields for turning it on is one-half of the total colour fields in one display frame. However, since the frequency of the green colour fields is twice that of the red colour field or the blue colour field, regardless of the light-up time of the green light source being shorter than the light-up time of any other colour light sources, colour breakup of green is not perceivable as well like the other colour.

Therefore, green saturation degrades. However, the red light source alone is turned on and off in the red colour field; the blue light source alone is turned on and off in the blue colour field; and the white balance is maintained approximately constant.

As a result, in a colour display device in which the red light source is turned on and off in addition to the green light source in the green colour field, red saturation and blue saturation do not degrade, but green saturation degrades. However, since human eyes are less sensitive to green saturation than to saturation of the other colour, the colour gamut sensed by human eyes does not degrade so much as compared with the case where saturation is decreased for the three colour evenly. As to the ratio of luminance for the red light source and the blue light source among respective colour fields in the novel driving scheme, its optimum value varies with the brightness of the colour display device, its frame frequency, environmental light, and so on. However, it would be easy for an ordinary person in the art to adjust the ratio of luminance among respective light sources in respective colour fields at an optimum value.

It is known that 'saccadic colour breakup' is not bothering so much when the brightness of the colour display device is low. The lighting ratios of the red light source and the blue right source relative to the green light source in the green colour field may be adjusted to enhance the green saturation of light in the green colour field higher as the brightness of the colour display device becomes lower if the display device is set in a low brightness level, that is, to enhance the green saturation by raising the ratio of quantity of green light in the green colour field.

14.8 LC response time requirement

For resolving problems in the FSC-LCD such as colour breakup, flickering and hue rotation at low temperature, it is commonly recognised that response time of LCD must be very short. However, we have resolved these problems by adopting our techniques mentioned above with TN-mode LCD without very fast response speed. Thus, we have done mass-production of our FSC-LCD easily.

Of course, response speed of existing TN-mode LCD is not fast enough to get high colour gamut at low temperature. In the future the improvement of the response time of LCD for FSC-LCD will be utilisation of FLCD as shown in Chapter 16 and nano-particle doped narrow gap TN LCD [15] which will be presented in the later section.

14.9 Implementation of a practical FSC-LCD: transparent FSC-LCD

One of the most promising practical usages of FSC-LCD will be transparent LCD. By taking its high optical transparency, FSC-LCD reaches 20 per cent transmittance due to the use without colour filter that absorbs light. Figure 14.12 shows a prototype of the transparent FSC-LCD whose specifications are: 15.6 inch, $1{,}600 \times 900$ pixels and a-Si TFT substrate. This technology will open a new era of signage and game machine.

14.10 Viewable under the sunshine

FSC-LCD has the virtue of good view ability under the direct sunshine; Figure 14.13 shows an experimental example of this demonstration. This FSC-LCD comprises 4 modules with the size of 10 cm × 10 cm and 24 × 24 pixels and narrow gap TN-LCD which is identical with those of FSC-LCD adopted in the airport indicator shown in Figure 14.1.

14.11 A further study for reducing LC response time of FSC-LCD at low temperature: doping of nano-particles

We examined several nano-particles such as this of MgO (solved in butyl alcohol, Ube Materials) [16], Pd metal, PγCyD-ZrO_2 (Toshima lab and Shiraishi lab) [17,18] and Aerosil (SiO_2, EVONIK). With an appropriate choice, we have obtained a good result as an example shown in Figure 14.14.

Figure 14.14 shows an experimental demonstration of how the doping of Nano-particles of MgO brings a finite response of a TN-LCD cell even at −10 °C;

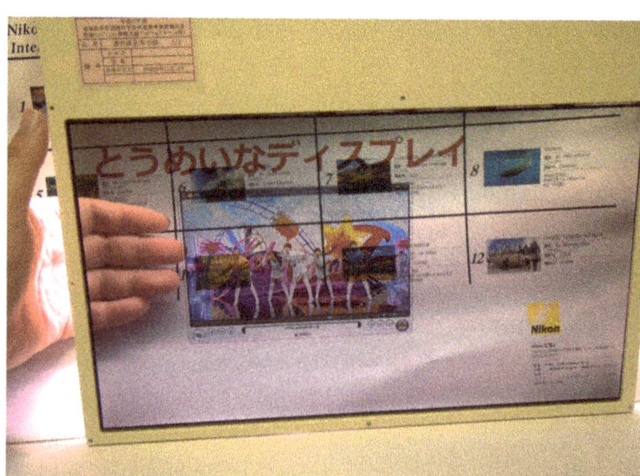

Figure 14.12 Transparent FSC-LCD

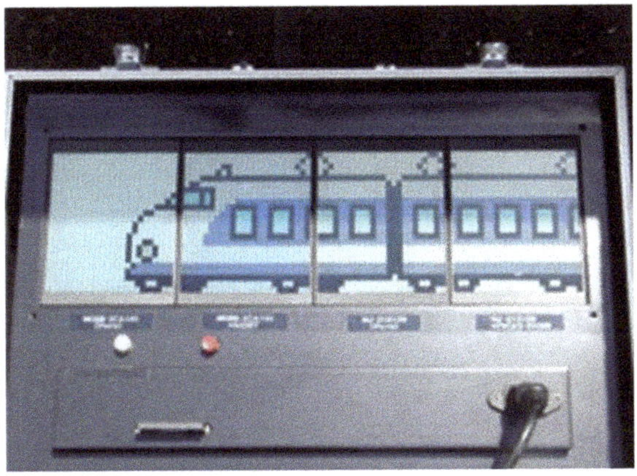

Figure 14.13 Experimental example of a good viewing ability of FSC-LCD under the direct sunshine (©SID)

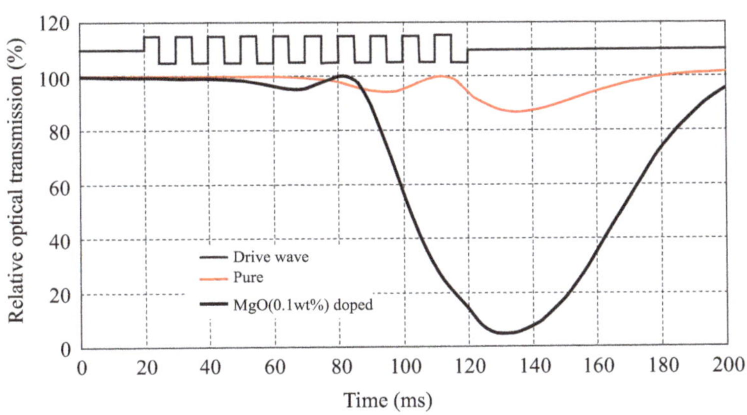

Figure 14.14 Temporal behaviours of TN-LCD with and without doping nano particles of MgO to a tone-burst voltage waveform at −10 °C

contrary to this without nano-particles almost no response to the same tone-burst voltage wave form. This kind of behaviour occurs also in a practical nematic LC. Now, shall go onto the explanation of this particular phenomenon in the followings as shown in Figure 14.15.

By choosing an appropriate combination of LC material and nano-particles, we shift the operating voltage towards low voltage as shown in Figure 14.15.

This kind of operating-voltage-reduction occurs also in TN-LCD cell. This voltage shift towards low voltage is reproduced by raising temperature by 10 °C (called Kobayashi temperature) and according to an independent experiment, this rising temperature brings a degrees of order parameter by 6 per cent. This means

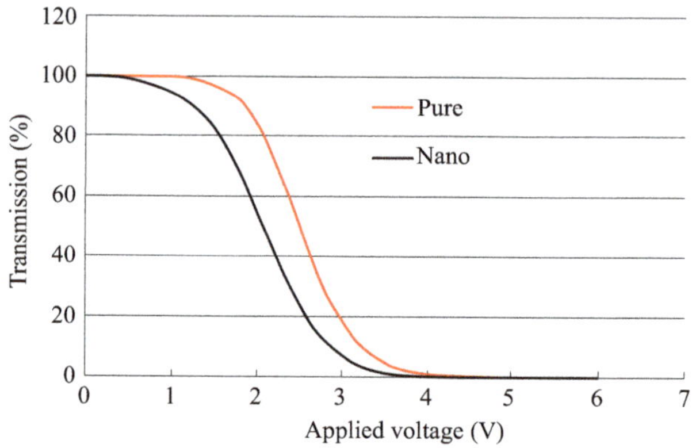

Figure 14.15 An example of reduction of operating voltage due to the existing of nano-particles of PγCyD-ZrO_2 in an ECB cell with NCL-NTN-02 (DIC)

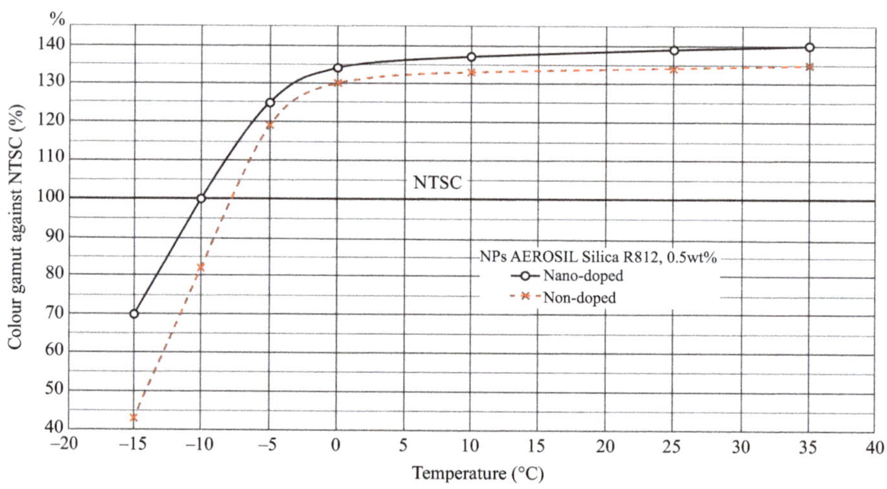

Figure 14.16 Temperature dependence of colour gamut of NTN-FSC-LCD without using colour filter with and without doping the nano-particles of PγCyD-ZrO_2

that the existence of nano-particles keeps a disordered state in an LC-Cell as if keep the LC-cell at an elevated temperature by 10 °C and thus this brings a finite temporal behaviour as shown in Figure 14.14 [19]. Thus, this effect brings a wide colour gamut even at low temperature as shown in Figure 14.16.

Figure 14.16 demonstrates that colour gamut reaches 140 per cent of the NTSC specification due to colour filter less and kept 100 per cent even at −10 °C.

14.12 Conclusions

FSC-LCD has superior characteristics compared to the existing LCD with colour filters in low power consumption, wide colour gamut, high optical transparency and high legibility under outdoor; thus, it has wide diverse applications and utilities such as transparent display and outdoor use. To realise practical FSC-LCD, we have resolved problems in this type LCD such as flickering, colour quality at low temperature and colour breakup by adopting RGBG four colour fields FSC method instead of conventional RBG three colour fields FSC method. In addition to these results, we presented a further research of low temperature operation of LCD in nano-particle-embedded LCD.

References

[1] *New York Times*, 4 September 1940.

[2] Clark M.G. and Shanks I.A. 'A field sequential color CRT using a liquid crystal switch', 1982 SID International Symposium, Digest of Technical papers, 18, pp. 172–76.

[3] Vatne R., Johnson P.A. and Bos P.J. 'An LC/CRT field sequential color display', 1983 SID International Symposium, Digest of Technical Papers, 19, pp. 28–32.

[4] Hasebe H. and Kobayashi S. 'A color field-sequential LCD using modulated backlight', 1985 SID International Symposium, Digest of Technical Papers, 1, pp. 81–84.

[5] Tanaka T., Hasebe H. and Kobayashi S. 'A flat panel field sequential LCD based on the combination of ferroelectric switch matrix and modulated LED backlight', Proc. of Japan Display, (October, Tokyo 1986) pp. 360–362.

[6] Okita M. US Patent No.: US 9,117, 393.

[7] http://www.hdtlcd.com HDT Inc. (Accessed on 1 July 2018).

[8] http://www.okayaelec.co.jp OKAYA ELECTRIC INDUSTRIES Co., Ltd (Accessed on 1 July, 2018).

[9] Kobayashi S., Shiraishi Y., Toshima N., *et al.* 'Further study of optical homogeneous effects in nanoparticle embedded liquid-crystal devices' *Journal of Molecular Liquid* 2018, 267, 303–307.

[10] Yamada F., Nakamura H., Sakaguchi Y. and Taira Y. 'Color Sequential LCD Based on OCB with an LED Backlight' 2000 SID International Symposium, Digest of Technical Papers, 30 1180–1183.

[11] Okita M. US Patent No.: US 7,170, 482.

[12] Bergquist J. and Wennstam C. 'Field-sequential-colour display with adaptive gamut', 2006 SID International Symposium, Digest of Technical Papers. 37, pp. 1594–97.

[13] Huang Y.P., Chen K.H., Lin F.C. and Shieh H.P.D. 'Adoptive LC/BL feedback control in field sequential color LCD technique for color breakup minimisation', *J .Display Technology*, 2008, 4/3, 290–295.

[14] Sekiya K., Miyashita T. and Uchida T. 'A simple and practical way to cope with color breakup on field sequential LCDs', 2006 SID International Symposium, Digest of Technical Papers 32, 717–720.

[15] Kobayashi S., Kundu B., Miyama T., *et al.* 'Narrow-gap field-sequential TN-LCD with and without nanoparticle doping' *Journal of the SID* 2011; 19/11,787.

[16] http://www.ubematerials.co.jp./production.http.MgO (Accessed on 1 July 2018).

[17] Shiraishi Y., Kobayashi S. and Toshima N. 'Improvement of the performance of liquid crystal displays by doping with super molecule-protected metal nanoparticles'. *Israel Journal of Chemistry,* 2013; 908–916.

[18] Sawai H., Shiraishi Y., Miyama T., Kobayashi S. and Toshima N. 'Zirconia nano-colloids having a nano-space of poly(cyclodextrin): Preparation and application to liquid crystal devices', *J. Nanoscience Nanotechnology* 2013; 13, 1–8.

[19] Kobayashi S., Akimoto M., Takatoh K., *et al.* 'Electro-optical properties of LCD doped with nanoparticles and with optical compensators: Ways for fast response' *MCLC* 2014; 594, 21–30.

Chapter 15
Color breakup reduction in field-sequential color LCDs

Han-Ping D. Shieh[1], *Yi-Pai Huang*[1], *and Fang-Cheng Lin*[1]

Climate change is now widely recognized as a major environmental problem. In the face of climate change, energy-efficient household electronics should be developed and popularized. In recent decades, thin-film transistor liquid crystal displays (TFT-LCDs) have become the most widespread display technology for applications ranging from small mobile devices, car navigation systems, laptops, and monitors to larger devices such as TVs and projectors. Although they have many advantages, LCDs still have some room for improvement. Their most notorious issues are an imperfect "dark" state, low color saturation, and low optical throughput. Thus, enhancing image quality and reducing power consumption are two of the main concerns in current LCD manufacturing.

This chapter first introduces the color synthesis mechanisms of LCDs and then illustrates the different types of eye movements and the reasons why eye movements cause a color breakup problem with high-optical-throughput color-filter-less field-sequential color LCDs (FSC-LCDs). Finally, an evaluation metric is introduced for color breakup solutions.

15.1 Colors in liquid crystal displays

LCDs have become the current dominant display technology. Their high market penetration level can be attributed to a host of advantageous features, including high resolution, high brightness, light weight, and a thin form factor. In an LCD, vivid color is produced through one of two possible mechanisms: spatial color synthesis by means of micro color filters (typically red (R), green (G), and blue (B)) or temporal color synthesis by means of the rapid, time-sequential display of three primary colors (R, G, and B), as shown in Figure 15.1.

[1]Department of Photonics, National Chiao Tung University, Taiwan

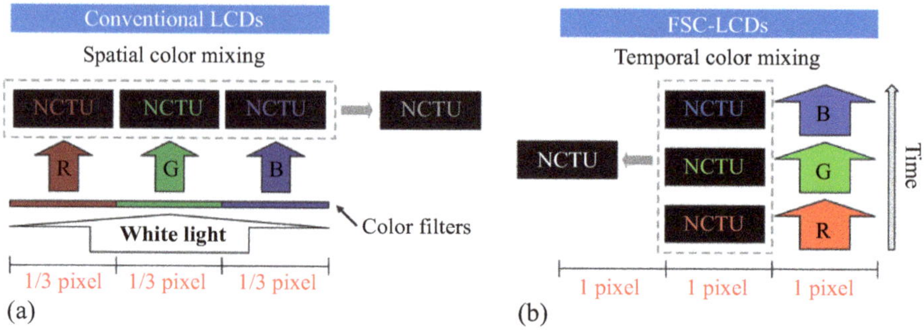

Figure 15.1 Display mechanisms of (a) a conventional spatial color synthesis LCD with color filters has a low optical throughput of 7% and (b) an FSC-LCD, with an ideal triple optical throughput of 21%. Removing color filters, the FSC-LCD by temporal color synthesis also triples horizontal panel resolution compared with the spatial color mixing type LCD

15.1.1 Spatial color synthesis

A traditional TFT-LCD is commonly illuminated using a constant all-on backlight that consists of conventional cold cathode fluorescence lamps (CCFLs) or white light-emitting diodes (W-LEDs). The nonpolarized backlight is transformed into the linear state after propagating through a polarizer on the TFT-array glass. Different voltages between the two glasses, a bottom TFT-array glass and a top color-filter-array glass, are used to control the liquid crystal (LC) orientations and create different phase retardations due to the birefringence property of LCs. When propagating through the LC cells, the linearly polarized light is modulated by the different LC orientations. Subsequently, the light from the different subpixels propagates into R, G, and B color filters at different intensities and becomes colorized. When the light passes through an analyzer on top of the color filter glass, a color image is finally created through spatial synthesis by the human eye. In a commercial 10-bit LCD, the intensity of each type of subpixel (i.e., R, G, and B subpixels) can range over 1,024 shades of gray. Thus, by combining different intensities of the R, G, and B subpixels, such a 10-bit LCD can produce a palette of 1.07 billion colors (1,024 shades of R × 1,024 shades of G × 1,024 shades of B).

LEDs have several advantages, such as having a small form factor and being free of mercury, high power efficiency, fast switching operation, a wide color gamut, and a long lifetime. Therefore, using LEDs in an LCD backlight system not only saves power but also enables the panel to be lighter and thinner. Due to the fast response of LEDs, it is simple to switch such an LED backlight on and off; thus, the advanced technologies of scanning backlight driving [1], high dynamic range (HDR) [2], and FSC [3] were swiftly developed following the introduction of LED backlight systems.

Although LCDs have been used in a wide variety of applications, they still have some drawbacks that have yet to be resolved, such as low optical throughput and a limited color gamut. The generated light propagates through an optical stack comprising a set of polarizers, color filters, and diffusers, and a substantial proportion of the light is absorbed or scattered by these optical films. Overall, only approximately 7% of the backlight reaches the front-of-screen image, with more than 90% of the backlight being consumed. Additionally, CCFLs and W-LEDs have broad spectra that cover all visible wavelengths (e.g., 400–700 nm), and the transmittance spectra of the color filters are not pure peaks at R, G, and B, as shown in Figure 15.2(a). Consequently, the primary colors perceived by the human eye from a conventional LCD are often made up of several different frequencies, resulting in color gamut shrinkage [4]. Typically, the color gamut of a traditional TFT-LCD using a CCFL or W-LED backlight is only 72% NTSC. Compared with a typical backlight module, an RGB-LED backlight has sharper and more well-defined emission spectra that better match the spectra of the color filters in an LCD. Thus, an LCD with an RGB-LED backlight yields a larger color gamut, typically as high as 120% NTSC, as shown in Figure 15.2(b).

The shortcomings of a limited color gamut and low light efficiency arise from the use of a wide-band CCFL/W-LED backlight and color filters that absorb approximately 70% of the backlight intensity. As the world's population becomes increasingly aware of its ecological footprint and the harm that is being caused to our planet, a pressing need to reduce the power dissipation of state-of-the-art LCDs

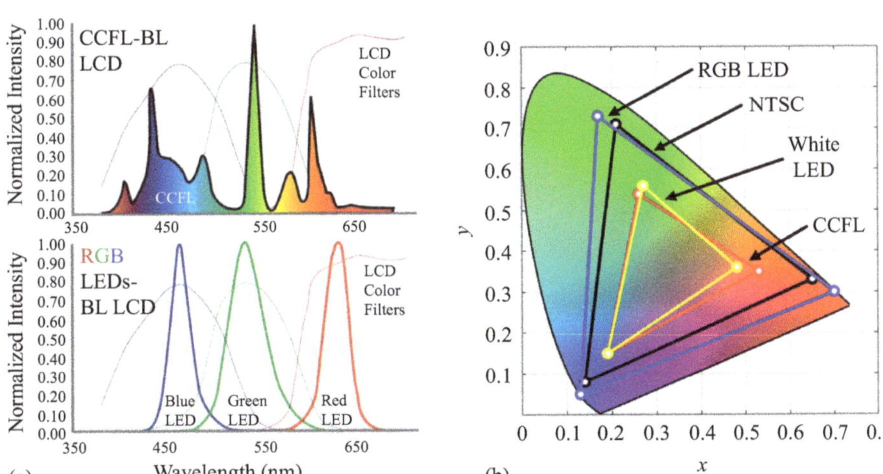

Figure 15.2 (a) Typical color filter transmittance spectra with illuminant spectra of a CCFL backlight (top) and an RGB-LED backlight (bottom); and (b) LCD color gamut with backlights employing the RGB LEDs, CCFLs and white LEDs [4]. An RGB-LED backlight has narrow emission spectra to fit the spectra of color filters of an LCD, resulting in less color crosstalk among different color filters. Thus, an RGB-LED LCD yields larger color gamut than that of using CCFL or W-LED backlight

is arising. In 1985, Kobayashi *et al.* proposed a method of temporal color synthesis, namely, FSC technology, which can produce a vivid color image without using color filters [3]. By using R, G, and B LEDs, which are near-point light sources with narrower spectra, to replace the conventional CCFL or W-LED backlight module, an FSC-LCD can be developed with a color gamut that can easily reach more than 120% NTSC.

15.1.2 Temporal color synthesis

The color-filter-less FSC technology greatly enhances the light efficiency and color gamut of LCDs. In contrast to the spatial color mixing mechanism of a conventional color-filter-type LCD, the R, G, and B fields for each frame in an FSC-LCD are sequentially and rapidly displayed. The three-primary system requires a minimum refresh rate of 180 Hz to prevent luminance flicker. The human eye can then perceive full-color images through a temporal color mixing mechanism. Because FSC-LCDs lack color filters, they have the advantages of a lower material cost, a larger color gamut, a three-times-higher possible screen resolution, and lower power consumption as higher light efficiency [5].

However, FSC-LCDs well-known suffer from color breakup, which causes the image quality to degrade [6]. As explained in Section 15.3, color breakup occurs when a relative velocity exists between a displayed object and the eyes of a viewer. In this case, the constituent color components do not fully overlap at the retina and may be perceived in the front-of-screen image. This color breakup phenomenon may occur during smooth pursuit motion but is particularly noticeable during saccadic eye movements, which may result in large retinal color displacements.

15.2 Types of eye movements

Since only the fovea, which is distributed over approximately 10° of the retina [7], is employed for distinct vision, the human eye tends to undergo either voluntary or spontaneous movements while perceiving scenery. Most primates use two types of voluntary eye movements, namely, saccades and smooth pursuit, in combination with fixation to track objects of interest.

15.2.1 Saccades and fixation

A saccade is a rapid, random eye movement exhibited while perceiving a static object. The eye moves around the object and locates interesting regions (fixation points) from which to gather visual information. Saccades in response to an unexpected stimulus normally take approximately 200 ms to initiate and last approximately 20 to 200 ms, depending on their amplitude. Figure 15.3 shows an example of saccadic movement. When a player taking the role of a police officer wants to shoot terrorists in a shooting game, he or she must rapidly move his or her eyes (saccades) and gaze at objects (fixation) to distinguish terrorists from fellow

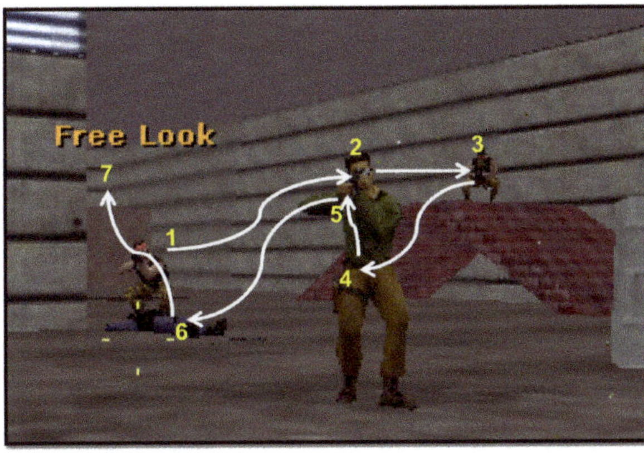

Figure 15.3 An example of saccadic movement while playing a shooting game with tracing directions (white lines) and fixation points (yellow numbers). (Adopted from the online game of Counter Strike)

Table 15.1 Summary of the two common types of eye movements for the color breakup discussion

Types of eye movement	Stimulus	Effect	Speed
Smooth pursuit	Slowly moving objects	Track slowly moving objects	Slow, usually only up to 90°/s
Saccade	Either peripherally detected motion or a decision to change fixation	Examine new objects, visual search	Very fast, up to 1,000°/s

officers. Generally, saccades occur spontaneously when a person is examining new targets, reaching a peak angular speed of 1,000°/s.

15.2.2 Smooth pursuit

Smooth pursuit is a predictable eye movement executed while following a moving object. When following a moving object, the human eye traces it by moving at the same velocity as that of the object, focusing the object on the fovea to capture clear images. Compared with the saccade velocity, the pursuit velocity is much slower; research has revealed the maximum pursuit velocity to be approximately 90°/s. If a target moves at a velocity faster than 90°/s, the eye cannot pursue it [8].

Eye movement is a natural response exhibited by human beings when attempting to perceive a whole scene (Table 15.1). However, unconscious eye movements result in color breakup on FSC-based displays and thus limit the applications of FSC technology.

15.3 Why the eye perceives color breakup

Color breakup is perceived while the eye is moving. Therefore, the color breakup phenomenon strongly depends on the two main types of eye movements: saccades and smooth pursuit. Accordingly, color breakup can be categorized into static and dynamic color breakup based on the characteristics of the relevant images and eye movements.

15.3.1 Static color breakup

The static color breakup phenomenon occurs when observing static images. While a human is viewing a static image, his or her eyes move around the image to gather details. During or after this saccadic movement, static color breakup is perceived. The mechanism of static color breakup is illustrated in Figure 15.4 [9]. Figure 15.4(a) shows a static white bar with a black background presented on an FSC display. When perceiving this image, the eyes gaze at and sweep across the white bar, exhibiting saccadic movement. In a conventional FSC-LCD with an RGB color sequence, the display shows the R, G, and B fields sequentially. When the eyes move in a saccade, the images corresponding to different color fields are projected onto the retina separately, and color breakup is perceived, as illustrated in Figure 15.4(b). The determining factors of static color breakup are the saccade velocity, the image color and luminance differences between the object and the background, and the display field rate.

15.3.2 Dynamic color breakup

The other type of color breakup, dynamic color breakup, occurs when perceiving dynamic images [10].

To focus the object being observed on the fovea, the human eye pursues a moving object at the same velocity as that of the object (Figure 15.5). For example, when a white bar is moving from left to right on an FSC-LCD that uses an RGB color

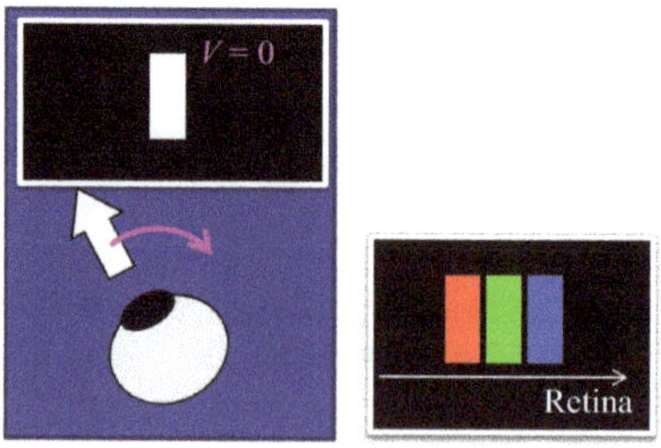

(a) (b)

Figure 15.4 Mechanism of the static color breakup: (a) white bars on an FSC display and a saccade path and (b) observed color breakup during or just after saccade

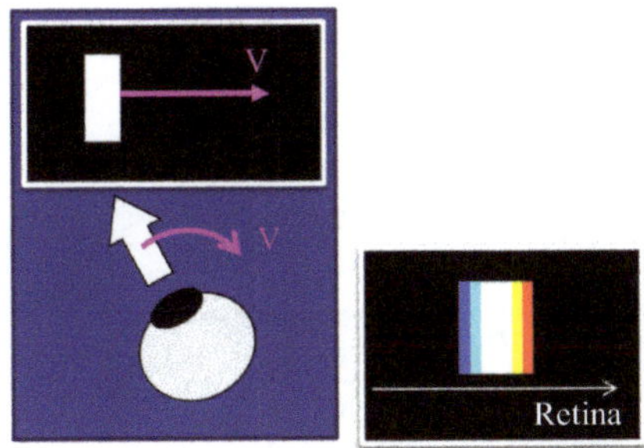

Figure 15.5 Mechanism of the dynamic color breakup: (a) human eyes synchronized with the moving image and (b) the eye-trace integration on the retina. The extra rainbow color will be observed at the bar edge

sequence, the observer perceives it through pursuit movement. The three primary colors are projected onto the retina separately, which causes colorful bands to appear at the edges of the white bar observed as a result of eye integration, as shown in Figure 15.5(b). As can be understood from the observation mechanisms of these two types of color breakup, static color breakup is more serious and troublesome than dynamic color breakup because of the faster associated eye velocity.

15.4 Prior color breakup solutions in LCDs

Color breakup artifacts are highly annoying, causing some viewers discomfort and, in some cases, even nausea. Color breakup suppression has been implemented in digital light processing projectors by inserting additional monocolor fields [11] or increasing the field rate to 540 Hz or higher [12]. Similarly, most traditional color breakup solutions for FSC-LCDs have involved either increasing the field rate or inserting additional monocolor fields (e.g., RGBY, RGBCY [13], and RGBD [14]). The insertion of a white field (W_{min} RGB [15]) enables effective color breakup suppression but fails for images containing complements of the three primary colors, such as cyan, magenta, and yellow. Alternatively, the black field insertion method (360 Hz RGBKKK) has been reported to narrow the color breakup width, making it less perceptible [16], as shown in Figure 15.6. Although RGB-LED backlights can be switched very rapidly, the response time of LCs prevents the implementation of many of the above methods (e.g., field rate > 300 Hz) for large-screen or high-resolution FSC-LCDs.

In the pursuit of fast response panel technologies, it is unlikely that affordable panels with frame rates higher than 240 Hz will be widely available in the foreseeable future. In recent years, Lin *et al.* have proposed the Stencil-FSC [17,18] and local primary desaturation (LPD) [19,20] methods to effectively suppress color breakup, making FSC technology potentially practical.

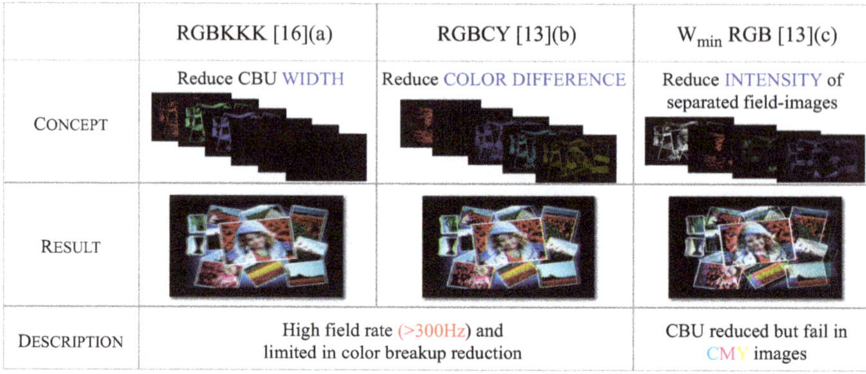

Figure 15.6 Prior color breakup suppression methods: (a) black field insertion; (b) mono-color field insertion; and (c) global minimum white-field insertion

15.5 Color breakup reduction with stencil-FSC

The "stencil" technique is a unique painting technique based on the concept of painting the base color first and then painting in the details. Sometimes, multiple stencil layers may be utilized for an image to add colors or to create the illusion of depth. Additionally, in the human visual system, the luminance contrast sensitivity function is significantly higher than the chromatic contrast sensitivity functions in both the temporal and spatial domains, indicating that the visual system is more sensitive to small changes in the luminance contrast of an image than it is to changes in chromatic contrast [21]. Based on the stencil concept and the properties of the human visual system, the Stencil-FSC method incorporates local color backlight dimming technology. The LC panel acts as a monochrome stencil mask to produce the outline of an image, and then the spatially modulated color LED backlight "paints" colors onto this mask to create a multicolor field, as shown in Figure 15.7(b). Based on the stencil concept and the properties of the human visual system, the Stencil-FSC method incorporates local color backlight dimming technology. The LC panel acts as a monochrome stencil mask to produce the outline of an image, and then the spatially modulated color LED backlight "paints" colors onto this mask to create a multicolor field, as shown in Figures 15.7(b)–(d).

In this way, the luminance is concentrated in the multicolor field, reducing the luminance of the residual field image(s) and successfully suppressing color breakup. For practical applications, the Stencil-FSC method can be driven at field rates of 240 Hz (four field images), 180 Hz (three field images) or 120 Hz (two field images) to effectively suppress color breakup for large-scale TFT-LCD applications. Currently, 180 Hz Stencil-FSC has been implemented in a 65-inch color-filter-less MVA LCD with a 32 × 24 RGB-LED backlight resolution. As evaluated using the IEC 62087 video on the basis of the average of the average picture level curves, 180 Hz

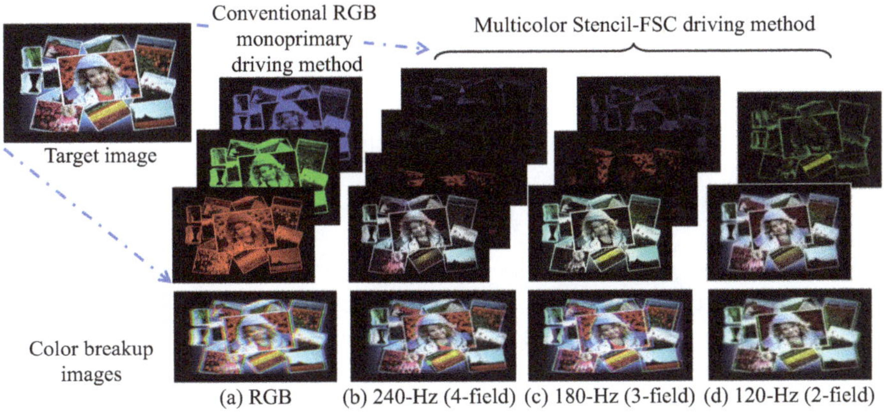

Figure 15.7 Schematic plot and corresponding simulated color breakup images of (a) the current RGB-FSC method and (b–d) the proposed Stencil-FSC method driven at (b) 240 Hz; (c) 180 Hz; and (d) 120 Hz. Stencil-FSC produces a multicolor field image that contains most of the image information, resulting in lower intensity of the remaining primary color fields. Thus, color breakup is suppressed

Figure 15.8 Reduction in green color saturation using 180 Hz Stencil-FSC: (a) the target image, Lotus (provided by Jacky Lee) and (b) the simulated image, with an average color difference of 2.9 in CIEDE2000

Stencil-FSC (52 W on average) achieves a reduction in display power consumption of 20% compared to that of a 65-inch conventional color filter LCD (250 W).

However, 180 Hz Stencil-FSC induces green desaturation because redundant red and blue colors are mixed into the green-based color field [22]. This issue is especially apparent in an image that contains abundant green information, as illustrated by the region indicated by the black oval in Figure 15.8. For the example image, *Lotus*, the

LC signals for the first field are taken from the green band; in this case, the blue and red LC signals are lower in intensity than the green LC signals are.

Therefore, even though the red and blue backlight intensities in the indicated region are low, blue and red light nevertheless propagates through the green-based field and contributes redundant luminance, resulting in a reduction in the green color saturation. Consequently, a modified algorithm has been proposed, namely, 180 Hz min-based Stencil-FSC, in which the average CIEDE2000 color difference of the *Lotus* image is decreased from 2.9 to 0.3. However, the intensity of the modified first-field image is not as high as that of the green-based first-field image. Thus, the color breakup suppression is slightly decreased.

To cope with the shortcomings of the previously discussed methods, a novel algorithm called LPD has been further proposed to redistribute the original three highly saturated primary color backlight fields, for example, R, G, and B, into less saturated fields. Color breakup is suppressed by means of the reduced color differences between the fields.

15.6 Color breakup suppression with local primary desaturation

Traditional FSC-LCDs use the maximum display gamut to sequentially show highly saturated R, G, and B field images at an 180 Hz field rate to create a vivid color image. Traditional driving does not consider the color contents of the target image, and color breakup is very severe, as shown on the left-hand side of Figure 15.9(c).

In contrast, when the local primary desaturation (LPD) method is used with a field rate of 180 Hz, the RGB-LED backlight module is segmented, and the

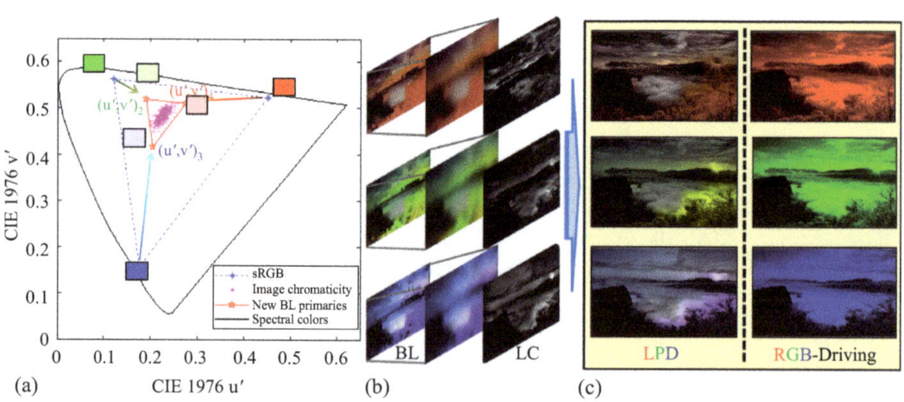

Figure 15.9 (a) The color triangles of sRGB and LPD backlight primary colors in the CIE 1976 u′v′ uniform chromaticity diagram; (b) the three backlight field images and LC signals using the LPD method; and (c) the three field images using the conventional RGB-driving (right) and the LPD method (left)

individual segments are independently controlled to provide a small but sufficient color gamut for each input image. First, the pixel chromaticity distribution of each backlight segment in the CIE 1976 u′v′ uniform color space is analyzed. Subsequently, the backlight primary colors encompassing all pixel chromaticities are dynamically desaturated, as illustrated in Figure 15.9(a). As a result, the colors of the three backlight fields are desaturated by mixing one or more other colors into the same image field (Figures 15.9(b) and (c) left). Thus, the color differences between the field image are reduced, and color breakup is significantly reduced.

15.7 Color breakup evaluation metric

In addition to efforts to suppress color breakup, many studies have also been dedicated to developing an evaluation metric that can objectively predict the subjective visibility of color breakup. For example, some studies have sought to define a threshold between visible and invisible color breakup [6,23–25] and to comprehensively quantify color breakup visibility using a scoring system [26,27]. Most of the earlier studies on color breakup evaluation were based on fixed patterns, for example, solid-colored blocks, whereas more recent studies are trying to extend the evaluation metrics designed for fixed patterns to natural images because practical FSC displays typically show natural images and the color breakup performance strongly depends on the image content [28,29].

The latest such study [30] has demonstrated the mechanism of color breakup perception from the perspective of human vision. First, visual saliency theory has clearly demonstrated that the content of a typical image unevenly attracts human visual attention and that a visual saliency map can be used to quantitatively represent this unevenness of attention [31]. Thus, it can be straightforwardly inferred that color breakup occurring in a more salient image region will contribute more to the overall color breakup visibility than breakup occurring in a less salient region will. Second, while an observer is viewing a still image, the general visual saliency map is strongly associated with the free saccadic movements between eye fixations within the image region [32,33], which are constrained to a certain extent, while the color breakup is induced by external factors. Therefore, the visual saliency map for color breakup should be accordingly constrained, i.e., the overall color breakup visibility should be determined by image regions with visual saliency values higher than a certain threshold, called the dominant visual saliency (DVS) regions [30].

A computational model has been presented on the basis of the theory proposed in [30], as illustrated in Figure 15.10. First, the visual saliency map of a retinal image with color breakup is calculated. Next, a threshold is used to constrain the visual saliency map to obtain the DVS regions. A color difference map between the retinal images without and with color breakup is then calculated pixel by pixel. Here, the color difference is expressed in terms of the chroma difference ΔC^*, which is the Euclidean distance between two points in the CIELAB color space. Finally, using the visual saliency values in the DVS regions as a weighting map, the chroma differences of all pixels are summed and normalized by the area of all DVS regions.

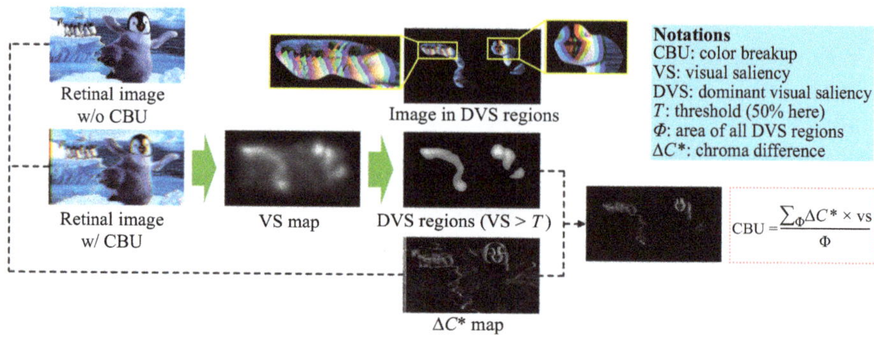

Figure 15.10 Calculation flow of the color breakup evaluation metric based on the DVS theory. The notations are explained in the figure

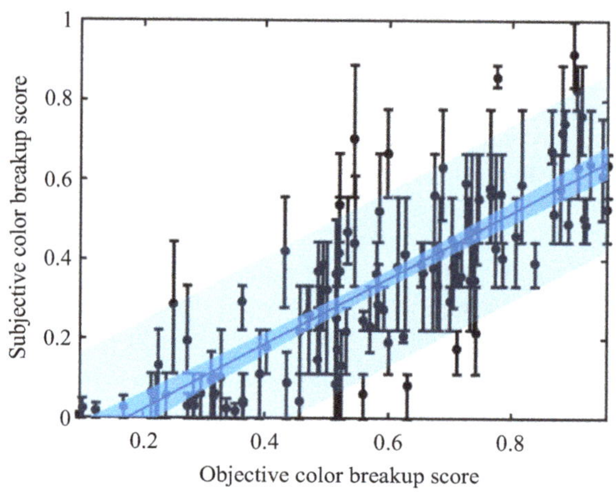

Figure 15.11 Objective color breakup visibility versus subjective visibility, where each error bar shows the mean, upper, and lower quartiles of the subjective data from the 18 subjects. The dark and light red bands denote confidence and prediction intervals, respectively, with a confidence level of 90%

The effectiveness of the color breakup evaluation metric introduced above has been verified based on an image database dedicated to color breakup evaluation, which contains 125 test images with various image complexities and color breakup performances, as well as subjective scores provided by 18 subjects for all the test images.

Figure 15.11 shows the objective scores for the 125 test images, as derived using the evaluation metric, versus the subjective scores; this plot reveals a high Pearson linear correlation coefficient (PLCC) of 0.82 between the objective scores and the mean subjective scores, a.k.a. the mean opinion scores (MOSs). This high

PLCC performance demonstrates that the color breakup evaluation metric is an effective predictor for natural images with varying contents.

15.8 Summary

The underlying color breakup reduction mechanism in Stencil-FSC relies on displaying a multicolor image that contains most of the image information in the first single field, resulting in low intensities of the residual field images. Then, even when the human eye moves at a fast velocity relative to an object on an FSC screen, the resulting separated color fields are dark; thus, color breakup is suppressed. The algorithm is simple, requiring only one convolution for backlight intensity evaluation. However, the 180 Hz Stencil-FSC method faces a trade-off between image fidelity and color breakup suppression. In contrast, the LPD method utilizes a small but sufficient color gamut to display three new desaturated field images that are generated depending on the target image to reduce the color differences between the field images. In this way, LPD maintains high image fidelity and suppresses color breakup. However, three convolutions are required to evaluate the backlight intensities for the three field images when using LPD. This computational complexity results in a high hardware cost for LPD execution. Despite these minor issues, the FSC technique is very viable for application in future display scenarios, such as ultralow-power LCDs, lightweight augmented reality and virtual reality displays, and low-heat and high-brightness vehicle heads-up displays.

References

[1] Sluyterman S. 'Dynamic Scanning Backlight Makes ClearLCD™ TV Come Alive'. *Information Display*. 2005, vol. 10(5), pp. 28–31.

[2] Seetzen H., Whitehead L. A., and Ward G. 'A High Dynamic Range Display Using Low and High Resolution Modulators'. in *Proceedings of SID Symposium Digest*. 2003, vol. 34, pp. 1450–53.

[3] Hasebe H. and Kobayashi S. 'A Full-Color Field Sequential LCD Using Modulated Backlight'. in *Proceedings of SID Symposium Digest*. 1985, vol. 16, pp. 81–83.

[4] Koskela T. 'LED Backlighting for LCDs Requires Unique Drivers, in Power Electronics Technology'. *Press in Power Electronics*. 2008.

[5] Huang Y. P., Lin F. C., and Shieh H. P. D. 'Eco-Displays: The Color LCD's Without Color Filters and Polarizers'. *J. Display Technol*. 2011, vol. 7(12), pp. 630–32.

[6] Jarvenpaa T. 'Measuring Color Breakup of Stationary Image in Field-Sequential-Color'. *J. Soc. Info. Display*. 2005, vol. 13, pp. 139–44.

[7] Osterberg G. A. 'Topography of the Layer of Rods and Cones in the Human Retina'. *Acta Ophthalmol Suppl*. 1935, vol. 13, pp. 1–103.

[8] Meyer C. H., Lasker A. G., and Robinson D. A. 'The Upper Limit of Human Smooth Pursuit Velocity'. *Vision Research*. 1985, vol. 25(4), pp. 561–63.

[9] Yohso A. and Ukai K. 'How Color Break-up Occurs in the Human-Visual System: The Mechanism of the Color Break-up Phenomenon'. *J. Soc. Info. Display*. 2006, vol. 14(12), pp. 1127–33.

[10] Kurita T. and Kondo T. 'Evaluation and Improvement of Picture Quality for Moving Images on Field-sequential Color Displays'. *International Display Workshop*. 2000, pp. 69–72.

[11] Langendijk E. H. A., Swinkels. S, Eliav D., and Ben-Chorin M. 'Suppression of Color Breakup in Color-Sequential Multi-Primary Projection Displays'. *J. Soc. Info. Display*. 2006, vol. 14, pp. 325–29.

[12] Mori M., Hatada T., Ishikawa K., et al . 'Mechanism of Color Breakup in Field-Sequential-Color Projectors'. *J. Soc. Info. Display*. 1999, vol. 7, pp. 257–59.

[13] Cheng H., Rao L., and Wu S. 'Color Breakup Suppression in Field-Sequential Five-Primary-Color LCDs'. *J. Display Technol.* 2010, vol. 6, pp. 229–34.

[14] Huang Y. P., Chen K. H., Chen C. H., Lin F. C., and Shieh H. P. D. 'Adaptive LC/BL Feedback Control in Field Sequential Color LCD Technique for Color Breakup Minimization'. *J. Display Technol.* 2008, vol. 4(3), pp. 290–95.

[15] Pan D., Chang S. D., Hu K. J., Huang J. F., and Tsai Y. S. 'A Novel Driving Method and Device to Reduce Color Breakup in Color-Sequential Displays'. *J. Soc. Inf. Display*. 2008, vol. 16, pp. 669–74.

[16] Sekiya K., Miyashita T., and Uchida T. 'A Simple and Practical Way to Cope with Color Breakup on Field Sequential Color LCDs'. in *Proceedings of SID Symposium Digest*. 2006, vol. 37, pp. 1661–64.

[17] Lin F. C., Huang Y. P., Wei C. M., and Shieh H. P. D. 'Color Breakup Suppression and Low Power Consumption by Stencil-FSC Method in Field-Sequential LCDs'. *J. Soc. Info. Display*. 2009, vol. 17(3), pp. 221–28.

[18] Lin F. C., Chang C. W., Huang Y. P., and Shieh H. P. D. 'High-Image Reproduction by the Low-Field-Rate Stencil-LPD Method for Field-Sequential-Color LCDs'. *J. Display Technol.* 2015, vol. 11(12), pp. 1069–75.

[19] Lin F. C., Zhang Y., and Langendijk E. H. A. 'Color Breakup Suppression by Local Primary Desaturation in Field-Sequential Color LCDs'. *J. Display Technol.* 2011, vol. 7(2), pp. 55–61.

[20] Zhang Y., Lin F. C., and Langendijk E. H. A. 'A Field Sequential Color Display with Local Primary Desaturation Backlight Scheme'. *J. Soc. Inf. Display*. 2011, vol. 17(3), pp. 242–48.

[21] Fairchild M. D. *Color Appearance Models*. Hoboken, NJ: John Wiley & Sons, 2nd Edition, 2005, pp. 26–29.

[22] Huang Y. P., Lin F. C., Tsai C. C., and Shieh H. P. D. 'Stencil Field-Sequential Color Method for Color Breakup Suppression on 180Hz LCD-TV'. *Eurodisplay Symp. Digest Tech Papers*. 2009, vol. 29.

[23] Langendijk E. H. A., Cennini G., and Belik O. 'Color-Breakup Evaluation of Spatio-Temporal Color Displays with Two- and Three-Color Field'. *J. Soc. Inf. Display*. 2009, vol. 17, pp. 933–40.

[24] Yoshida A., Kobayashi M., and Yoshida Y. '24.4: Subjective and Objective Assessments of Color Break-up on Field Sequential Color Display Devices', in *Proceedings of SID Symposium Digest*. 2011, vol. 42, pp. 313–16.

[25] Wang J., Zhang Y., Li X., and Weng Y. 'Color Breakup Visibility Thresholds for 2-Field Sequential Colors'. *Color Research & Application*. 2017, vol. 42, pp. 580–90.

[26] Cheng Y. K. and Shieh H. P. D. 'Relative Contrast Sensitivity for Color Break-up Evaluation in Field-Sequential-Color LCDs'. *J. Display Technol.* 2009, vol. 5, pp. 379–84.

[27] Zhang Y., Langendijk E. H. A., Hammer M., and Hinnen K. 'A New Color Breakup Measure Based on Color Difference Between Fields and Contrast to the Surrounding'. *J. Display Technol.* 2012, vol. 8, pp. 145–53.

[28] Kim J. U., Yang C. M., Bae J. H., Kim C. W., Lee H. S., and Kim D. 'Evaluation of Static Color Breakup for Natural Images on Field Sequential Displays'. *Proc. SPIE*, 2014, #9015, 90150L-1.

[29] Yang C. M., Bae J. H., Kim C. W., Kim D. S., Cho S. P., and Lee H. S. 'Content-Dependent Reduction of Static Color Breakup on Field Sequential Color LCDs'. *J. Display Technol.* 2016, vol. 12, pp. 673–80.

[30] Qin Z., Lin Y. J., Lin F. C., *et al.* 'Image Content Adaptive Color Breakup Index for Field Sequential Color Displays Using a Dominant Visual Saliency Method'. *J. Soc. Inf. Display*. 2018, vol. 26(2), pp. 85–97.

[31] Harel J., Koch C., and Perona P., 'Graph-Based Visual Saliency,' *Advances in Neural Information Processing Systems*. 2007, vol. 19, pp. 545–52.

[32] Foulsham T. and Underwood G., 'What Can Saliency Models Predict About Eye Movements? Spatial and Sequential Aspects of Fixations during Encoding and Recognition,' *Journal of Vision*. 2008, vol. 8, pp. 1–17.

[33] Sun X., Yao H., and Ji R., 'What Are We Looking For: Towards Statistical Modelling of Saccadic Eye Movements and Visual Saliency'. *IEEE Conference on Computer Vision and Pattern Recognition (CVPR)*, pp. 1552–59.

Chapter 16

Field-sequential full colour displays using polymer-stabilised ferroelectric LCD

Toru Fujisawa[1]

The effect of polymer stabilisation of liquid crystal alignment is an attractive method for modifying the electro-optical properties of liquid crystals. The polymer stabilisation in ferroelectric liquid crystals allows a bistable switching of a biaxial alignment, which is surface-stabilised ferroelectric liquid crystal display (SS-FLCDs), to alter into uniaxial alignment of V-shaped switching to enable AC driving. We demonstrated an excellent performance to be suitable for a field sequential full-colour LCD exhibiting a fast switching with response time of less than 500 μs.

16.1 Introduction

Ferroelectric liquid-crystal (FLC) displays, which are well known as surface-stabilised ferroelectric liquid crystal display possessing fast switching response and bistability [1], are attracted to research interest. However, there are several issues to meet the requirements of practical uses in the application of displays. Inherent issues regarding orientation of FLCs utilized smectic C* phase are as follows:

1) Non-recovering to original orientation by mechanically deforming a layer structure of smectic C* phase.
2) Zigzag defects occurring at transition from smectic A phase to smectic C* phase deteriorate black level causing contrast reduction in the displays.

Aligning uniformly liquid crystals of smectic C* phase along rubbing direction of alignment layer require a sequence of transition from N* phase with non-twisting alignment to smectic A phase prior to transition of smectic C* phase under precise temperature control.

In addition, the inherent issues of electro-optical properties with respect to full-colour displaying are caused by bistable switching as well as hysteresis phenomenon based on ferroelectric properties. Biaxial orientation to be switching between a light state and a dark state of surface-stabilised ferroelectric liquid crystal displays is impediment to displaying grey scale produced with voltage-modulation to

[1]DIC Corporation, Fine Synthesis Technical Group 4, Saitama, Japan

be necessary to displaying full colour. The biaxial alignment is not suitable for producing grey-scale owing to no existence of intermediate state between a dark state and a bright state. Moreover, the biaxial orientation which is possible to switch between the two states depending on polarity of applied voltage by direct-current driving causes a possibility of an image sticking with an ion impurity of FLCs.

A suppression of helical structures by using thin cell gap as well as controlling layer structures of smectic C* phase by inducing pre-tilt angle at the surface of alignment layer are required to obtain uniform of biaxial orientation. In order to unwind the helical structures, it is necessary that the cell gap <2 μm which is smaller than the helical pitch of smectic C* phase. In a point view of productivity in LCDs, the manufacturers demand a cell gap >3 μm to yield a high productivity. It is important issue to solve such a narrow cell gap toward mass productions of FLCDs. A birefringence Δn lowering than approximately 0.16 of conventional ferroelectric liquid crystals, and adjusting appropriate helical pitch capable of unwinding the helical structure with a thick cell gap are necessary to manufacture FLCDs.

The solution for these issues that we have adapted is the polymer stabilised V-shaped switching ferroelectric liquid crystal displays exhibiting a continuous grey scale capable of AC-driving with suppression of occurring zigzag defects and with a high-speed response below 400 μs, called PSV-FLCDs introduced by Kobayashi *et al.* [2].

The advantages for PSV-FLCDs are as follows:

1) A continuous grey-scale capability for full-colour displaying by disappearing the bistability
2) Optical uniaxiality state causing a dark state
3) Enabling AC driving based on polymer-stabilised V-shaped switching
4) Fast switching speed
5) Capability of unwinding helical structure in smectic C* phases by polymer stabilisation.

Polymer stabilisation using UV-curable liquid crystals causing change of electro-optical properties in LCDs is an interest research topic to progress to the performance of LCDs.

16.2 Polymer stabilisation in FLCs

PSV-FLC materials, which are composed of a ferroelectric liquid crystal mixture and several photo-curable monomer mixtures including newly synthesised monomers, are used.

PSV-FLCDs are fabricated with a unique process combined with a UV-light exposure to form polymer network for stabilising orientation of FLCs and an AC voltage applied to create uniaxial orientation without zigzag defects.

For preparing defect free test cells and prototype FSC-LCD, the utilisation of polyimide RN-1199 (Nissan Chem. Corp., Tokyo, Japan) film, which is featured by a very high surface-smoothness, is very effective for eliminating the appearance of zig-zag defect (refer to Chapter 1 in Volume 1). The prepared mixture was injected into an empty cell with 1.4–3 μm of cell gap at a temperature above isotropic phase via capillary action or vacuum filling. After the injection, the cell was cooled gradually at a cooling rate of 2 °C/min to room temperature at which the smectic C* phase is observed. And then, the FLC mixture was photo-cured with a 5 mW/cm^2 of UV light source at 365 nm for 10 min to form polymer network in the smectic C* phase; a square wave voltage of ±5–10 $V_{0\text{-}p}$ at a frequency of 1 kHz was applied simultaneously during the UV exposure in order to obtain V-shaped switching. After that by performing photo-curing, PSV-FLCDs showing a V-shaped switching are attained.

By this process, biaxial orientation switching based on FLCs changes into uniaxial orientation of V-shaped switching to be possible to produce a continuous grey scale [3,4].

Polarising micrograph of PSV-FLCDs fabricated with test cell is shown in Figure 16.1(a). The polarising micrograph shows both switching area and non-switching static area during UV exposure. The switching area sandwiched with electrodes exhibits a dark state in uniaxiality which direction of polarising light is accordance with the direction of layer normal in smectic C* phase. In contrast, the non-switching area without electrodes exhibits biaxial orientation including two states composed of a bright state and a slightly dark state as shown in left side of Figure 16.1(a). PSV-FLCDs illustrates that the original biaxial orientation is not changed after UV exposure without applied voltage, whereas the biaxial orientation before UV exposure is drastically changed by UV exposure accompanied with applied voltage.

Observation with polarizing microscope equipped with a high sensitivity camera was carried out to research how the formation of polymer network was differed from both the biaxial orientation and the orientation in uniaxiality. A clear image of the polymer networks as shown in Figure 16.1(b) could be observed by the elimination of optical influences in FLCs surrounding polymer networks with heating up to isotropic transition temperature in FLCs. As the result, a distribution of refractive index differences at boundary between polymer network and isotropic phase in FLCs was detected as an image.

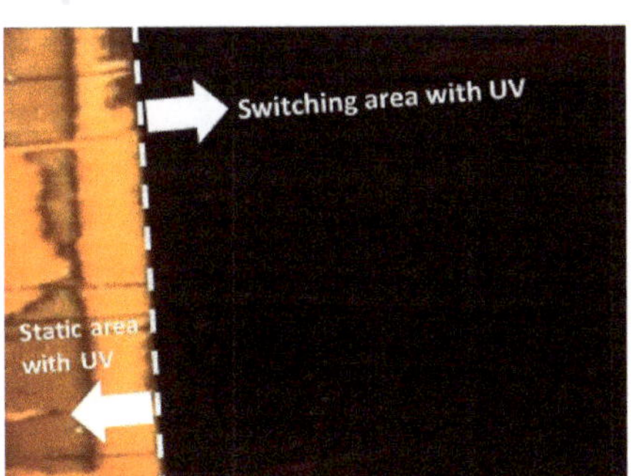

(a) Alignment in PSV-FLCs

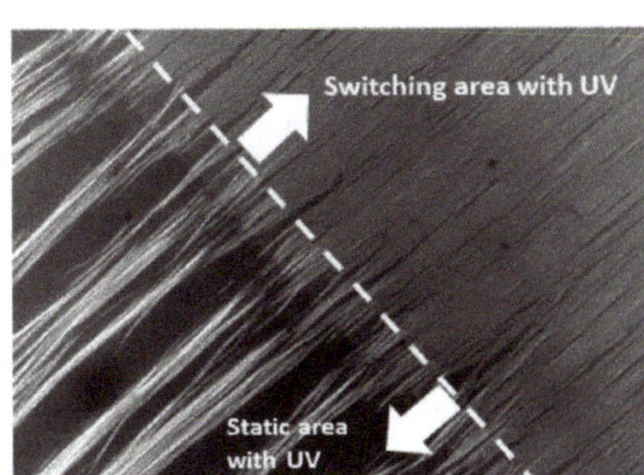

(b) Structure of polymer networks formed in the bulk of FLCs

Figure 16.1 Polarising micrographs

Figure 16.1(b) shows a polarising micrograph of polymer networks forming inside the PSV-FLCs fabricated with above description. As shown in Figure 16.1(b), a lot of fine polymeric fibres are densely aligned along rubbing direction of alignment layer.

The sparse and dense of polymeric fibres at random are formed in the non-switching area where alignment of liquid crystals is stabilised without applied voltage by photo-polymerisation as shown in static area of Figure 16.1(b). On the other hand, the high dense of aligned polymeric fibres is spreading to a switching area sandwiched between a pair of electrodes as shown in switching area of Figure 16.1(b). It is found out that the dense of polymeric fibres is relevant to a dark state of optical uniaxiality in PSV-FLCDs through observation of polarising micrograph at a temperature above isotropic transition in host FLCs.

From the comparison of two cases of polymer stabilisation with applying electric field or without electric field, it is suggested that dynamic behaviour of molecules in FLCs impacts on the density of polymeric fibres distributed in the switching area. In other words, it is understood that appearance of optical uniaxiality producing a dark level in display is depended strongly on the structure of polymer network formed inside a FLC layer.

16.3 Switching properties of V-shaped FLCs

Comparing with transmittance of other operational mode by nematic liquid crystals such as twisted nematic mode, the issue of transmittance in PSV-FLCs has been arisen. PSV-FLC requires ideally a large switching angle at 90° to satisfying the $\lambda/2$ of retardation in smectic C* phase. In contrast, the switching angle to be required in SS-FLCDs is dispensed with an only half switching angle compared with PSV-FLCs needing a large switching angle. These differences concerning switching angle are dependent on the relevance of arrangement between optical axis of FLCs at off-state and the direction of polarising light. As the direction of alignment in the uniaxiality of PSV-FLCs at off-state is positioned at symmetrical centre of switching angle in smectic C* phase, it is necessary that the director of FLCs is tilted at 45° to both side from the symmetrical centre of switching angle for ensuring $\lambda/2$ of retardation. Aiming at the increase of switching angle in smectic C* phase, an attempt to widen operational temperature range in smectic C* phase was performed to attain a high performance of PSV-FLCs possessing high transmittance. In addition, decreasing birefringence in smectic C* phase is required with increasing a narrow cell gap suitable for conventional SS-FLCDs. Therefore, the investigation of ferroelectric liquid crystals exhibiting a low birefringence with a widening temperature range was carried out in terms of thermal stability of FLCDs, an increase of switching angle, and extending cell gap.

The electro-optical properties of the test cells were evaluated with a polarising optical microscope equipped with a hot stage to control temperature of a sample and a photomultiplier to detect light intensity. A 60 Hz of square waveform from 0 to 10 $V_{0\text{-}p}$ is applied to measure E-O property of the test cell. A 0 per cent of transmittance is defined as an intensity of transmitted light through a pair of crossed polariser. A 100 per cent of transmittance is defined as intensity of transmitted light through a pair of parallel polariser. A maximum transmittance is a transmittance measured at which a 10 $V_{0\text{-}p}$ of square waveform is applied to avoid the damage of orientation in smectic C* phase by overload of applied voltage.

The maximum transmittance as a function of cell gap in PSV-FLCDs was investigated as shown in Figure 16.2(a). It is found out that the maximum transmittance increases as cell gap increases from 1.4 to 1.8 μm, reaching a transmittance at 60 per cent at a 2.0–2.5 μm. The maximum transmittance sharply decreases at the cell gap above 2.5 μm owing to an increase of retardation. It is expected that the transmittance at the thickness over 2.5 μm is improved by a decreasing birefringence in ferroelectric liquid crystals. On the other hand, driving voltage of V90 increases monotonously as cell gap increases, as shown in Figure 16.2(b).

Figure 16.3 shows Δn as a function of wavelength in smectic C* phase of previous FLCs and developed FLCs. 1.7 μm of cell gap satisfying $\lambda/2$ of retardation is a necessary to maximise transmittance in case that the birefringence Δn in

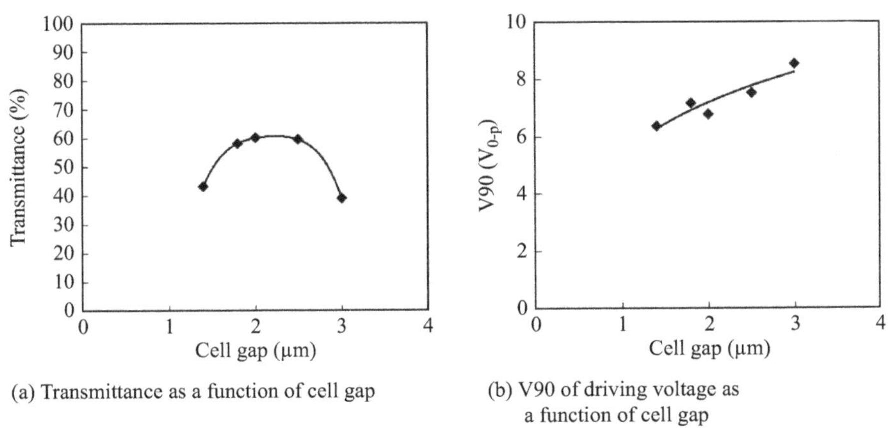

Figure 16.2 Cell gap dependence of transmittance as well as V90

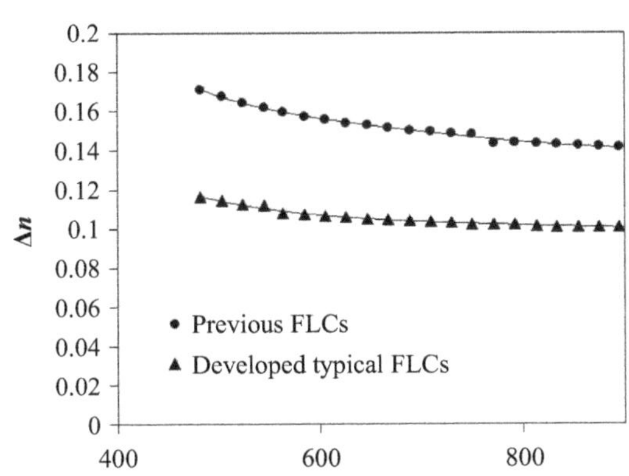

Figure 16.3 Wavelength dispersion of Δn in FLCs

Table 16.1 Physical properties of FLCs

Sample no.	Δn	Switching angle	Max. transmittance	Phase transition temperature/°C				
				Cry.	SmC*	SmA	N^*	Iso
FL1609	0.12	70	75%	−27	112.6	113.6	129.3	
FL1735	0.11	61.2	77%	−20	103.9	106.1	119.3	

previous FLC indicates 0.16 at 550 nm. In contrast, newly developed FLC, which birefringence Δn indicates 0.11 at 550 nm, is possible to extend to 2.5 μm of cell gap to necessary to maximising transmittance.

We have succeeded in realising the first ferroelectric liquid crystals exhibiting a low birefringence with a wide temperature range from −27 °C to 112.6 °C in smectic C* phase. The physical properties are shown in Table 16.1.

Although a birefringence in realising the mixture is decreased by adding a low birefringence substance, the issues to narrow a temperature range in smectic C* phase by adding the substance into the mixture are arisen. In addition, most smectic liquid crystals exhibiting a low birefringence tending strongly to extend readily the temperature range in SmA phase or nematic phase lead to narrowing the temperature range in smectic C* phase.

By using substance exhibiting isotropic-nematic-SmC phase in sequences to ensure a stability of smectic C* phase, temperature range of smectic C* phase is widened without expanding SmA phase. It is revealed that SmA phase of narrow temperature range within 2 °C of SmA phase to be appearance between N* phase and smectic C* phase and a high transition temperature of smectic C* phase over 100 °C are important parameters increasing the switching angle at a room temperature. Consequently, the switching angle >60° is attained as shown in Table 16.1. However, although a high maximum transmittance more than 75 per cent even at a 2.5 μm of cell gap is reached by using a 0.11–0.12 of birefringence Δn in smectic C* phase accompanied with switching angle beyond 60°, the maximum transmittance by biaxial orientation in smectic C* phase is dropped to transmittance at 60 per cent in orientation of uniaxiality by the fabrication process of PSV-FLCDs. The transmittance-reduction is caused by the switching angle decreasing approximately from 60° to 53° with polymer stabilisation by UV exposure.

Figure 16.4 shows the performance of our PSV-FLCDs exhibiting a high contrast ratio of 900:1 with reaching a 60 per cent of transmittance at a 2.5 μm of cell gap. The figure of this improvement was achieved by using new FLC with optimising the composition.

16.4 Switching properties

Formation of three-dimensional high-dense-aligned polymer networks in PSV-FLCs as shown in Figure 16.1(b) affects the change of the polarisation reversal current and switching properties to be relevant to motion of liquid crystal molecules based on

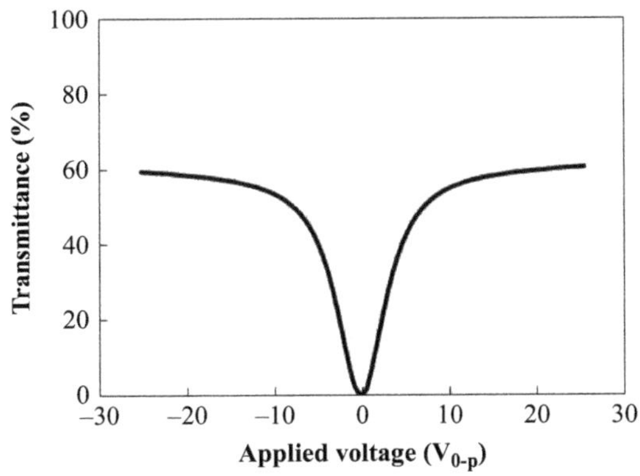

Figure 16.4 V-shaped E-O property

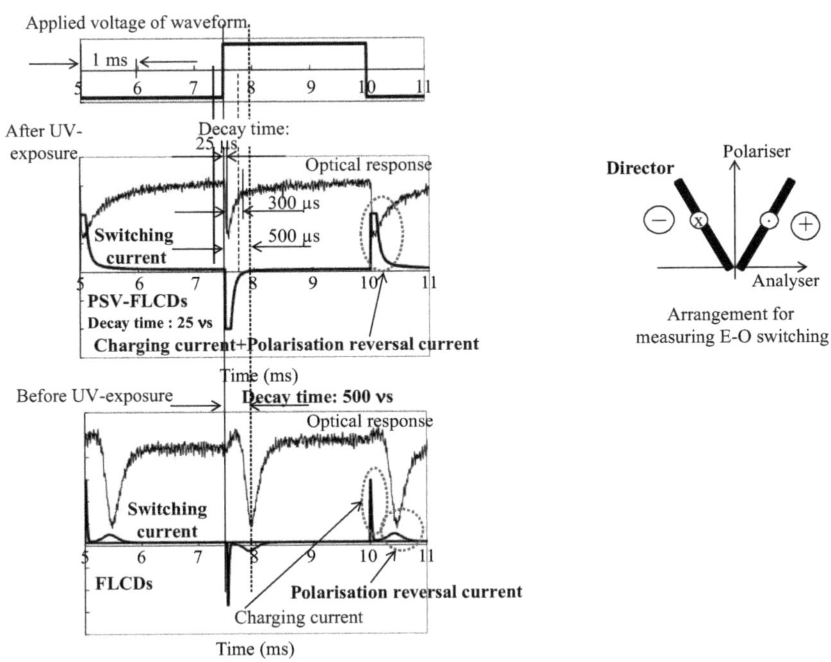

Figure 16.5 Optical response and polarisation reversal current before and after UV-exposure

ferroelectricity in smectic C* phase. The change of switching properties in terms of both optical response and polarisation reversal current, which bistable switching properties before UV-exposure are changed into exhibiting V-shaped switching properties after UV exposure, are shown in Figure 16.5. Comparing the optical response waveform and current waveform with before and after UV exposure, the differences of polarisation reversal current before and after UV exposure can clearly be seen in Figure 16.5. The peak of polarisation reversal current before the UV exposure has appeared at the end of the charging current of capacitor to be occurred at a point of time in the polarity inversion of applied rectangular waveform. In contrast, the peak of polarisation reversal current after UV exposure overlaps the duration in the charging current of capacitor to be occurred after the polarity inversion in rectangular waveform. These different switching processes of polarisation reversal current before and after UV exposure lead to different optical response in each. The decay time in optical response in V-shaped switching after UV exposure, which the polarisation reversal current overlaps the charging current, is accelerated to the 25 μs in comparison of the decay time of 500 μs in bistable switching before UV exposure. The reason for change of optical response is that the effect of anchoring force by the polymer network formed inside smectic C* phase causes the polarisation reversal current occurring immediately after the polarity inversion in rectangular waveform without the delay time as shown in Figure 16.5. The switching of such a mechanism is a major feature of the PSV-FLCD.

16.5 Prototype of a TFT driving panel

We fabricated a field-sequential full-colour (FS-FC) LCD with 4″ diagonal and SVGA (800 × 600, 254 ppi, 1.8 μm of cell gap) specification that makes it possible to display high-quality motion pictures without blurring and without significant colour breakup due to the fast response time less than 350 μs ($\tau_{rise} + \tau_{decay}$) of our PSV-FLCDs [5,6] as shown in Figure 16.6. Materials of PSV-LCDs for a TFT driving used a ferroelectric liquid crystal composition containing a UV-polymerisable liquid crystalline monomer with spontaneous polarisation 33 nC/cm^2, switching angle 54°, and phase sequence: Iso. · 76.3 °C N* 70.6 °C SmA · 66.2 °C · smectic C* · −12 °C Crystal. The prepared mixture was injected into an empty cell at a temperature showing a

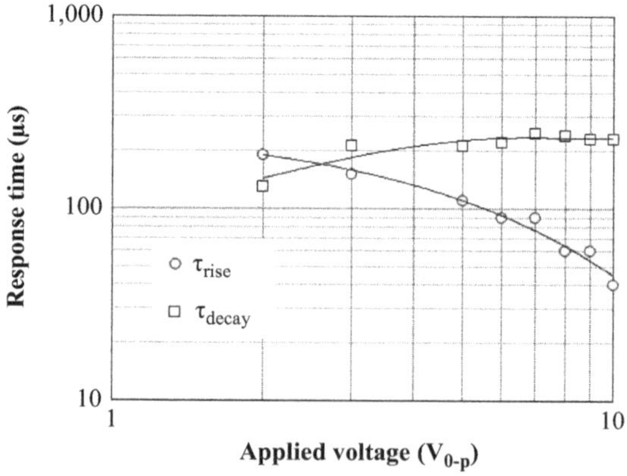

Figure 16.6 Response time as a function of applied voltage in fabricated TFT driving PSV-LCDs

Figure 16.7 A photograph of displayed images on the screen of the prototype field sequential full-colour LCDs using PSV-FLCD (Specification: 4″ diagonal and SVGA (800 × 600 pixels))

nematic phase via vacuum filling, because the viscosity of the smectic liquid crystal is too high to fill by vacuum. After the injection, the cell was cooled gradually at a cooling rate of 2 °C/min to room temperature at which the smectic C* phase is observed. And then, the FLC mixture was irradiated with a 5 mW/cm^2 of UV light source at 365 nm for 10 min with simultaneously applying AC voltage to form optical uniaxiality alignments at the area of electrode in the smectic C* phase. A square wave voltage of $\pm 9\ V_{0\text{-}p}$ at a frequency of 1 kHz was applied simultaneously to the data line of TFT array during the UV exposure in order to change bistable switching into V-shaped switching at the area of pixels.

The panel is driven by voltage modulation which is possible to display 256 grey scales at a voltage of $\pm 9\ V_{0\text{-}p}$ or less. Images are displayed with the frame frequency as 60 Hz as a default frequency, and a flashing LED array backlight system is used to emit three monochrome lights in the sequence of a red, green, and blue light at intervals of 1/180 s, where the frame period of 1/60 s is divided into three sub-frames: R, G and B of monochrome images.

These emissions of monochrome lights are synchronised with the addressing period of images for PSV-FLCDs. The black frame is inserted as erasing period to prevent a generating overlap of colours between the sub-frames. Thus, we could demonstrate the potential of polymer-stabilised ferroelectric liquid crystal displays for the application as full-colour displays and analogue grey-scale displays as shown in Figure 16.7. If it is possible to further improve the uniaxial orientation and the transmittance in the future, such displays could be used for excellent motion picture performance.

16.6 Conclusions

We have succeeded in obtaining ferroelectric liquid crystals exhibiting low birefringence of 0.11 with a wide temperature range from −27 °C to 112.6 °C in smectic C* phase.

It has been revealed that a temperature range within 2 °C of SmA phase to appear between N* phase and smectic C* phase and a high transition temperature of smectic C* phase over 100 °C are important parameters influencing the increase of the switching angle over 60° C at a room temperature.

It is estimated that a dark level in uniaxiality depends strongly on the structure of polymer networks formed in the bulk of FLCs. Even in a 2.5 μm of cell gap, PSV-FLCDs by using FLCs possessing a low birefringence of FLCs make it possible to form a uniaxial alignment exhibiting contrast ratio of 900:1 and to achieve a reaching 60 per cent of maximum transmittance. We have succeeded in demonstrating a field-sequential full-colour LCDs with the specification of 4″ diagonal and SVGA (800×600 pixels) that is capable of displaying full-colour images of not only high resolution still images but also fast-moving video images without motion blurring.

References

[1] Clark A. N. and Lagerwall S. T. 'Submicrosecond Bistable Electro – optic Switching in Liquid Crystals'. *Appl. Phys. Lett.* Vol. 36, p. 899(1980).

[2] Taguchi Y. S., Iimura Y., Kobayashi S., Hasebe H. and Takatsu H. 'Liquid Crystalline Polymer Stabilized FLCDs with Conventional Rubbed Polyimide Films or with Photo Alignment Films of Poly (vinyl Cinnamate)'. *Mol. Cryst. and Liq. Cryst.* 1997, Vol. 292, pp. 333–43.

[3] Miyazaki, Y., Furue, H., Takahashi, T., Shikada, M. and Kobayashi, S. 'Mesogenic Polymer-Stabilized FLCDs Exhibiting Asymmetric and Symmetric (V-Shape) Electrooptic Characteristics'. *Mol. Cryst. and Liq. Crst.*, 2001, Vol. 364, pp. 491–99.

[4] Kawamoto S., Oh-kochi M., Kundu S., Hasebe H., Takatsu H. and Kobayashi S. 'Polymer-Stabilized V-Mode FLCDs and Their Application to Color Sequential Full Color LCDs'. *DISPLAYS*, Vol. 25, issue 1, pp. 45–47 (2004).

[5] Fujisawa T., Hayashi M., Hasebe H., Takeuchi K., Takatsu H. and Kobayashi S. 'Novel PSV-FLCDs with High Response Speed, High Optical Throughput, and High Contrast Ratio with Small Voltage Shift by Temperature: Application to Field Sequential Full Color LCDs'. SID '07 Digest, pp. 633–36 (2007).

[6] Fujisawa T., Nishiyama I., Hatsusaka K., Takeuchi K., Takatsu H. and Kobayashi S. 'Field Sequential Full Color LCDs Using Polymer-stabilized V-shaped Ferroelectric Liquid Crystals'. *Ferroelectrics*, Vol. 364, pp.78–85 (2008).

Chapter 17

Optical security devices using UV-curable nematic liquid crystals

Keizo Nakayama[1]

This chapter begins by reviewing current counterfeit-proofing concepts and the security applications of liquid crystal (LC) materials that have been proposed to date. Next, security devices that use intrinsic random features for applications in a variety of fields are reviewed and the core concept of use of random features is discussed. The random optical textures that occur in LC phases, which have generally been considered to be defects for the purposes of most engineering applications, can then be regarded as favourable characteristics for security devices; the use of the unique patterns of these optical textures as identification information, which was previously proposed by the author, is then explained. Because security measures that use these intrinsic random features appear to be an unfamiliar concept for most LC researchers, a more detailed explanation is provided that traces the concept of security measures that use the random features in LC materials, and real data from actual fabricated devices is then provided.

17.1 Introduction

It is becoming increasingly important to verify the authenticity of many artificial materials and products, including banknotes, identification (ID) cards, and cash cards. In addition, any keys to private equipment or facilities, such as room keys or car keys, are also becoming targets that are worthy of consideration. To enhance security in general, security devices that use digital technologies are widely used today. One specific example is a smart card or chip card that contains an embedded integrated circuit (IC). However, the risk that new counterfeiting techniques will be developed to overcome this security measure must always be a concern. There is also the option of repeating and updating the existing security measures using more recently developed technology in the same technical field. However, because it is easy for the counterfeiters to narrow down the technology that is required for counterfeiting in this case, the development cycles of anticounterfeiting technology

[1]Department of Electric and Electronic Engineering, Kindai University, Japan

and counterfeit technology can become something of a vicious circle. To break this circle, some diversification of the ideas used to provide the security measures is effective. In such a case, two or more security measures based on technologies from different fields must be embedded in the artificial item that requires authentication, which can also reduce the worry that counterfeiters would want to forge that particular item. This is because counterfeiters would need to collect and master quite a wide variety of knowledge and different types of counterfeiting equipment. Therefore, it is expected that the barrier to counterfeiting of a target object that has these diverse security techniques will become very high. Security measures that use analogue features at opposite poles of digital technologies are probably a most suitable application for this diversity idea.

The holographic seals and optically variable inks that are currently widely used may also be classified as analogue technology. Additionally, optical technology is compatible with analogue technology and is suitable for use in security technology, and many security technologies have been developed on this basis [1]. LC materials, which have unique optical properties, have drawn attention as promising materials for security technology applications. The cholesteric LC pigments available from Chelix Technologies Corp. (California, USA) use circularly-polarised-selective and wavelength-selective reflections to provide optically variable security inks [2]. The linearly photopolymerisable polymer/LC polymer (LPP/LCP) technology produced by Rolic Technologies Ltd. (Allschwil, Switzerland) combines LCs with photo-alignment layers that can be patterned via irradiation by linearly polarised ultraviolet (UV) light through a photomask to realise polarised latent images [3]. This technology was improved using a stack composed of other functional layers and through use of dye-doped cholesteric LC materials [4,5]. In addition, security devices that can store dual or plural latent images have also been reported [6,7]. These devices are based on a polymer surface photo-modification procedure, which is performed by surface irradiation using unpolarised UV light through a photomask combined with a subsequent rubbing process to form patterned substrates. Using this technique and a guest-host LC, plural latent images that can be switched via the arrangement of the polarisers are realised.

As mentioned above, LC materials are highly promising as base materials for support of analogue security technologies based on control of their orientation, treatment or other properties. Furthermore, security technology using randomness is attracting considerable attention in the analogue security field [8,9].

17.2 Intrinsic random features for security devices

Intrinsic natural randomness, which denotes uncontrollable variations in the features of certain natural materials or artificial products, can be described as being non-uniform and uncontrollable, depending on the types of features being considered. While natural randomness occurs frequently in certain features of materials, it is typically unsuitable for industrial applications because homogeneous products with consistent uniformity are generally desired. Therefore, these instances of natural randomness, from this unfavourable viewpoint, have been considered to be defects in most engineering applications.

In contrast, in the security field, a random nature is strongly related to provision of uniqueness or a counterfeit-proof nature, based on processes like fingerprint authentication. Therefore, intrinsic randomness is a very attractive property in this field. In fact, security measures using the intrinsic random features of various materials have been studied for use in anticounterfeiting technology for currency and banknotes [8,9]. In addition, in the military sphere, to enhance the supervisory weapons monitoring systems operating between countries with arms agreements, this type of security measure has been studied in the form of unduplicatable tags or unique identifiers that could be embedded on each item of equipment [10].

Many ideas for anti-counterfeiting media of this type, which are considered to be analogue technology, have been individually proposed using various materials in various scientific fields, and range from electronic circuit characteristics [11] to the surface shapes of the products themselves, which can be made from paper or plastic [12,13]. The following concepts are also relevant to optical technology: three-dimensional random arrangements of polymer fibres in nonwoven fabrics [14], speckle patterns from transparent inhomogeneous media [15,16], and the inhomogeneous emission patterns of polymer light-emitting devices [17]. These proposed materials, which show intrinsic natural randomness or variations in types of features that are difficult to tamper with or copy, provide unique information with high security, such as that required for ID tags. Crucial artificial products such as credit cards, on which these features are embedded, can be authenticated by simply reading and verifying the unique information provided by these features in each product.

More specifically, if these devices or ID tags using this concept are embedded into every product and a secure network of verification terminals is then built, where each terminal comprises a reader and an authentication server that stores information about each tag, any product's authenticity can then be confirmed by verifying the information read from the tag using information from the server.

17.3 Concept of the LC optical ID tag

The characteristics that make materials suitable for application to security measures that use their natural random features are the ease with which they can be fabricated and the difficulty involved in cloning them. The author believes that LCs are extremely suitable for this type of application and have wide potential applicability. This section describes the concept and the advantages of using LCs for security measures based on use of random features.

There are many LC phases, in which various molecular orientations occur. Under polarisation-based observation, the textures that reflect on specific molecular orientations can be observed. The molecular orientation is dependent on a combination of the LC phase and the glass surface conditions, and thus the textures must also be dependent on the same combination. For example, the non-uniform optical texture that can be observed in a planar nematic (N) phase cell is the schlieren texture, while that in a homeotropic chiral nematic (N*) phase cell is the fingerprint texture. The natural random textures in LC phases have previously been used to identify the type of LC phase because it is easy to fabricate an LC sample cell or device that indicates these natural optical textures. However, the detailed patterns obtained at each LC cell are different and this non-uniformity between the fabricated cells makes it difficult to apply these cells for practical use in normal optical device fields. These characteristics, however, correspond well with the characteristics required for security technology that uses natural random features.

The fluidity of the LC materials, which indicates that the orientations of the molecules can be changed easily using external factors, is a very attractive characteristic for many applications. In applications to security technologies using natural random features, this highly sensitive fluidity makes it easy to obtain random features. However, after the required random pattern is obtained, this fluidity becomes extremely troublesome. This is because the identification information, i.e., the random pattern obtained, must become permanent regardless of any external effects, such as mechanical pressure or temperature changes. Therefore, when the permanent unique pattern has been decided, it is necessary to fix this pattern. To realise this, we can use UV-curable LC materials, in which the molecules can be fixed by photopolymerisation induced by UV light irradiation.

In short, a security tag that can offer permanent ID information can be realised easily by making the most of the orientation flexibility of LC molecules when using UV-curable LC materials [18–21]. When the authenticity of an item that has been embedded with the proposed tag using LCs is critical, the pattern of the tag in the item can be read optically, and it can then be confirmed that the pattern is registered in the database. Therefore, for simplicity, the proposed tag will be referred to as the liquid-crystalline optical ID (LCOID) tag from here onwards, and means an optically read ID tag using LC materials. In addition, because this method has realised fixation of a random texture using photo-induced polymerisation, it can be classified using an eye point with a counterfeit-proofing device produced using a chemical approach [22].

Section 17.3 focuses on the essentials of the LCOID tag. Section 17.4 will describe the actual results obtained from fabricated LCOID tags.

17.3.1 Fabrication steps

When producing multiple tags that only show the identification information with very precisely controlled technology, a process that deliberately produces unique information is required for individual tag creation. However, because the natural random pattern of the LC's optical texture should be acquired automatically for every tag, a special process that guarantees their uniqueness is not needed. While some molecular alignment controls are needed to acquire the natural random patterns, these controls are less rigid than the alignment control technology required in the manufacturing processes of displays or ordinary optical elements. The barrier to be overcome for practical realisation of the manufacturing technology required for LCOID tags should thus be very low.

When embedding an LCOID tag into a card or other carrier, it is necessary to coat the tag with an outer film. The coating film used will normally have some birefringence properties, with the result that the pattern that is optically read from the LCOID tag coated with the film can have lower contrast than the pattern that would be obtained from the LC cell itself without the coating film. While this effect would be an extremely important issue in ordinary applications, it will not be an issue in the case of a security application as long as a random pattern is observed. In addition, if the birefringence properties of the coating film were to be uneven, this would represent a very serious glitch in ordinary optical applications, but in the proposed security application of the film, even this unevenness can be included as identity information.

17.3.2 Enrolment step and authentication step

Depending on the type of functionality that is mounted in the scanner that is used to read the information from the LCOID tags, we can select the information category that can be used for authentication, and this can be related to a security level. Here, we consider the case where the information is read from tags in the form of a two-dimensional light intensity pattern using polarisers. When the tag with two inserted polarisers is embedded into a card, for example, the scanner or verification terminal that is placed in each shop or other location can be made using a simple architecture that consists only of the functions required to read the light intensity. In this case, when the complete light intensity pattern information has been obtained by a counterfeiter in any way, it becomes theoretically possible to print a film with the same pattern using microprinting technology. However, this does not represent a limitation of the security level of the proposed LCOID tag, because this approach does not use the complexity of the 3D orientation of the LC molecules thoroughly.

Alternatively, when the polarisers are not included in the tag itself but are mounted in the scanner, the security level becomes extremely high. Because a light intensity pattern that reflects the disorder of the three-dimensional orientation of the LC molecules can be obtained using the polarisers, the tag itself is usually transparent. When the layout of the two polarisers and the tag is controlled for optical reading using the scanner, different light intensity patterns can be obtained from just one tag. This means that this type of tag introduces the additional parameter of the angle of the polarisers into the authentication procedure, in addition to the two-dimensional optical intensity information. To pass through this security checking process, counterfeiters must clone the tag such that it has the same optical anisotropy or birefringence in three dimensions. It will probably still be difficult to produce an LC film that simply has the same molecular orientation pattern at present, even if the LCOID tag was to be investigated in detail and the molecular orientation pattern was to be fully grasped. Because this security system is based on optical information that is dependent on the optical anisotropy property of LCs, it is possible to select the properties of the light, e.g., the magnitude and wavelength (i.e., light intensity and colour), and the layout (e.g., the angles of polarisers or incidence) used for the optical reading system in a manner similar to passwords.

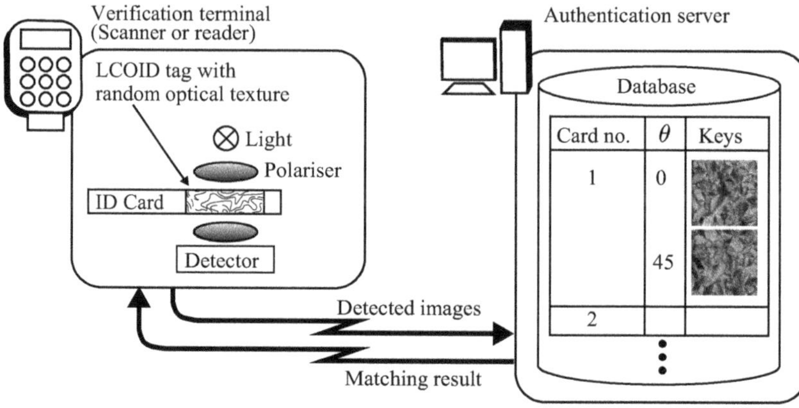

Figure 17.1 Schematic diagram of basic security system using the proposed liquid-crystal optical ID (LCOID) tag. The verification terminal has the function of automatic rotation of the two polarisers and the angle is denoted by theta in the database of the authentication server. The number of times that the check is performed, which involves rotation of polarisers, can be controlled based on the desired security level

Needless to say, this choice will be directly related to factors such as the authentication accuracy, security level and cost requirements.

Another more precise procedure can also be used, in which the light intensity pattern obtained under crossed Nicols observation is used to provide the identification information. The LCOID tags are embedded in this case in products for which authenticity is important. Prior to the distribution of these products, unique information or keys, which are simply images that can be detected optically from the embedded tags or some information that is generated from these images, are stored with an associated identification number in a database. In addition, the patterns are registered using the angle between the crossed polarisers and the tag. If more advanced authentication is required, the number of angles can be increased by rotating the crossed polarisers. When the product is to be used, the keys that are derived directly from the product must then be matched to the keys that are stored in the database before use of the product is accepted. It is possible to multiplex the checking of the patterns by changing the angle of the crossed polarisers in accordance with the desired security level. Figure 17.1 shows a basic security system using the proposed LCOID tag with two-angle multiplexing; it is assumed that the products for which the authenticity is important are ID cards.

17.3.3 Addition of indicator function using a phase transition phenomenon

LC materials show phase transition phenomena that are caused by external factors, for example, by external electric fields or temperature changes. If a phase transition occurs, then the orientation of the LC molecules will change, and the observed texture will also change as a result. Because UV-curable LC materials are used in this research, patterning is easily realised using partial photo-induced polymerisation. This means that LCOID tags can be produced in which different domains show different textures. For example, when a thermal phase transition is used, after the partial photo-induced polymerisation process is performed using a photomask, the LC cell is then heated to the temperature at which the sample shows an isotropic phase, and the entire cell is then irradiated using UV light. This process realises a cell that has a domain with random texture in the shape of the photomask and an isotropic domain without an optical texture. This indicator function, which can record not only the random identification information but also the desired controlled pattern, is only one of the merits of the LCOID tag.

This desired pattern indicator function can be used in two ways. The first way involves a method in which many tags are produced with the same recorded pattern, such as the logo mark of a company. While each tag appears to have the same mark, which cannot be used for identification, all of these marks consist of unique optical texture patterns that act as hidden identification information. Therefore, rigid discrimination of each tag is made possible by management of the hidden micropatterns. The second way involves use of a tag with double identification information. Each tag is fabricated using a mask in the shape of unique information, such as a serial number. When checking this tag initially, a pattern written using the photomask is investigated, and if is necessary to guarantee the genuine nature of the information further, it will be possible to check whether a random pattern on the tag is the same as that stored in the database.

17.4 Practical LCOID tag using nematic LCs

This section focuses on the fabricated LCOID tags and evaluates their performance. Although various categories exist in the LC phases, in which the various textures are observed, based on the current availability of LC materials, research that focused on nematic LC materials has mainly been performed to date. When a material based on a nematic LC is used, the schlieren texture of the nematic phase [21] and the fingerprint texture of the chiral nematic (cholesteric) phase [18–20] can both be used. Because the fingerprint texture is finer than the schlieren texture, an LCOID tag using the fingerprint texture has the potential to realise a higher security level. However, because there is no difference between the essential properties of the two tag types, an LCOID tag using the fingerprint texture of the cholesteric phase is mainly discussed here.

17.4.1 Experimental materials

In this study, a UV-curable LC material (UCL-001-K1, DIC Corp., Tokyo, Japan) was used and was doped using a chiral dopant (S-811, Merck Ltd., Tokyo, Japan). This mixture exhibited an isotropic (Iso)-chiral nematic (N*) phase sequence. The phase transition temperature (T_{NI}) from the Iso to the N* phase of the mixture with 1.8 weight per cent concentration was 40 °C. A commercial cell (E.H.C. Co. Ltd.,

Tokyo, Japan) was used with surfaces that had been treated to ensure homeotropic molecular alignment.

First, the effects of the chiral dopant concentration on the optical textures of the N* phase in the LC are discussed. The mixture with 1.8 weight per cent of chiral dopant showed the so-called fingerprint texture in the LC phase, while the mixture with 9.3 weight per cent dopant concentration showed a fan-shaped texture in the LC phase. In the mixture with a 3.1 weight per cent concentration, the fingerprint and fan-shaped textures coexisted. A texture is thus essentially applicable to the LCOID tag if the texture has some randomness. Every mixture could be used in the LCOID tag research because they all seemed to have some randomness. From the viewpoints of texture complexity and ease of analysis, we decided to use the fingerprint texture that was obtained in the mixture with the 1.8 weight per cent dopant concentration.

17.4.2 Thermal stability of polymerised pattern

The permanence of the fixed molecular orientation is important for practical use of the tag. An LCOID tag using the fingerprint texture that was made using the above mixture was checked by thermal testing at 150 °C for 30 min, and no marked changes were found under crossed Nicols observation. Furthermore, an LCOID tag using the schlieren texture, which was made from the UV-curable nematic LC material only, showed no marked changes under crossed Nicols observation after cooling at −30 °C for 6 h and heating at 150 °C for 3 h.

When ordinary security usage, e.g., in a credit card, is taken into consideration, thermal stability up to approximately 80 °C, which is the temperature on a car dashboard during summer, is required. While the actual thermal stability is dependent on the materials used, existing commercial UV-curable LC materials can satisfy this requirement. Therefore, these results indicate that it will not be necessary to worry greatly about the heat resistance of the proposed LCOID tags for most security device applications.

17.4.3 Uniqueness

For the application to security devices for the use in identification, the fingerprint texture pattern in the N* phase requires uniqueness, in the same manner that no two human fingerprint patterns are the same. When numerous LCOID tags are fabricated on a laboratory level, there is the possibility that the properties of the LCOID tags may be influenced by drift in their implementation conditions. To eliminate the drift as far as possible and to enable evaluation of more essential randomness from the LC material, pattern generation with reorientation in the same LC cell that had not been photopolymerised was used.

Ten fingerprint texture patterns were observed repeatedly in the LC cell with the UV-curable LC material at the same place in the same cell after every re-orientation process, which was performed by heat treatment. When the cell is heated to the Iso phase, the fingerprint texture disappears, but cooling the cell to the N* phase again causes the fingerprint texture to reappear. Figure 17.2 shows two of

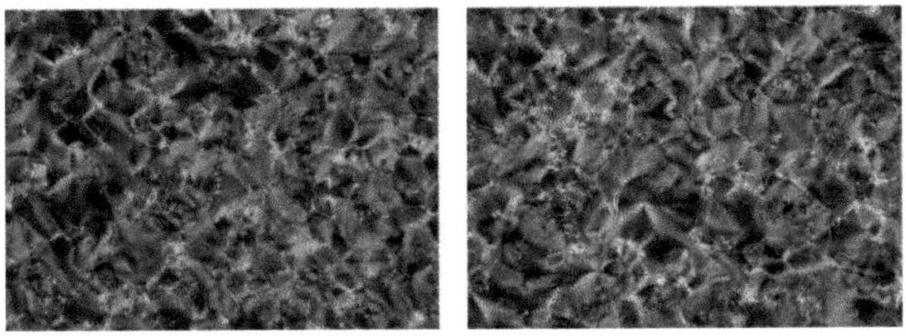

(a) (b)

Figure 17.2 Two polarising micrographs of fingerprint texture patterns acquired at the same place in the same LC cell. Part (b) was acquired after reorientation. The two patterns are not same, which indicates that the fingerprint texture in the LC has intrinsic natural randomness

Table 17.1 Normalised cross-correlation coefficients of ten patterns of fingerprint textures observed in the same LC cell

Pattern no.	**1**	**2**	**3**	**4**	**5**	**6**	**7**	**8**	**9**	**10**
1		0.03	0.02	0.03	0.01	−0.03	0.05	−0.01	−0.01	−0.01
2			0.03	0.08	0.09	0.03	0.07	0.04	0.01	0.02
3				0.07	0.05	0.03	0.09	0.02	0.07	0.03
4					0.05	0.06	0.06	0.05	−0.02	0.05
5						0.04	0.08	0.07	0.08	0.06
6							0.06	0.05	0.04	0.03
7								0.04	0.01	0.03
8									0.11	0.07
9										0.04
10										

the ten polarising micrographs that were captured using a digital camera. These micrographs had an area of approximately 360 μm × 270 μm and an image size of 1,600 × 1,200 pixels.

The randomness of the fingerprint texture was estimated quantitatively via normalised cross-correlation between all the fingerprint texture patterns. Table 17.1 shows the 45 correlation coefficients required for all combinations of the ten different patterns. The maximum value of these 45 correlation coefficients was 0.11. The values were low enough to confirm that the ten fingerprint texture patterns were different. Of course, in the case of a simpler texture, for example, the schlieren texture, a larger area must be used for the evaluation. Moreover, in further research for practical use, the number of different patterns for evaluation of uniqueness should be more than ten.

These results indicate that different patterns were observed, even though the same cell was used, and further indicates that the patterns of the different LC cells or LCOID tags should not be the same. Therefore, unique and random patterns can be obtained easily using the fingerprint texture of the N* phases, even if the procedure for provision of variation is deliberately not introduced at all.

17.4.4 Patterned LCOID tag

The characteristic feature of the LCOID tag is its ability to record a designed pattern. To fabricate a tag that is to be written using a designed pattern, the photopolymerisation process must be performed using a patterned mask. The fabrication process used in this work consisted of the following steps, as shown in Figure 17.3. The temperature of the cell that was filled with the LC mixture was kept below T_{NI}, thus allowing the mixture to exhibit random fingerprint textures over the whole area of the cell. Then, part of the cell was irradiated using UV light through the mask; only the mixture in this irradiated area was photopolymerised, and its fingerprint texture was then fixed (Figure 17.3(a) and (b)). Next, the cell temperature was increased to more than T_{NI} to make the mixture in the non-irradiated areas change phase from N* to Iso; the fixed molecular orientations in the previously irradiated areas remained unchanged. Finally, the whole cell area was irradiated without the mask and the Iso phase areas were then polymerised (Figure 17.3(c) and (d)). After these steps, we obtained a patterned LCOID tag, in which the entire cell was polymerised to give permanence to the molecular orientation.

Figure 17.4(a) shows a polarising micrograph of an LCOID tag that was fabricated using the N* LC material when using a mask in the shape of a '6'. The bright region is the area with polymerised N* orientation, which exhibits the fingerprint texture, while the dark region is the polymerised Iso orientation area that exhibits no optical texture. As described in Section 17.3.3, this tag stores two types of information, as illustrated in Figure 17.4(b) and (c).

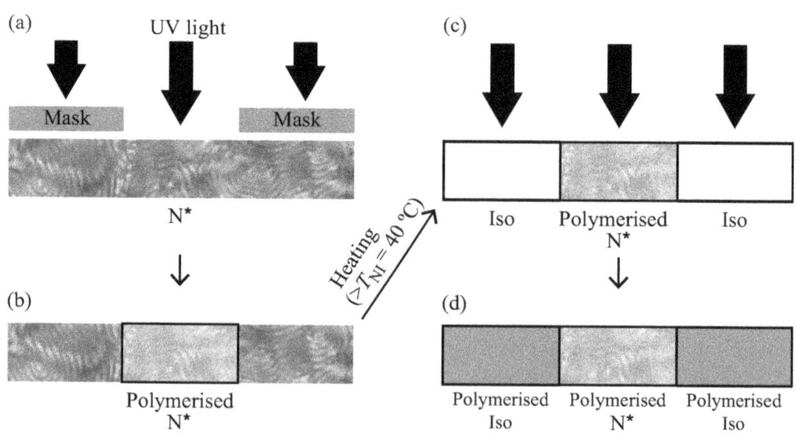

Figure 17.3 Photopolymerisation process for patterned LCOID tag: (a), (b) steps for partial photopolymerisation, and (c), (d) steps for complete photopolymerisation

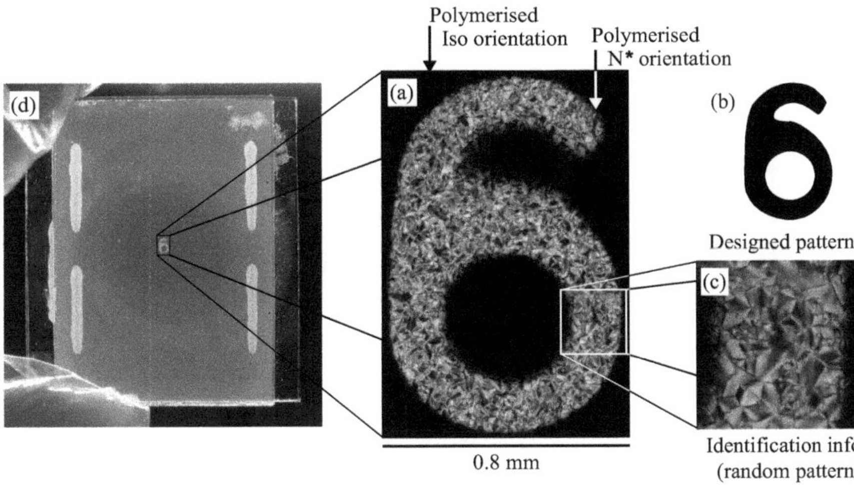

Figure 17.4 Fabricated LCOID tag using N LC materials, which provides both the designed pattern and identification information: (a) polarising micrograph of the tag; (b) designed pattern stored in the tag; and (c) unique identification information; and (d) photograph of the tag taken using a digital camera without polarisers*

If an observer looks at Figure 17.4(a), their eyes will first be drawn to the artificially designed pattern, which is the pattern of the number six, as shown in Figure 17.4(b). If the logo mark of a company or similar is desired, the required tag can only be produced using a photomask of the desired shape. The visualisation image, which has high contrast and is shown in Figure 17.4(a), was obtained using polarisers based on the optical anisotropy property of the LC. Because human eyes do not contain polarisers, we cannot observe the differences in the optical anisotropy directly. Fortunately, the fingerprint texture in the polymerised N* orientation area caused strong light scattering, while the lack of optical texture in the polymerised Iso orientation area caused almost zero light scattering. If we make the most of this characteristic, then the artificial information can be made visible to the naked eye without use of polarisers. Figure 17.4(d) shows a photograph of this tag that was taken using a digital camera without polarisers to imitate human eye observation against a black background. The pattern of the '6' was clearly visible because of the strong scattering from the polymerised fingerprint texture. While it is not necessary to use this characteristic to provide high security levels, LCOID tags have potential for use as simple information indicators for end users.

The second type of information, as shown in Figure 17.4(c) is harder to see, and is desired to be virtually invisible to the end users, including counterfeiters. This is why naturally or automatically generated unique or random patterns are essential information for security applications. The randomness of this pattern is discussed in Section 17.4.3; our attention must therefore now turn to the cloning of LCOID tags in detail. The counterfeiting technology that is assumed to be most

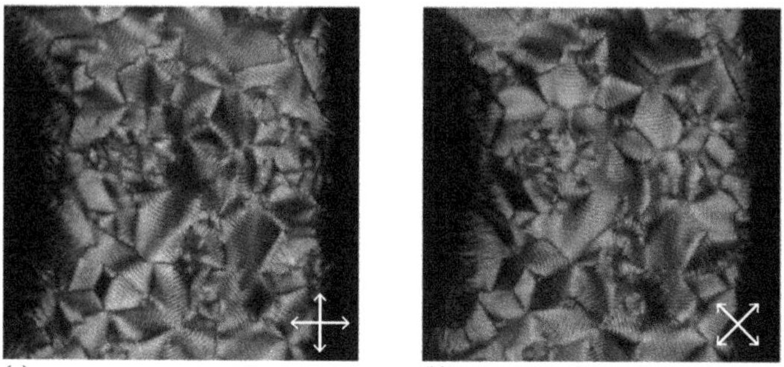

(a) (b)

Figure 17.5 Polarising micrographs taken at the same area of the same LCOID tag under two different crossed-polariser configurations. The crosses represent the directions of the crossed polarisers in each case. The two patterns in the photographs differ in optical intensity and it should not be possible to realise this property using microprinting technologies

commonly used for LCOID tags is a method based on microprinting technology. Because microprinting technologies control the colours and densities of the inks that are to be distributed on films, but not their anisotropy, the anisotropic features of the LCOID tag should be used in the authentication process to the utmost limit. The most simple and powerful method to use the optical anisotropy of the tag in the identification information is to check two or more light intensity patterns that were obtained by varying the angles of the polarisers. In other words, this is the strategy in which the level of security is improved by multiplexing of the polarisation information to reflect the anisotropy of the genuine LCOID tag. Figure 17.5(a) and (b) shows polarising micrographs that were taken at the same region of the same LCOID tag; however, there are differences in optical intensity between these two photographs. This is because the angles of the crossed polarisers that were used in the two photographs were different, and the optical intensity under polarising observation is dependent on the configurations of the LC molecular orientations and the polarisers. By introducing this method into the authentication process, it becomes impossible for tags that have been cloned using microphotography and microprinting technologies to pass through the authentication process. One remaining possibility for cloning of this tag involves use of the LC material itself. While improper use of photo-alignment technology and related processes can be considered, perfect cloning of the complex 3D molecular orientation at this level of fineness is likely to remain a very difficult task.

17.5 Conclusion

The LCOID tag described here uses the randomness of the optical texture of the LC, which is rarely considered to be an industrial advantage. LCs have various phases, and this means that various LCOID tags using various textures, in which both the style and the fineness differ, can be produced easily. Therefore, the introduction of diversity to the LCOID tag is also easy, and it must also be quite disagreeable to have to deal with all these diversities from the viewpoint of a counterfeiter.

Many ideas based on LC materials for anticounterfeiting technology, including our proposed LCOID tag, have been proposed as described in the 'Introduction' section. These ideas will cover various security levels, where the LC materials can provide security elements that range from features that can be checked using the human eye to those that must be checked using a machine and can range from a colour-shifting effect to the generation of unique information. The potential for integration of various security measures using techniques based on LC materials only is anticipated. In addition, LCs, which are soft materials, still have many unique characteristics, and the idea, which represents a novel security application, may be buried.

Acknowledgements

The author thanks Emer. Prof. J. Ohtsubo from the Shizuoka University for providing the opportunity to perform the research and offering valuable comments. This work was supported in part by JSPS KAKENHI under Grant number 25630309.

References

[1] van Renesse R. L. *Optical Document Security*. 3rd edn. Boston: Artech House; 2005.

[2] Jiang Y., Wilson R., Hochbaum A. and Carter J. 'Novel pigment approaches in optically variable security inks including polarizing cholesteric liquid crystal (CLC) polymers'. Proc. SPIE 2002, vol. 4677, pp. 247–54.

[3] Moia F., Seiberle H. and Schadt M. 'Optical LPP/LCP devices: A new generation of optical security elements'. Proc. SPIE 2000, vol. 3973, pp. 196–203.

[4] Moia F. 'New colored optical security elements using Rolic's LPP/LCP technology: Devices for first- to third-level inspection'. Proc. SPIE 2002, vol. 4677, pp. 194–202.

[5] Moia F. 'New color-shifting security devices'. Proc. SPIE 2004, vol. 5310, pp. 312–20.

[6] Yamaguchi R., Mizutori R., Kawamura T. and Sato S. 'Polarized latent image forming in liquid crystal devices using polymer-surface photo-modification technique'. Proc. SPIE 2004, vol. 5618, pp. 166–73.

[7] Yamaguchi R. 'Security applications of liquid crystal film with plural latent images'. Proc. SPIE 2009, vol. 7232, pp. 72320Q-1–8.

[8] National Research Council. *Counterfeit Deterrent Features for the Next-Generation Currency Design*. Washington, D.C.: National Academy Press; 1993. pp. 74–75, pp. 117–20.

[9] National Research Council, *A Path to the Next Generation of U.S. Banknotes: Keeping Them Real*. Washington, D.C.: National Academy Press; 2007. pp. 90–91.

[10] Graybeal, S. N. and McFate, P. B. 'Getting out of the STARTing block'. *Scientific American*. 1989, vol. 261(6), pp. 23–29.

[11] Lofstrom, K., Daasch, W. R. and Taylor, D. 'IC identification circuit using device mismatch'. IEEE International Solid-State Circuits Conference. Digest of Technical Papers, 2000, pp. 372–73.

[12] Haist, T. and Tiziani, H. J. 'Optical detection of random features for high security applications'. *Optics Communications*, 1998, vol. 147(1–3), pp. 173–79.

[13] Buchanan J. D., Cowburn R. P., Jausovec A. V., *et al.* 'Forgery: 'fingerprinting' documents and packaging', *Nature*. 2005, vol. 436, p. 475.

[14] van Renesse, R.L. '3DAS: a 3-dimensional-structure authentication system'. European Convention on Security and Detection; Brighton, UK, May 1995 (Institution of Electrical Engineers, 1995), pp. 45–49.

[15] Pappu, R., Recht, B., Taylor, J. and Gershenfeld, N. 'Physical one-way functions'. *Science*, 2002, vol. 297(5589), pp. 2026–30.

[16] Matoba, O., Sawasaki, T. and Nitta, K. 'Optical authentication method using a three-dimensional phase object with various wavelength readouts'. *Appl. Opt.* 2008, vol. 47(24), pp. 4400–04.

[17] Tada, K. and Onoda, M. 'Polymer light-emitting devices for artificial fingerprints'. *Jpn. J. Appl. Phys.* 2003, vol. 42(Part 2, No.9A/B), pp. L1093–95.

[18] Nakayama, K. and Ohtsubo, J. 'Application of random texture in cholesteric liquid crystal for security devices'. *Mol. Cryst. Liq. Cryst.* 2010, vol. 516(1), pp. 253–59.

[19] Nakayama, K. and Ohtsubo, J. 'Randomness of fingerprint texture pattern in chiral nematic phase and its application'. *J. Soc. Elect. Mat. Eng.* 2011, vol. 20(2), pp. 165–68.

[20] Nakayama, K. and Ohtsubo, J. 'Optical security device providing fingerprint and designed pattern indicator using fingerprint texture in liquid crystal'. *Opt. Eng.* 2012, vol. 51(4), p. 040506-1–040506-3.

[21] Nakayama, K. and Ohtsubo, J. 'Optical security devices using nonuniform schlieren texture of UV-curable nematic liquid crystal'. *Appl. Opt.* 2016, vol. 55(5), pp. 1012–16.

[22] Arppe, R. and Sørensen, T. J. 'Physical unclonable functions generated through chemical methods for anti-counterfeiting'. *Nat. Rev. Chem.* 2017, vol. 1(4), pp. 0031-1–13.

Chapter 18
Application of liquid crystals to MEMS

Tomohiro Tsuji[1] *and Shigeomi Chono*[1]

In this chapter, the applications of liquid crystals (LCs) to microactuators which are suitable for MEMS usage are introduced. The major source of the actuator drive is the LC flows induced by the applied electric fields, and thus this chapter begins with the reviews of the experimental and theoretical aspects of the backflow effect of LCs between parallel plates. Then, the experimental and simulation results on the actuator drive under the pulsed electric fields are presented and the effects of the electric field on the driving performance of the LC actuator are considered. These results suggest that the LC materials have the potential of application to MEMS devices.

18.1 Introduction

Since the discovery of the applicability of LCs to display devices in the 1960s, many studies on LCs have been carried out both fundamentally and practically. As a result, LC displays have established a strong position as a substitute for cathode ray tube displays. The manufacturing of such displays has grown into a large industry with a global scale and further development is expected. However, the technical development has reached a plateau except for a few trivial improvements.

A research field, such as 'mechanics of liquid crystals', should exist and various studies concerning LCs should be performed from a mechanical viewpoint, because in mechanical engineering, solid mechanics and fluid mechanics are established in the fields of solid, and gas and liquid, respectively. However, there is no such research field, while there are some studies where an LC is regarded as a fluid [1–4]. Figure 18.1 shows the conceptual applications of LCs; that is, we input some factors, such as electromagnetic field, flow, heat, and optic, into LCs and output different factors by utilising the easy change in orientation state, which is the most representative feature of LCs. An example of such applications is an LC display in which an electric field is input and a change in transmitted light is output. We can regard an LC as a medium of converting energy by combining input and output factors appropriately. The applications of LCs are assumed to increase by changing the combination of input and output factors or by proposing other factors.

[1]School of Systems Engineering, Kochi University of Technology, Japan

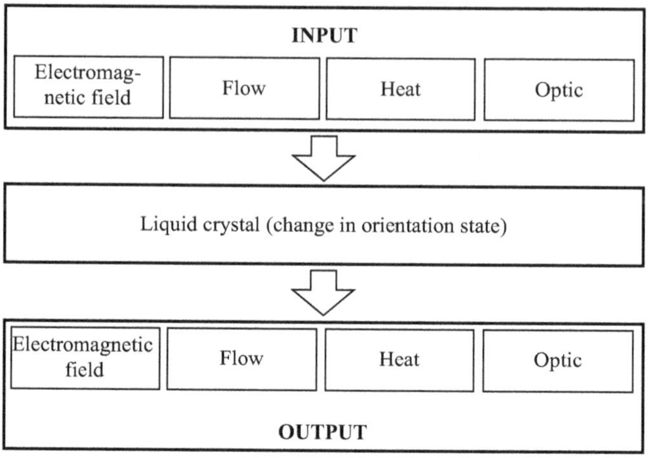

Figure 18.1 Application of LC

Although it is well known that the application of a velocity gradient makes LC molecules rotate and reorient to a different direction, its reverse phenomenon, such that the rotation of the molecules (by applying, for example, an electric field) induces a velocity gradient, is not clearly understood. Since this phenomenon corresponds to the energy conversion from electrical to kinetic energies, to utilise such induced flow leads to the development of microactuators. They have various advantages in comparison with other types of microactuators: simple structure, high shape adaptability, easy downsizing, and low driving voltages. Therefore, LC actuators are often preferred to other MEMS or microactuators. Some studies [5–7] were made for the generation of a flow, which was called 'backflow', however, in these studies, the mechanical applications of the backflow were not considered, although its effect on molecular orientation was investigated.

In this chapter, we present the experimental and computational results of the backflow, and introduce the applications of the backflow to microactuator devices. Hereafter, the LC used is 4-cyano-4′-pentylbiphenyl (5CB), which has a molecular weight of approximately 250 and exhibits a nematic state in the temperature range of 22.5–35 °C.

18.2 Backflow effect of nematic LCs

18.2.1 Visualisation of backflow

We focus on measuring velocity profiles for various applied voltages, gap size between the plates, and boundary conditions for director orientation, when an electric field is applied to an LC confined between two parallel plates.

Figure 18.2 shows an experimental cell for the visualisation experiment. The procedure for manufacturing the cell is as follows. We prepared two flat glass plates on which a transparent electrode layer (ITO layer) is formed by sputtering. The surface of the ITO layer is coated with an alignment layer, which helps align

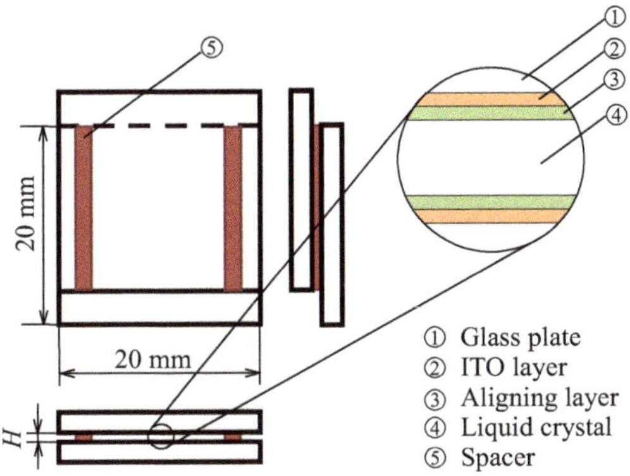

Figure 18.2 An experimental cell for visualisation

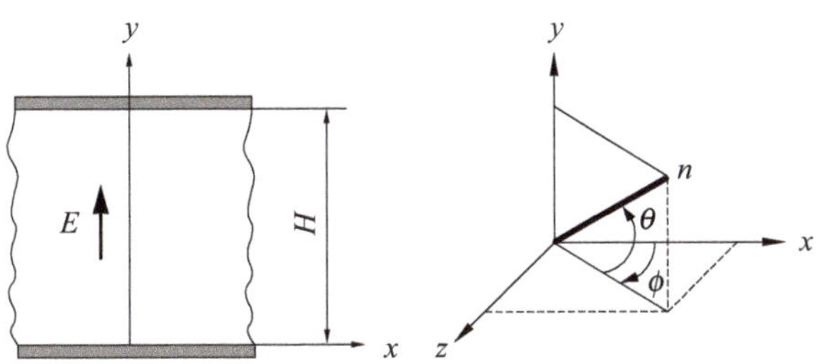

Figure 18.3 Coordinate of the director

the LC molecules. We used two liquid materials for alignment; one is for the planar alignment to induce a small pre-tilt angle, and the other is for homeotropic alignment. As shown in Figure 18.2, two thin spacers of thicknesses 25 and 50 μm are sandwiched between the two glass plates. We slightly staggered one glass plate relative to the other to apply the electric field more easily. When the polar coordinates of the director are expressed in terms of two angles $\theta(y)$ and $\phi(y)$, as shown in Figure 18.3, we define the pre-tilt angle θ_w as $\theta(0)$ [$=\theta(H)$] and the twist angle ϕ_w as $\phi(H)$–$\phi(0)$, where H is the gap between the upper and lower plates.

We set up a two-dimensional traversing device, on which the experimental cell (Figure 18.2) was mounted and placed on the stage of the polarisation microscope. We then took microscopic images at a rate of 30 frames/s using a CCD camera, and stored the images on computer. Figure 18.4 depicts the procedure for measuring the velocity profiles. First, we selected the focus position of the microscope between the plates. Second, using the traversing device, we looked for a particle on which the focus was adjusted. Finally, we applied a step-function voltage to the LC using a DC stabilised power supply to record images of the particle in motion. We repeated the procedure at 11 different points between the plates to obtain the velocity profiles. At every point, we performed the measurement 10 times under the same condition. The experimental parameters are the cell gap $H = 25$ and 50 μm, the applied voltage $V = 5$

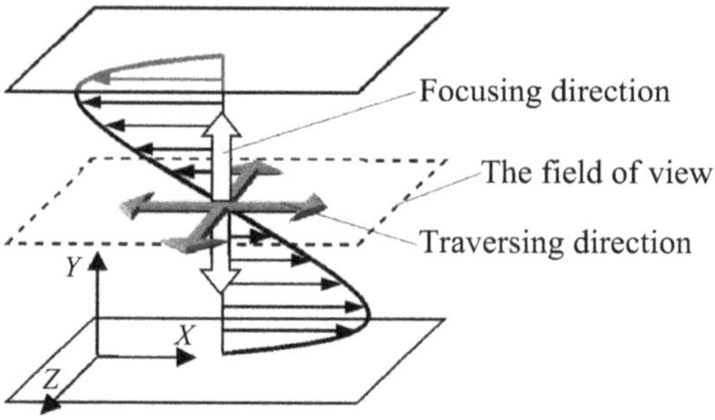

Figure 18.4 Procedure for measuring velocity profile between parallel plates

and 10 V, the molecular configuration mode, i.e., planar and twist modes, and the twist angle $\phi_w = 0°$, 90°, and 180°. The twist mode with $\phi_w = 0°$ corresponds to the planar mode. The experiment was performed at a room temperature of 25 °C. To visualise the induced backflow between plates, we dispersed SiO_2 particles with a diameter of 2.5 μm in 5CB when the gap H was 50 μm. We determined the volume fraction of the particles to be approximately 0.01 per cent in a preliminary experiment. When H was 25 μm, SiO_2 particles with a diameter of 1 μm were used.

Figure 18.5 shows the velocity profiles for $\phi_w = 0°$, 90° and 180° when $V = 5$V and $H = 50$ μm. The averages of 10 measured data per set are plotted with error bars; the curves are numerical predictions which are explained in the following section. The values at the plates ($y = 0$, 50 μm) are not measured but are plotted assuming a no-slip boundary condition. The velocity profile for $\phi_w = 0°$ shows an antisymmetric S-shaped profile with respect to $y = 25$ μm that roughly follows the numerical prediction. As ϕ_w increases, the region of negative value enlarges whereas that with positive value diminishes; that is, the velocity profile breaks its antisymmetry, and gradually becomes unidirectional at $\phi_w = 180°$.

18.2.2 Continuum simulation of backflow

In this section, we report computational results for the backflow using the Leslie–Ericksen theory [8–10], which has been widely used as a constitutive equation for low-molar-mass nematic LCs.

The isothermal flow of LCs under an electric field is governed by continuity, momentum, angular momentum, and constitutive equations, which are written as

$$\Delta \cdot \mathbf{v} = 0 \tag{18.1}$$

$$\rho\left\{\frac{\partial \mathbf{v}}{\partial t} + (\mathbf{v}\cdot\nabla)\mathbf{v}\right\} = (\varepsilon_\perp E + \Delta\varepsilon \mathbf{n}\cdot\mathbf{E}\mathbf{n})\cdot(\nabla\mathbf{E}) - \nabla p + \nabla\cdot\tau \tag{18.2}$$

$$\mathbf{n}\times\left\{\Delta\varepsilon\mathbf{n}\cdot\mathbf{E}\mathbf{E} - \frac{\partial F}{\partial \mathbf{n}} + \nabla\cdot\left(\frac{\partial F}{\partial \nabla\mathbf{n}}\right) + (\alpha_3 - \alpha_2)\mathbf{N} + (\alpha_6 - \alpha_5)\mathbf{A}\cdot\mathbf{n}\right\} = \mathbf{0} \tag{18.3}$$

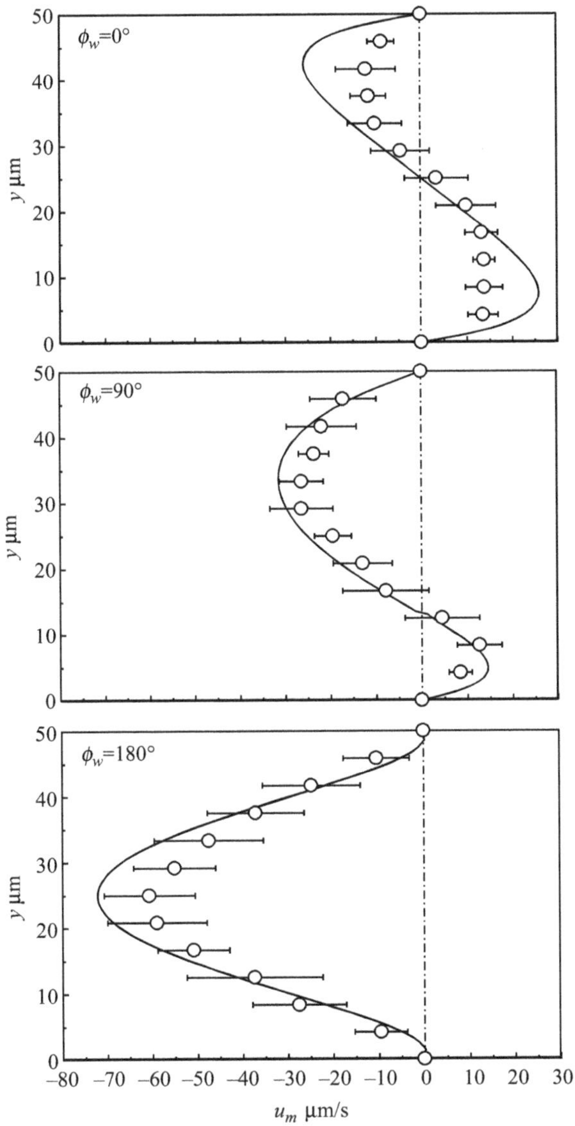

Figure 18.5 Effect of twist angle on the backflow at V = 5 V and H = 50 µm

$$\begin{aligned}\tau &= \alpha_1 \mathbf{nnn} \cdot \mathbf{A} \cdot \mathbf{n} + \alpha_2 \mathbf{nN} + \alpha_3 \mathbf{Nn} + \alpha_4 \mathbf{A} \\ &\quad + \alpha_5 \mathbf{nn} \cdot \mathbf{A} + \alpha_6 \mathbf{A} \cdot \mathbf{nn} - \frac{\partial F}{\partial \nabla \mathbf{n}} \cdot (\nabla \mathbf{n})^T\end{aligned} \tag{18.4}$$

In these equations, $\mathbf{v}$ is the velocity vector, ρ the fluid density, p the pressure, $\mathbf{n}$ the director indicating the average of the directions of LC molecules, $\mathbf{E}$ the electric field vector, $\varepsilon_\perp$ and $\varepsilon_{//}$ the dielectric constants perpendicular and parallel to

Table 18.1 Material constants of 5CB at 25 °C

Leslie viscosities [×10^{-2} Pa s]						Elastic constants [×10^{-12} N]			Dielectric constants [×10^{-11} F/m]	
α_1	α_2	α_3	α_4	α_5	α_6	K_1	K_1	K_1	$\varepsilon_{//}$	$\varepsilon_{\perp}$
0	−8.6	−0.4	8.9	5.9	−3.1	6.4	3.8	8.6	15.7	5.7

the director, respectively, $\Delta\varepsilon = \varepsilon_{//} - \varepsilon_{\perp}$ the anisotropy of the dielectric constants, **A** the rate of the deformation tensor, **N** the angular velocity of the director relative to that of the surrounding fluid, $\alpha_1 - \alpha_6$ the Leslie viscosities, and F the free-energy density due to the spatial distortion of the director field expressed as

$$F = \frac{1}{2}\left\{k_1(\nabla \cdot \mathbf{n})^2 + k_2(\mathbf{n} \cdot \nabla \times \mathbf{n})^2 + k_3|\mathbf{n} \times \nabla \times \mathbf{n}|^2\right\} \tag{18.5}$$

Here, K_1, K_2 and K_3 are the elastic constants representing the splay, twist, and bend deformations of the director, respectively. The material constants of 5CB at 25 °C are listed in Table 18.1 [11–13].

An LC was placed between two parallel plates with the gap H, as shown in Figure 18.3, and at time $t = 0$, a uniform electric field with the intensity E (=constant) was imposed in the y-direction. A no-slip condition for the velocity field was applied to the plates. On the other hand, for the orientation field, the tilt angle was fixed at 1° and the twist angles were set at 0° and 180° in the present computation, where the tilt angle is θ at the plates and the twist angle is defined as the difference in ϕ between both plates when we express the posture of the director with θ and ϕ, as shown in Figure 18.3. For initial values, we set $\mathbf{v} = 0$ in the entire domain and chose a one-dimensional steady-state solution without an electric field for director orientation. A finite difference method and a second-order Runge–Kutta method were used to discretise the governing equations. From a preliminary computation, we selected $\Delta y = H/100$ and $\Delta t = 10^{-9}$ s.

Figure 18.6 shows the transient velocity profiles for $H = 50$ μm and $\phi = 0°$ when we applied a direct-current voltage of 10 V at $t = 0$ and kept the value constant. Figure 18.6(a) and (b) shows the change from the start to the maximum and that from the maximum to the cessation, respectively. The induced velocity is negative in the upper half region and positive in the lower half region; thus, the profile is S-shaped and the bulk flow rate is zero. A series of such changes occur in 0.3 s with an acceleration time of about 0.1 s and a deceleration time due to viscosity of about 0.2 s. The difference in velocity profile between the acceleration and deceleration periods is small. Almost no flow occurs in the early stage of the acceleration period $t = 0$–0.06 s, and the flow is rapidly accelerated for $t = 0.06$–0.09 s. Similarly, a rapid deceleration occurs for $t = 0.09$–0.12 s, and afterward, the flow is decelerated very slowly for $t = 0.18$–0.30 s. The maximum induced velocity is approximately 80 μm/s.

On applying a step-function voltage, a visualisation particle accelerates to a peak speed, and then decelerates, as just explained. In this transient process, the

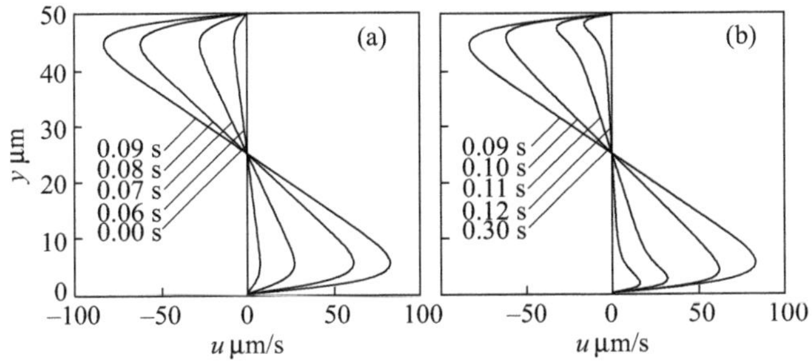

Figure 18.6 Change in velocity profile with time (twist angle = 0°): (a) from start to maximum and (b) from maximum to cessation

peak value is denoted u_m. We plot the maximum value at each point in the y-direction between the plates in Figure 18.5. Note that the maximum velocities in the y-direction do not necessarily correspond to the same time. The experimental results and the numerical predictions agree qualitatively in regard to the velocity profiles and breaking of antisymmetry with increasing ϕ_w. Quantitative agreement was obtained at $\phi_w = 90°$ and 180°, but the differences are large at $\phi_w = 0°$. The experimental values are lower than the predictions in all instances, because plate edge effects were ignored in computations.

18.2.3 Molecular dynamics simulation of backflow

The backflow has been studied mainly in terms of the Leslie–Ericksen macroscopic continuum theory. Here, we study the mechanism of backflow generation under the application of an electric field by molecular dynamics simulation, and investigate the roles of intermolecular interaction in the generation of bulk velocity. We consider the dynamics of a nematic LC confined between parallel plates. The computational domain and molecular model are shown in Figure 18.7. The LC molecules are represented as ellipsoids. Without the electric field effect, the molecules interact through the Gay–Berne potential [14]:

$$\phi_{ij}^{\mathrm{GB}} = 4\varepsilon\left(\mathbf{u}_i, \mathbf{u}_j, \hat{\mathbf{r}}_{ij}\right)\left\{\left[\frac{\sigma_0}{r_{ij} - \sigma\left(\mathbf{u}_i, \mathbf{u}_j, \hat{\mathbf{r}}_{ij}\right) + \sigma_0}\right]^{12} - \left[\frac{\sigma_0}{r_{ij} - \sigma\left(\mathbf{u}_i, \mathbf{u}_j, \hat{\mathbf{r}}_{ij}\right) + \sigma_0}\right]^{6}\right\} \tag{18.6}$$

Here, $\mathbf{u}_i$ and $\mathbf{u}_j$ are unit vectors representing the orientation of molecules i and j, respectively, and $\hat{\mathbf{r}}_{ij}$ is the unit vector of vector $\mathbf{r}_{ij}$ ($\hat{\mathbf{r}}_{ij} = \mathbf{r}_{ij} / r_{ij}$) that connects the centres of mass of molecules i and j. $\sigma(\mathbf{u}_i,\mathbf{u}_j, \hat{\mathbf{r}}_{ij})$ represents the contribution of molecular orientation on the intermolecular distance, while $\varepsilon(\mathbf{u}_i,\mathbf{u}_j, \hat{\mathbf{r}}_{ij})$ defines the potential well. The parameters of the potential (related to $\sigma(\mathbf{u}_i,\mathbf{u}_j, \hat{\mathbf{r}}_{ij})$ and $\varepsilon(\mathbf{u}_i,\mathbf{u}_j, \hat{\mathbf{r}}_{ij})$) are molecular length scale σ_0, energy scale ε_0, molecular aspect ratio σ_r, energy ratio ε_r, and the constants for the potential well μ and ν.

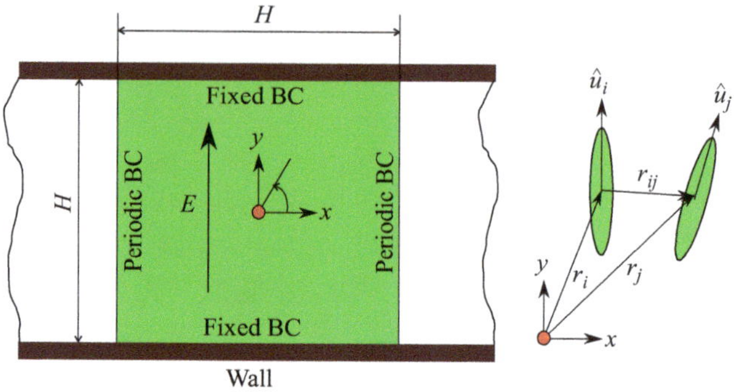

Figure 18.7 Computational domain and molecular model

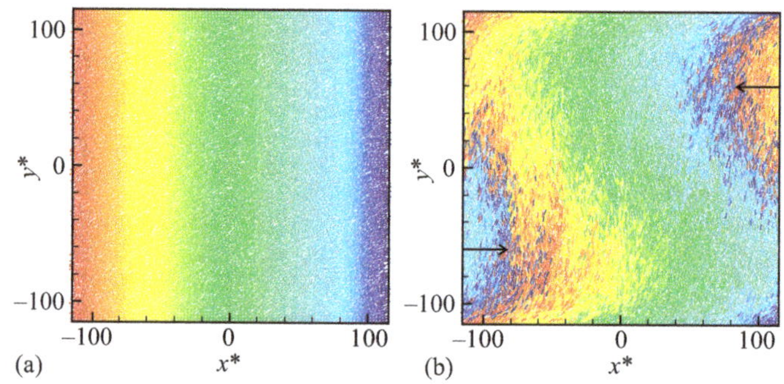

Figure 18.8 Snapshot of molecules at equilibrium (a) and after application of electric field for time t = 1,000 (b)*

To account for the effect of electric field, we use a similar approach as used in references [15] and [16], in which the electric field induced molecular torque is considered as the contribution of the induced polarisation. The application of an electric field $\mathbf{E} = E\mathbf{e}$ induces an electric dipole $\mathbf{p}$ on each molecule, $\mathbf{p} = \alpha E(\mathbf{e}\cdot\mathbf{u})\mathbf{u}$, where α is the polarisability, and E and $\mathbf{e}$ are the strength and unit vector of the electric field, respectively. The induced dipole contributes to the molecular interaction through a dipole–dipole interaction potential [17]:

$$\phi_{ij}^{\mathrm{DP}} = \frac{1}{r_{ij}^3}\left\{\mathbf{p}_i \cdot \mathbf{p}_j - 3(\mathbf{p}_i \cdot \hat{\mathbf{r}}_{ij})(\mathbf{p}_j \cdot \hat{\mathbf{r}}_{ij})\right\} \tag{18.7}$$

Therefore, the total potential ϕ_{ij} is the sum of the Gay–Berne potential and the dipole–dipole potential, $\phi_{ij} = \phi_{ij}^{\mathrm{GB}} + \phi_{ij}^{\mathrm{DP}}$.

Snapshots of molecules at equilibrium and after the application of an electric field are shown in Figure 18.8. The colours represent the initial horizontal position x^* of each molecule. It can be observed that after the application of the electric field, the molecules in the upper region move to the left side, while the molecules in the lower region move to the right side. This confirms that the application of an electric field induces an S-shaped backflow obtained in the calculation using continuum mechanics.

18.3 Development of LC actuators

18.3.1 Experiment

In the previous section, the focus has been on backflow in cells with fixed upper and lower plates (Figure 18.9(a)). Because the induced backflow exerts shear stresses on both plates, the upper plate can be driven by the shear stresses if it was free to move in its plane (Figure 18.9(b)). This is the principle behind LC actuators.

The induced backflow vanishes in a short time because of its viscosity, even though we keep applying an electric field (Figure 18.10(a)). Therefore, to use LC actuators, in practice, a repeated application of an electric field over an appropriate duration is necessary; that is, we need to return the molecules along the electric field direction to their initial positions by releasing the electric field, and then reapply the electric field (Figure 18.10(b)). In this circumstance, frequency and duty ratio also become input parameters. Here, we describe experiments where, by applying a rectangular wave voltage to an LC confined between two plates, we can drive the free upper plate in its plane.

Figure 18.11 shows an experimental cell used in this study. Whereas the lower glass plate is 20 mm × 20 mm in size, the upper plate is 10 mm × 10 mm in size with a thickness of 0.15 mm and mass of 47 mg. For the lower plate, we sputtered an indium-tin-oxide (ITO) layer on its upper surface (the side in contact with the LC), and formed an alignment layer on the ITO layer. For the upper plate, we sputtered an ITO layer on all surfaces to enable a voltage to be applied easily using a steel rod. An alignment layer was formed only on the lower surface (the LC side). A number of polystyrene particles were sprayed on the lower plate to keep the two plates parallel and the gap constant.

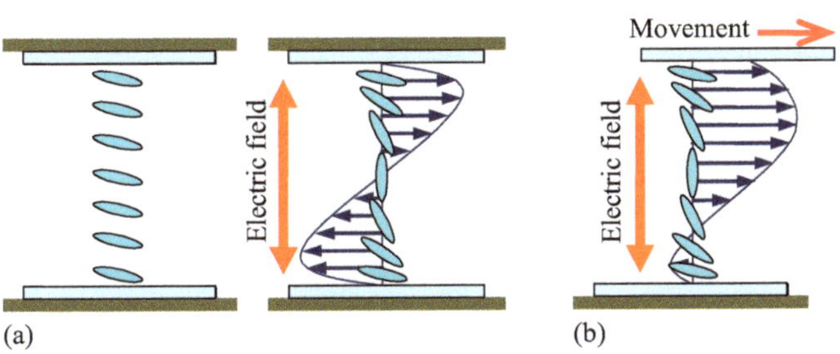

Figure 18.9 Generation of backflow between two parallel plates and principle of LC actuators: (a) with the upper and lower plates fixed, the molecules rotate and induce an S-shaped backflow profile, on applying an electric field to an LC and (b) with a free upper plate, an induced backflow exerts a shear stress driving the plate horizontally

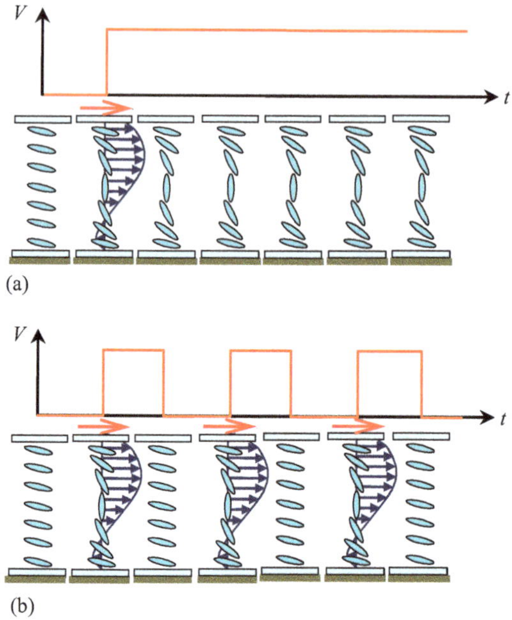

Figure 18.10 Application of an electric field and molecular orientation profiles: (a) a single step-like change in voltage is applied to an LC. The induced backflow vanishes in a short time because of the viscosity of the LC, even though we keep applying an electric field. (b) A repetitive application of an electric field is necessary for the continuous movement of the upper plate; the molecules along the electric field direction return to their initial positions by releasing the electric field, and then reapply the electric field

Under a polarising microscope, we imaged a small point marked on the upper surface of the upper plate using either a CCD camera (640 × 480 pixels and 0.75 μm/pixel in space resolution) at 30 frames/s or a high-speed camera (512 × 512 pixels and 0.80 μm/pixel in space resolution) at 1,000–4,000 frames/s, saving the video recording to a computer for analysis. The experimental parameters are the applied voltage V = 0–10 V, the duty ratio of D = 1–50 per cent, the frequency f = 1–1,000 Hz, and the cell gap (the diameter of the sprayed polystyrene particles) H = 1–50 μm.

Figure 18.12(a) shows the displacement l of the upper plate as a function of the time t when $V = 10$ V, $D = 5$ per cent, $H = 10$ μm, and $f = 1$, 10, and 100 Hz. The upper plate moves as the rectangular wave voltage rises and comes to rest after a short distance. By repeating the voltage signal, the upper plate moves forward in a step-like motion (at $f = 100$ Hz, this motion is not observed in this figure, but is confirmed in video footage using a high-speed camera). For each f, the microscopic displacements l are macroscopically linear with elapsed time t, hence producing macroscopic upper plate speeds that are constant. Similar results were obtained for different frequencies. We take this constant speed as the average speed denoted

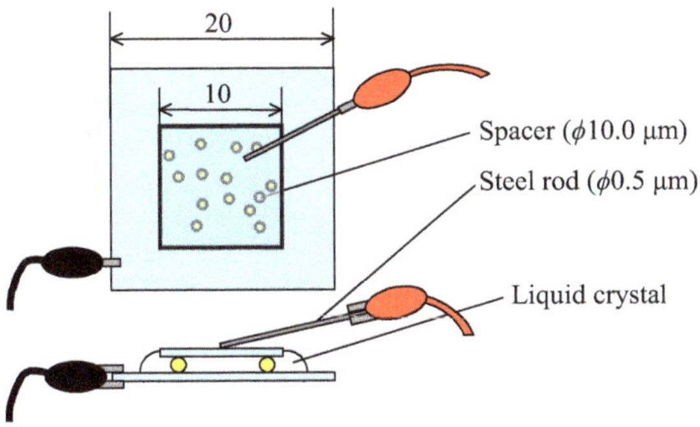

Figure 18.11 *Experimental cell. The lower glass plate is 20 mm × 20 mm in size, and the upper one is 10 mm × 10 mm in size with 0.15 mm in thickness and 47 mg in mass. For the lower plate, we evaporated an ITO layer on its upper surface (the side in contact with the LC), and formed an alignment layer on the ITO layer. For the upper plate, we evaporated an ITO layer over all surfaces so a voltage can be easily applied using a steel rod. An alignment layer was formed on only the lower surface (the LC side). A number of polystyrene particles were sprayed on the lower plate to keep the two plates parallel and the gap constant*

by $\bar{U}$. Figure 18.12(b) shows the relationship between $\bar{U}$ and f. At each f, 10 measurements were taken to plot the average value along with its standard deviation. With increasing f, $\bar{U}$ increases sharply, attains a maximum value of 120 μm/s at $f = 175$ Hz, and then decreases gradually. For lower f, $\bar{U}$ becomes large with increasing f, because the molecules can respond to the repeating voltage signal. Hence, increasing f corresponds simply to increasing the number of repetitions per unit of time. However, for higher f, the molecules are unable to respond to the change in voltage, so that their rotation cannot induce a satisfactory backflow, and $\bar{U}$ exhibits a gradual decrease with f.

18.3.2 Numerical simulation

An interesting result was also obtained that the maximum speed of the upper plate was attained at the frequency of the applied voltage of 175 Hz, which is quite high compared with the intrinsic response frequency (approximately 10 Hz) of the LC and the cell used in this experiment. Here, we propose a simple model to predict numerically the performance of LC actuators by combining the flow of an LC and the motion of the upper plate of an LC cell, and compare the predictions with the experimental results. Moreover, we elucidate the mechanism of LC actuators by analysing the behaviour of molecules and the induced velocity profiles in detail.

The equation of motion of the upper plate is expressed as

$$m\frac{dU}{dt} = \tau_w a - \text{sgn}(U)\mu_k mg \tag{18.8}$$

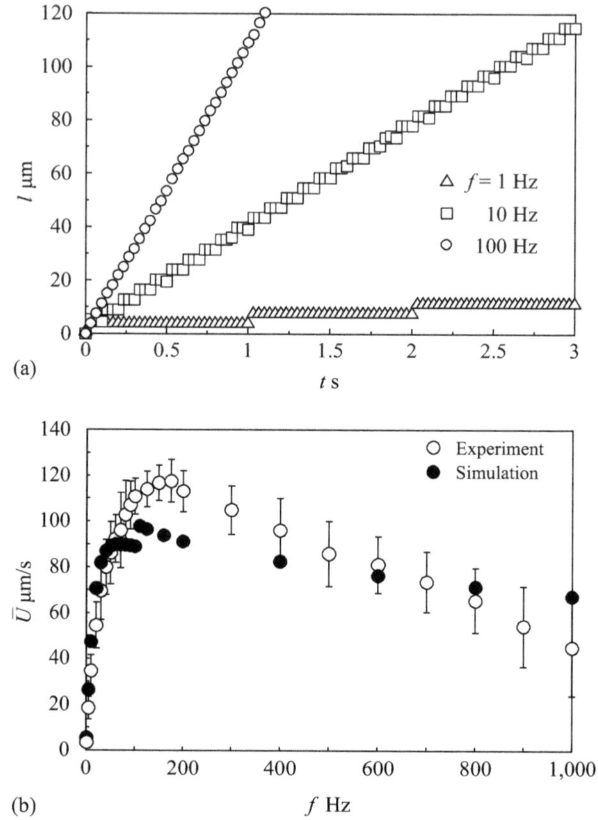

Figure 18.12 Experimental and computational results when $V = 10$ V, $D = 5$ per cent, and $H = 10$ μm: (a) displacement of the upper plate at $f = 1$, 10, 100 Hz. The upper plate moves by repeating the step-like motion (although the step-like motion is unobservable at $f = 100$ Hz, but is confirmed in video images using a high-speed camera). The displacement l with a microscopically step-like change is macroscopically linear to the time t, and hence, the upper plate moves macroscopically at a constant speed and (b) dependence of the speed of the upper plate on frequency of the applied voltage. The average values and the standard deviations of 10 measurements at each f are plotted. $\bar{U}$ exhibits a sharp increase, attains a maximum value of 120 μm/s at $f = 175$ Hz, and then decreases gradually with increasing f

where m and U are the mass and speed of the upper plate, respectively, a the area of contact between the upper plate and the LC, μ_k the coefficient of kinetic friction between the upper plate and the particles used to keep the two plates parallel and the gap constant, and τ_w the shear stress acting on the upper plate. It should be

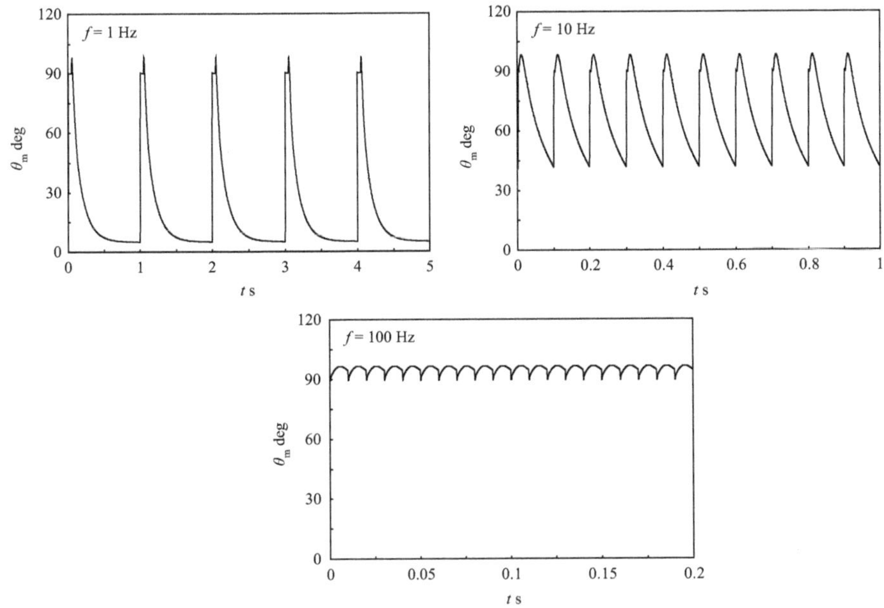

Figure 18.13 Director orientation angle at the middle point between two plates: (a) f = 1 Hz; (b) f = 10 Hz; and (c) f = 100 Hz

noted that (18.8) is not applicable when the upper plate is at rest and also the shear force $\tau_w a$ is less than the static frictional force $\mu_s mg$ (μ_s is the coefficient of static friction), since the plate does not move. By applying the shear stress of (18.4) at the upper plate, $\tau_{yx}(H)$, to τ_w, and by using $u(H) = U$ as a no-slip condition, where $u(y)$ is the x component of the velocity vector, we obtain a coupled problem of the flow of LCs and the motion of the upper plate.

In Figure 18.12(b), we superpose the numerical results for the relationship between the average speed $\bar{U}$ and the frequency f. The numerical results are almost the same as the experimental results. However, if we compare both results in detail, agreement is not satisfactory. The quantitative disagreement is assumed to stem mainly from approximating to a unidirectional flow in a Hele–Shaw cell, hence neglecting the effects of the surface tension and the drag force at the edge of the upper plate. However, the present simple model is useful at least for the schematic design of an LC actuator.

Figure 18.13(a)–(c) shows the orientation angle θ_m of the director located at the middle point between two plates. At $f = 1$ Hz, the director orients in the electric field direction at the moment when an electric field is applied, and afterwards returns to the initial orientation angle, 5° (a pre-tilt angle). When the field is released, θ_m is over 90°, which is due to the *kickback effect* [18]. At $f = 10$ Hz, the behaviour of the director just after the application of the electric field is similar to that at $f = 1$ Hz, but the director does not return to the initial orientation angle completely but starts the next cyclic motion at about 40°. At $f = 100$ Hz, surprisingly, the director oscillates with a small amplitude between 90° and 96°, leading to the conclusion that not backflow but kickback effect is dominant.

Figure 18.14 shows the velocity profile in the x-direction u and the profile of the molecular orientation θ in the early stage of a period after a complete cyclic motion is developed. f is 100 Hz, and t^* is the time non-dimensionalised by the period T $(=1/f)$. Double-S shaped velocity profiles are generated, and the velocity gradient and resultantly shear stress at the plate are large. This is the reason why the upper plate moves fast at a frequency beyond the intrinsic frequency of 5CB, 10 Hz.

We consider the mechanism by which double-S-shaped velocity profiles were induced, on the basis of the profiles of the molecular orientation angle shown in the right side of Figure 18.14. At $f = 100$ Hz, the orientation angle of molecules in the middle region between plates is over 90° because of the kickback effect, as stated before, so that application of an electric field to this orientation state makes the molecules in the middle region rotate in the opposite direction to surrounding molecules and induces opposite velocity gradient as shown in Figure 18.15(a). Then, we obtain a double-S shaped bulk velocity profile shown in Figure 18.15(b)

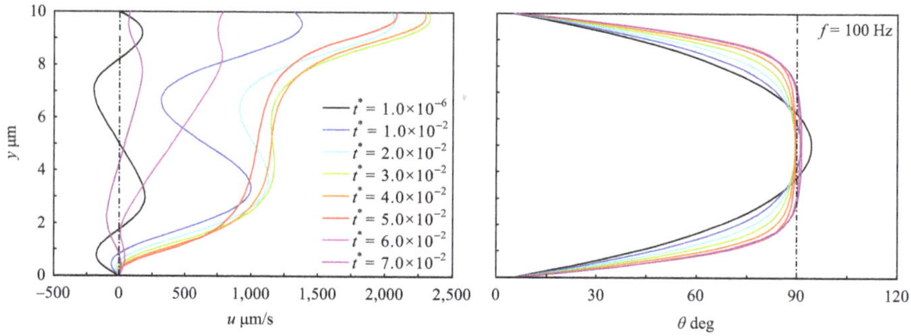

Figure 18.14 Velocity and orientation profiles between two plates at $f = 100$ Hz

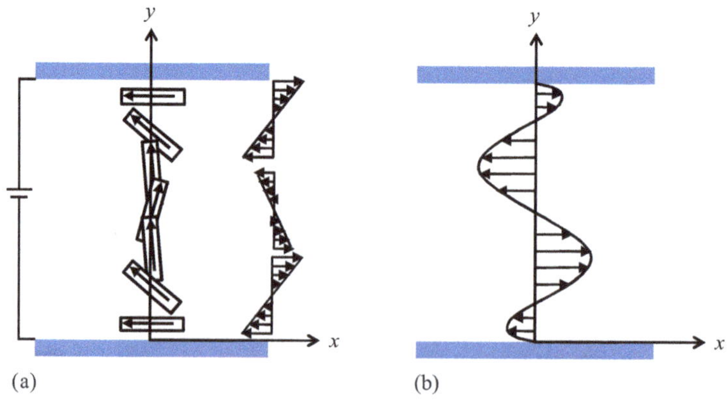

Figure 18.15 Velocity and orientation profiles between two plates

by integrating these velocity gradients in the y direction and by incorporating a no-slip boundary condition at the plates.

18.4 Conclusions

We have developed microactuators driven by a backflow which is induced by applying an electric field to LCs. Because of liquidity of LCs, there is no restriction on the shape of an object to be driven, and the driving force (or shear stress) acts along the tangential direction of the contact face between the object and the LC. Therefore, if the object has a round shape, the force causes the rotational motion of the object, which comes up with the development of LC micromotors.

Figure 18.16 shows a microscope image of an LC micromotor placed on a grain of rice. The outer and inner diameters of the outer cylinder are 0.1 and 0.08 mm, respectively, the outer diameter of the inner cylinder is 0.07 mm, and the gap is 5μm. The length and the mass of the inner cylinder are 1 mm and 0.004 mg, respectively. Both cylinders are made of glass. Figure 18.17 shows the relationship between the rotation speed and the frequency of the applied voltage. The voltage and the duty ratio are $V = 10$ V and $D = 5$ per cent, respectively. With increasing the frequency, the rotation speed increases sharply for $f < 300$ Hz, increases gradually until $f \approx 1{,}000$ Hz, and then decreases gradually. The frequency at which the rotation speed reaches a maximum value is about 1,000 Hz, which is much higher than that for the flat plate actuator ($\approx$175 Hz in Figure 18.12(b)). This is due to the difference in the molecular orientation profile at an equilibrium state; a hybrid molecular orientation was employed for this micromotor, whereas the planar molecular orientation was employed for the flat plate actuators.

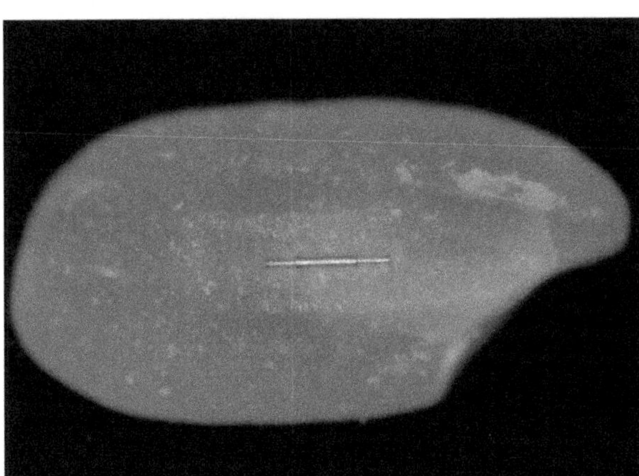

Figure 18.16 Microscope image of an LC micromotor placed on a grain of rice. The diameters of the inner and outer cylinders are 0.07 and 0.1 mm, respectively

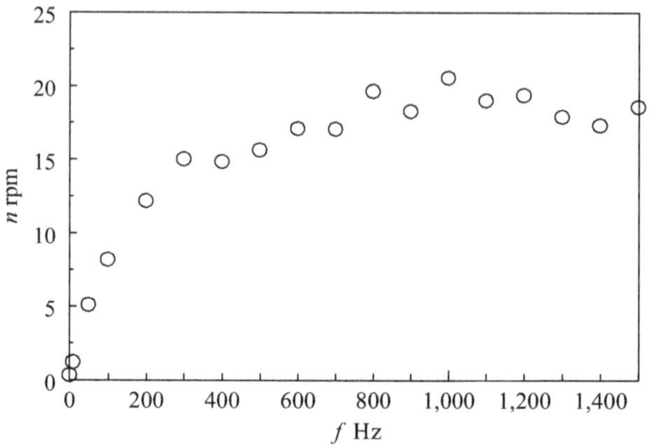

Figure 18.17 Effect of the frequency of the electric pulse on the circumferential speed of the inner cylinder surface of LC motor

The actuators including motors mentioned here are only a few examples and we expect various studies on the mechanics or dynamics of LCs.

References

[1] Rey A.D. and Denn M.M. 'Jeffrey-Hamel flow of Leslie-Ericksen nematic liquids'. *J. Non-Newtonian Fluid Mech.* 1988, Vol. 27, pp. 375–401.

[2] Chono S., Tsuji T. and Denn M.M. 'Spatial development of director orientation of tumbling nematic liquid crystals in pressure-driven channel flow'. *J. Non-Newtonian Fluid Mech.* 1998, Vol. 79, pp. 515–27.

[3] K. Koyama, *Development of Electro-Rheological (ER) Fluids*, Tokyo: CMC; 1999, *in Japanese.*

[4] Rey A.D. and Denn M.M. 'Dynamical phenomena in liquid-crystalline materials'. *Annu. Rev. Fluid Mech.* 2002, Vol. 34, pp. 233–66.

[5] Brochard F. 'Backflow effects in nematic liquid crystals'. *Mol. Cryst. Liq. Cryst.* 1973, Vol. 23, pp. 51–58.

[6] van Doorn C.Z. 'Dynamic behavior of twisted nematic liquid–crystal layers in switched fields'. *J. Appl. Phys.* 1975, Vol. 46, pp. 3738–45.

[7] Berreman D.W. 'Liquid–crystal twist cell dynamics with backflow'. *J. Appl. Phys.* 1975, Vol. 46, pp. 3746–51.

[8] Ericksen J.L. 'Anisotropic fluids'. *Arch. Ration. Mech. Anal.* 1960, Vol. 4, pp. 231–37.

[9] Ericksen J.L. 'Conservation laws for liquid crystals'. *Trans. Soc. Rheol.* 1961, Vol. 5, pp. 23–34.

[10] Leslie F.M., 'Some constitutive equations for liquid crystals'. *Arch. Ration. Mech. Anal.* 1968, Vol. 28, pp. 265–83.

[11] Kneppe H., Schneider F. and Sharma N.K. 'Rotational viscosity γ_1 of nematic liquid crystals'. *J. Chem. Phys.* 1982, Vol. 77, pp. 3203–08.

[12] Kneppe H., Schneider F. and Sharma N.K. 'A comparative study of the viscosity coefficients of some nematic liquid crystals'. *Ber. Bunsenges. Phys. Chem.* 1981, Vol. 85, pp. 784–89.
[13] Karat P.P. and Madhusudana N.V. 'Elasticity and orientational order in some 4′-n-alkyl-4-cyanobiphenyls: Part II'. *Mol. Cryst. Liq. Cryst.* 1977, Vol. 40, pp. 239–45.
[14] Gay J.G. and Berne B.J. 'Modification of the overlap potential to mimic a linear site–site potential'. *J. Chem. Phys.* 1981, Vol. 74, pp. 3316–19.
[15] Tian P., Bedrov D., Smith G.D., Glaser M., Maclennan J.E. 'A molecular-dynamics simulation study of the switching dynamics of a nematic liquid crystal under an applied electrical field'. *J. Chem. Phys.* 2002, Vol. 117, pp. 9452–59.
[16] Berardi R., Muccioli L. and Zannoni C. 'Field response and switching times in biaxial nematics'. *J. Chem. Phys.* 2008, Vol. 128, pp. 024905–1–024905–12.
[17] Houssa M., Rull L.F. and McGrother S.C. 'Effect of dipolar interactions on the phase behavior of the Gay–Berne liquid crystal model'. *J. Chem. Phys.* 1998, Vol. 109, pp. 9529–42.
[18] Chandrasekhar S., *Liquid Crystal*, Cambridge: Cambridge Univ. Press; 1992.

Chapter 19

Detection of volatile organic chemicals by using liquid crystals

Shoichi Ishihara[1], Jovia Jia Zhen Lee[2], and Khai Jun Kek[3]

The detection of volatile organic compounds (VOCs) is drawing much attention in numerous aspects. It is not only highly utilised in the field of detection of toxic and flammable gases but also in the field of disease diagnosis and healthcare. To date, much research has been done by using liquid crystals (LCs). In this chapter, in addition to introducing various prior arts, we investigated the structural dependence of LC materials which have been hardly studied so far. In particular we focused on the use of chiral nematic LCs which show selective reflection and clarified the relationship between these molecular structures and the shift in their reflection spectrum as a result from contact with VOCs. We also propose a mechanism to explain the red- or blue-shift in the peak wavelength of the selective reflection spectrum. Furthermore, enhancement of the sensitivity and the selectivity of this VOC detection method are discussed.

19.1 Introduction

As the ageing society becomes imminent, interests in own health and the environmental issues are increasing. The Ministry of Health, Labour and Welfare of Japan is also channelling resources for the development of new early diagnostic methods so as to reduce medical costs for the people. Efforts to detect early cancer in particular, which account for the top cause of death, are actively promoted [1,2], and citizens are strongly encouraged to undergo cancer screenings. However, high costs and painful methods of diagnosis are setbacks that cause such efforts to be less attractive for the people.

Recently, it has been reported that the expiration of air and urine of cancer patients contained a distinctive set of gases compared with those of healthy persons,

[1]Department of Electrical and Electronic Engineering, Graduate School of Engineering, Osaka Institute of Technology, Japan
[2]Panasonic Industrial Devices Singapore, Singapore (Currently: Centre for Innovation, Changi General Hospital, Singapore)
[3]Panasonic Industrial Devices Singapore, Singapore

Table 19.1 The biogas components and the significance of measurement

Component	Location	Clinical meaning/clinical application	Detection concentration
Ammonia	Breath	Screening for hepatic encephalopathy	A few ppm
Mercaptan (methanethiol)	Intro-oral	Parameter for bad breath (periodontitis, periodontal disease)	Tens of ppb
Acetone	Breath	Diagnosis of insulin deficiency disease (diabetes); Parameter for effect of dieting	~ 1 ppm
Isoprene	Breath	Parameter for cholesterol generation	~ 0.1 ppm
Isoprene	Breath	Measure grow rate of oral bacteria	
Carbon monoxide	Intro-oral	Parameter for CO poisoning; Diagnostic of hypoplastic anaemia	
Hydrogen	Breath	Parameter for digestive activity	
Hydrogen	Fart	Diagnosis of Lactose intolerance	A few ppm
Methane	Breath	Measure intestinal flora	~ 1 ppm
Acetaldehyde	Breath	Measure alcohol intoxication; Parameter for metabolic activity	
Ethanol	Breath	Parameter for drunkenness (check amount of drinking)	100 ~ 200 ppm
Carbon dioxide	Breath	Parameter for gastric infection by *Helicobacter pylori*	

and if those specific gases can be detected, the possibility that the diagnosis of cancer via these non-invasive methods could be viable. [3–5]. Analysis of such biogas components (VOC gas) is non-invasive and it can increase the quality of healthcare for the patient as such method can be used not only for medical examinations but also for daily healthcare administration. Table 19.1 shows the biogas components and the significance of measurement.

Gas sensors for detecting toxic gases, such as carbon monoxide (CO), hydrogen disulfide and flammable gases, such as propane and butane, have been used frequently. Gas sensors using semiconductors are small and inexpensive. Moreover, they can detect trace amounts of gas at the ppb level, and are used in many occasions. In order to detect cancer or lethal gases, extremely sensitive levels of detection ranging from 1 ppt to 1 ppb level are required. Research on detection of gases using nanotechnology such as micro electro mechanical systems (MEMS) and quartz crystal microbalance (QCM) are becoming more prevalent. In general, most sensors require a power supply, and in some cases, they are often used in a state where the sensing part is heated to a high temperature. Therefore, portability and convenience of use are some issues that need to be tackled. Recent detection method for gas sensors are described in detail in [6].

Applications of gas sensors include not only in disease diagnosis and health management, but also flammable gas alarms, alcohol sensors, food freshness checker, various toxic gas sensors, such as hydrogen sulphide, environmental protections, and for other use-cases such as biological or chemical terrorism. The overall gas sensors market is expected to be valued at USD 1,297.6 million by 2023 [7]. In addition, recent developments of gas sensors are focusing on portability, low power consumption and ease-of-use where one can easily read the changes using naked eyes.

19.2 Use of LCs as gas sensors

The LC has a property of self-organisation, and its orientation varies depending on temperature, pressure, external atmosphere, etc. [8]. Therefore, it can be regarded as a material suitable for a sensor material. The unique characteristic of the gas sensor using the LCs is that power supply is unnecessary and that change can be identified with the naked eye easily [9]. There are three types of detection methods: (i) detection by change of the transmitted light intensity, (ii) detection by degree of the light scattering and (iii) detection by change in colour of selective reflected light. In method type (i), a uniformly aligned LC layer is sandwiched between two crossed polarisers and the light transmittance T in this configuration is expressed by the following equation.

$$T = \sin^2(2\varphi)\sin^2\left(\frac{\pi\Delta nd}{\lambda}\right) \tag{19.1}$$

Here, φ is the angle between the transmission axis of the polariser on the incident light side and the director direction of LCs, and Δn is the refractive anisotropy of LCs for the incident light wavelength λ. d is the thickness of the LC layer. When a change occurs in the LC alignment due to the invasion of the gas, Δn and/or the angle φ at the relevant region changes and the transmitted light intensity changes according to (19.1). That is, as apparent from this equation, the larger the Δn of the LC materials used, the larger the change in the transmitted light intensity becomes which thus results in a sensor with higher sensitivity. The LCs used in method type (i) are mainly nematic LCs. Method type (ii) utilises a change in light scattering. The refractive index of LCs in a small region is changed upon contact with a gas. Light scattering thus occurs based on the difference of the refractive index between LCs in the region and LCs around the region like a polymer-dispersed LC display. This causes the sensor to become opaque. The LCs used in type (ii) are nematic LCs or lyotropic LCs. Method type (iii) utilises colour change due to selective reflection which is observed in cholesteric liquid crystals (CLCs) with a planar orientation. Figure 19.1 shows the state of the selective reflection in a CLC layer.

Much research from the era of Fergason [10] showed that CLCs have their selective reflection spectra changed by external field (temperature, pressure, atmospheric gas, etc.). Initiatives aimed for detecting specific gases are just beginning around 2010. Originally, the CLCs have a planar orientation whereby the LC molecules are aligned horizontally at the substrate interface. These rotate so that the orientation direction shows a spiral as it goes away from the substrate, thus reflecting the circularly polarised light selectively. The wavelength λ of the

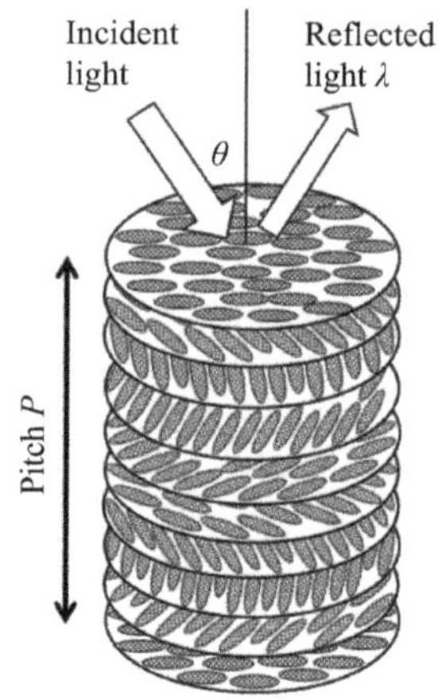

Figure 19.1 Selective reflection of CLCs

reflected light is expressed when the natural light hits incident at an angle θ with respect to the normal direction:

$$\lambda = p \cdot n \cdot \cos\theta \tag{19.2}$$

where p is a helical pitch of CLCs and n is the refractive index of CLCs. The sense of the reflected circularly polarised light is the same as that of CLCs and the theoretical maximum reflectance is 0.5.

When a VOC enters the CLC layer from the outside, the pitch p changes, and as a result the selective reflection wavelength λ changes, thus the intrusion of VOC is recognised as a change in colour. A power source is not needed and we can determine whether the gas is present with the naked eye. Hence, CLC has great potential to be a simple and highly accurate VOC sensor. The types of analytes include gases such as breath, scattered pesticides, toxic gases generated from building materials, and aqueous solutions containing excreted urine, viruses, bacteria, pathogenic bacteria, antibodies and so on. Early detection of infectious diseases using CLC is based on the change of the LC alignment with use of antigen-antibody reaction or specific reaction. The observation of the LC orientation change is usually done by immersing the LC sensor in an aqueous solution such as a buffer solution. For a gas sensor aimed at detecting pathogenic bacteria, it is necessary to pay attention to the LC material to be used so that the LC does not have the toxicity to living cells. McCamley *et al.* show that LCs which have no functional group, ether functional group/ester group, and cyanobiphenyl group are toxic, whereas LCs which contain fluorophenyl group are found to be non-toxic to living cells [11]. In general, LC materials are roughly divided into two categories: lyotropic LCs and thermotropic LCs, and the latter are classified into smectic LC, nematic LC and cholesteric LC. The prior arts of the gas sensors using each LC material are summarised in the next section.

19.2.1 Lyotropic LCs

Lyotropic LCs are formed when amphiphilic molecules are dissolved in an aqueous solution and exist in the form of droplets or lipid bilayer. A typical configuration is shown in Figure 19.2 [11]. The orientation of the LC layer or the orientation of the adjacent LC layer is disturbed upon contact with the analyte. Using lyotropic LCs, highly accurate detection of bacteria has been realised [12–16]. The structure of the lyotropic LC is similar to the structure of the biological membrane, and this type of the sensor is suitable for detecting pathogens invading the living body.

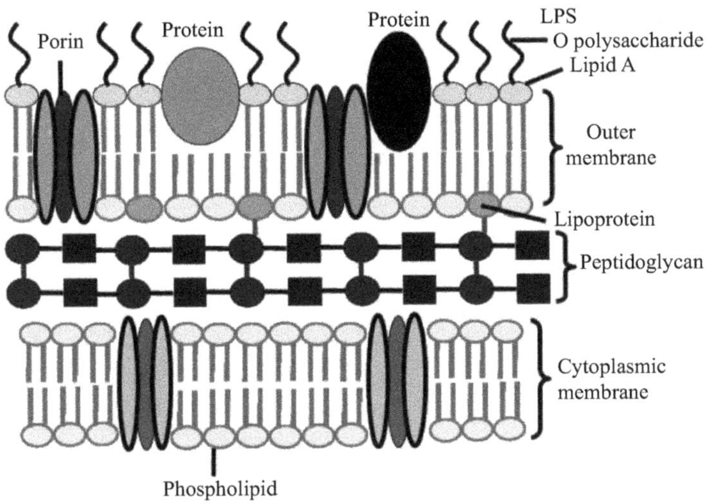

Figure 19.2 The cell membrane of a gram-negative bacterium. The phospholipids form a lyotropic LC layer which is permeable to only certain sized molecules. Reprinted with permission from Reference [11]. Copyright © 2007 by World Scientific Publishing Co. Pte. Ltd

19.2.2 Nematic LCs

In the uniformly aligned nematic LCs, the alignment of each molecule is oriented in the same direction. When the alignment is disturbed by contact with the analyte invading from the outside, it can be recognised as the change of the transmittance under crossed polarising plates (Figure 19.3). This disorder of the LC alignment upon contact with an analyte is characteristic of a LC gas sensor and the type of disorder can be categorised into two groups. One is based on the anchoring change [9,17–19] of LCs on the surface of the substrate, caused by the adsorption of the analyte on the surface of the substrate or by the chemical bonding of the analyte with the surface of the substrate. The other disorder is caused by a change in orientation [17,20–24] of LC molecules in the bulk of LC layer. In addition, when the degree of change in orientation is large, we can detect gas intrusion even by a change in electrostatic capacity [25].

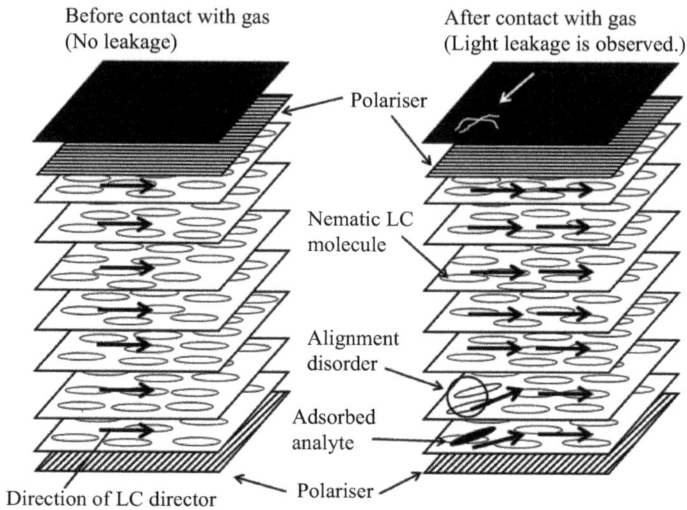

Figure 19.3 Light leakage caused by the alignment disorder when LCs make contact with an analyte

Recently, studies where carbon nanotubes (CNTs) are doped in LCs are being conducted to detect the gas as a change in conductivity [26–32]. Prior arts of gas sensors using nematic LCs are summarised in Table 19.2.

In Reference [20], detection of ppb level has been reported by Adgate *et al.* In addition, He *et al.* reported that sensitivity increases by about one digit with use of phenylcyclohexyl (PCH) LC [19]. The nematic LC materials used are mostly cyano-based LCs such as 5CB or E7 which are comparatively easy to obtain.

19.2.3 Cholesteric LCs

As apparent from (19.2) of the selective reflection, the selective reflection wavelength λ depends on the pitch p, the refractive index n, and the observation angle θ. In the case where the gas concentration is a trace amount, λ is almost determined by the change of p, as the refractive index n hardly changes even if the gas comes into contact with CLCs. The parameters which affect the p value are the above-mentioned pressure, temperature, atmospheric gas and so on. Some other specific parameters include the physical properties (phase transition temperature T_{NI}, elastic constants k_{11}, k_{22}, k_{33}) of CLC, the intermolecular interaction between the LC molecules and the VOC molecules, and furthermore the change in the LC alignment (order parameter, pre-tilt angle (molecular tilt angle)) caused by entry of the VOC gas into the LC layer affect p. All changes are interactive and eventually results in a state where the total free energy of the system becomes the minimum [25]. (For details, see Section 19.3)

A type of CLCs include LCs (which molecule itself has chirality) such as cholesterol derivatives and chiral nematic LCs (hereinafter referred to as Nm*LCs).

Table 19.2 Gas sensors using nematic LCs

LC	Substrate	Sensing mechanism	Analyte	Ref. no.
5CB	Surface–immobilised COOH/Au corrugated $H_3C(CH_2)_{14}SH$ and $HOOC(CH_2)_{10}SH$	Change of the direction of directors	hexylamine (ppb level)	[9]
		Anchoring transition: Homogeneous → Twist or Homeotropic	$(CH_3O)_2POCH_3$	
5CB	$AlClO_4$/11-mercaptoundecanoic acid (MUA)/gold	Change of the direction of directors	dimethyl methylphosphonate (DMMP) vapour (125 ppb), vesicular stomatitis virus (VSV)	[17]
	Partially biotin-$[(CH_2)_2[(CH_2)]_2O]_2$-$NHCO(CH_2)_{11}SH$ on gold	Homogeneous → Homeotropic	aviden (protein)	
		Order → disorder		
5CB	TEM grid slitting on top of octadecyltrichlorosilane (OTS)-coated glass	Change of the direction of directors Anchoring transition: Homogeneous → Homeotropic	lipid L-α-dialauroyl phosphatitidyl choline (L-DLPC)	[18]
nCB, 5PCH, 5CCH	TEM grids are placed on a polyimide-coated glass	Anchoring transition: Homeotropic →	bile acid which is produced by decomposition of cholesterol	[19]
	Sodium dodecyl sulphate (SDS) is adsorbed on LCs	Homogeneous alignment		
E7	Gallium^{3+} perchlorate/ 11-MUA/ gold	Order → disorder	organophosphate pesticides (diazinon), cholinesterase inhibition	[20]

(Continues)

Table 19.2 (Continued)

LC	Substrate	Sensing mechanism	Analyte	Ref. no.
E7	CNT and LC dispersed polymer is spin-coated on Au-comb electrode on a Si-substrate	Change of resistivity caused by the orientational ordering transition	100 ppm of acetone	[21]
5CB	Micrometre-sized droplets/ emulsions	Ordering transition Bipolar → radial	endotoxin which is a lipid-polysaccharide (glycophospholipid)	[22]
	octadecyltrichlorosilane (OTS)-coated glass	Homogeneous ← → Homeotropic		
4-decyloxy benzaldehyde (DBA) doped 5CB	dimethyloctadecyl[3-(trimethoxysilyl)-propyl] ammonium chloride (DMOAP) treated glass	Order (homeotropic) → disorder	Hydrazine reacts with DBA to become diimine.	[23]
Lauric aldehyde-doped 5CB	Planar orientation on a cleaned glass (homeotropic at the air-5CB interface)	LC orientation at the air interface and the glass surface changes according to analyte concentration.	butylamine	[24]
E7	Polymer-encapsulated droplet: poly(ethyleneimine) (PEI) and poly (4,4-dimethylazlactone) (PVDMA) bilayer	Bipolar → radial	bacterial endotoxin (lipopolysaccharide), cationic amphiphile	[33]

'Homogeneous' means homogeneous alignment and LC molecules are aligned horizontally on a substrate. 'Homeotropic' means homeotropic alignment and LC molecules are aligned vertically on a substrate.

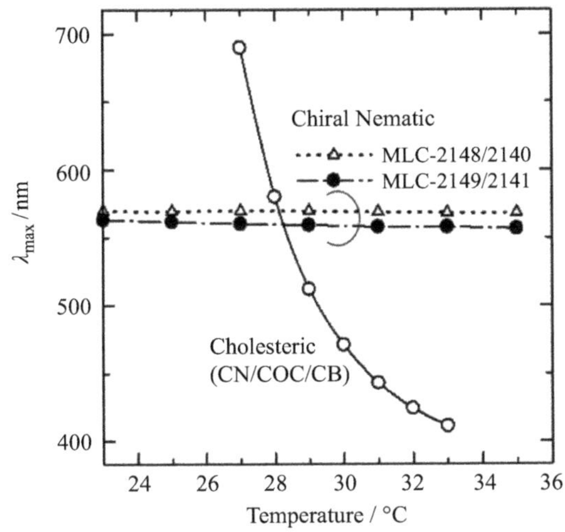

Figure 19.4 Temperature dependence of the selective reflection wavelength of spin-coated CLC layers

Nm*LCs are obtained by adding a chiral dopant to nematic LCs. In the former, the temperature dependency of the selective reflection peak wavelength is large, and a LC thermometer utilising this feature has been commercialised. On the other hand, the latter Nm*LC is suitable for gas sensor materials because the temperature dependency of the selective reflection peak wavelength is very small [34] and the degree of freedom of material selection is large. Figure 19.4 shows an example of the temperature dependency of each selective reflection peak wavelength λ_{max}. This figure compares the temperature dependency of λ_{max} for the mixture of cholesteryl nonanoate (CN), cholesteryl oleyl carbonate (COC) and cholesteryl benzoate (CB) with that of two kinds of Nm*LCs .

Table 19.3 summarises gas sensors using CLCs composed of cholesterol derivatives.

19.2.4 Chiral nematic LCs

Table 19.4 shows representative reports of gas sensors using Nm*LCs. In a gas sensor using LCs, the area in contact with gas is a liquid phase and has a problem of poor usability. Chang *et al.* stabilised the LC layer with polymer and improved usability [41,42]. In addition, Stumpel *et al.* investigated the inkjet method as a process of depositing a LC layer and made it easy to fabricate sensors arranged in a matrix [43].

Although many investigations on gas sensors using LCs have been done so far, LC materials and chiral materials to be used are limited, and their chemical structure and behaviour of selective reflection upon contact with gas have been hardly studied (except for specific reactions such as hydrogen bonding). In Section 19.3, the influence of the molecular structure and physical properties

Table 19.3 Gas sensors using cholesteric LCs

LC	Configuration and Preparation	Analyte	Shift direction of the selective reflection spectra	Remarks	Ref. no.
CB CN COC	Polymer stabilised CLC	heptane, trimethylamine, diisopropyl amine, hexylamine	Shorter shift (Hydrogen bonding is formed.)	High sensitivity is obtained in a primary amine	[35]
CB CN COC	Small droplets of CLC are dispersed in PDMS. Free CLC layers separated by multi-layered PDMS films	ethanol produced from fermentation	Shorter shift Dissolution of ethanol in CLC can create a net torque around the helical axis.	Micro-fluidic devices	[36]
CC CN COC	Spin-coated on a styrene plate	aromatic and halogenated hydrocarbons, tetrachloroethylene, methanol, THF, chloroform	Longer shift	Influence of the humidity is discussed.	[37]
CB CN COC	Glass/dodecyl amine doped CLC/ glass, glass/ PDMS/CLC/glass	aldehyde vapour, pentyl aldehyde	Shorter shift Formation of an imine bond between amine and aldehyde.	Amine, alcohol and acetone vapour don't react with amine.	[38]
CN CC	Cast by pipetting. After mixing PVC and silicone rubber with CC/ CN., CLC are immobilised.	acetone, ethanol, benzene, pyridine, hexane	Longer shift for pyridine, acetone. Colourless for benzene, hexane. No change for ethanol.		[39]

CB: cholesteryl benzoate, CC: cholesteryl chloride, CN: cholesteryl nonanoate, COC: cholesteryl oleyl carbonate, PDMS: poly (dimethylsiloxane), PVC: polyvinyl chloride

of the LC material used on the selective reflection wavelength was examined in detail.

19.3 Molecular structure of chiral nematic LC gas sensors

19.3.1 Chemical structure of LCs

Nm*LCs with selective reflection wavelength in the visible region have been prepared by adding chiral materials to the nematic LCs composed only of homologous LCs. The nematic LC is a mixture of LC compounds which have the same chemical structures except different terminal alkyl chain length. After diluting with cyclohexane, Nm*LC is dropped on the rubbed surface of the polyimide alignment film formed on the glass substrate and applied to the entire surface by spin coating.

Table 19.4 Gas sensors using chiral nematic LCs

LC	Chiral dopant	Preparation of a LC film	Analyte	Remarks	Ref. No
TM1013		LC-coating Embedded into a polystyrene matrix. Divinylbenzene was added to a styrene monomer.	methanol ethanol, THF chloroform, tet-rachloroethylene	Shorter shift for THF, Longer shift for bulky chloride atoms	[40]
5CB	Chiral diacrylate (LC756)	Reactive mesogen (RM82, RM105) or monoacrylate with a carboxylic acid unit (6OBAB, 6OBA) is mixed and polymerised after ink jet printing.	ethanol methanol	Hydrogen bonding Detection of counterfeit wine (5CB is extracted by evaporation after polymerisation)	[41,42]
4-(4-(6-acryloyloxy hexyloxy) benzoyloxy) methoxy benzene	R(+)-3-methyl adipic acid	Mono and diacrylate LCs or benzoic acid derivatives are mixed and polymerised after ink jet printing.	gaseous trinethyl amine (TMA)	Hydrogen bonding Detection of putrefactive odour.	[43]
E7	a chiral binaphthyl dithiol derivative, diamine derivative	Wire-bar coating	O_2 with TMA (ir-reversible), CO_2 (reversible)	Administration of CO_2/O_2 in food preservation	[44]
CNT doped E7	S1011 and DBD	Fill a cell with LCs by a capillary action.	acetone		[45]
E7	S1011 and DBD	Fill a cell with LCs by a capillary action.	acetone toluene	The higher the molecular polarity of the vapour, the greater the shift of the reflection spectra of CLCs.	[46]
E7	Bis-[binaphthylen-(2.20)-dioxy]-silan	Wire-bar coating	humidity	A dopant is hydrolysed.	[47]

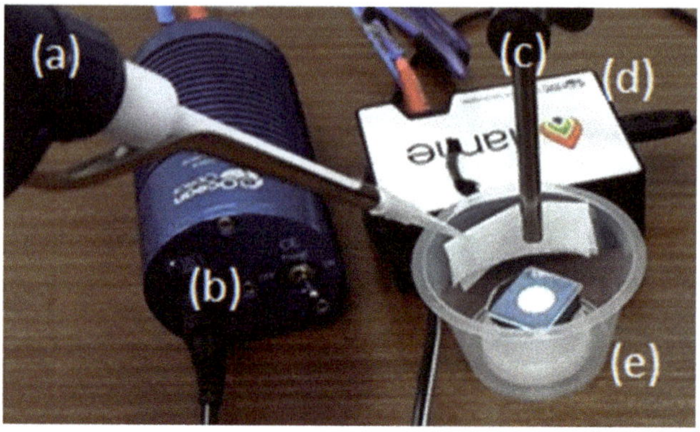

Figure 19.5 Experimental setup to measure selective reflection spectra: (a) micropipette; (b) halogen lamp; (c) optical probe; (d) spectrophotometer; and (e) sample. Reprinted with permission from Reference [25], Copyright (2017) John Wiley & Sons publications

Thereafter, when cyclohexane is removed by heating, a CLC substrate having an Nm*LC layer with a thickness of 15 μm to 20 μm is obtained.

The prepared Nm*LC layer is in contact with two different interfaces of the rubbed alignment film interface and the free air interface. Although LC molecules are aligned homogeneously by the anchoring strength of the rubbed polyimide alignment film at the interface of the alignment film, LC molecules at the air interface show nearly homeotropic alignment, since the anchoring strength of the alignment film on the glass substrate does not influence the opposite air interface. Such a hybrid LC configuration has less viewing angle dependency of the displayed colour as compared with the planar CLC configuration.

Thereafter, the prepared CLC substrate is placed in a 90 ml closed vessel, VOC is introduced using a 5 μl micropipette, and the colour change of the CLC layer is measured using a fibre type multi-channel spectroscopic system for reflection measurement (Figure 19.5).

Table 19.5 shows the experimental results of colour shifts at the time of VOC introduction such as methyl ethyl ketone (MEK), toluene, cyclohexane, and tert-butyl alcohol. LC-1 is a Schiff base LC (mixture of equal amounts of MBBA and EBBA), LC-2 is a Demus ester LC (mixed LC of D-402, D-501, D-502), LC-3 is a diphenylacetylene LC (mixture of equal amounts of PTP-304, PTP-402 and PTP-501), LC-4 is a n-type phenylcyclohexane LC (mixture of equal amounts of PCH302 and PCH502), LC-5 is a p-type phenylcyclohexane LC (mixture of equal amounts of PCH3, PCH5, and PCH7), LC-6 is a mixed LC E-7 composed of biphenyl type LCs and terphenyl type LCs, LC-7 is an alkenyl type LC and LC-8 is a pyrimidine type LC.

Conventionally, it has been well known that when an ester compound forms a hydrogen bond, the selective reflection spectrum shows a blue shift, but it was also found that a tolane or phenyl pyrimidine LC, which does not form a hydrogen bond, also shows a blue shift. The changes of the selective reflection spectra every 15s

Table 19.5 Correlation between molecular structures of the compounds of nematic LC (1) to (8) and direction of λ_{max} shift. R and R′ denote alkyl groups, respectively. Five microlitres of MEK, toluene, tert-butyl alcohol or cyclohexane was added into each LC. The original λ_{max} was around 550 nm. Reprinted with permission from Reference [25], Copyright (2017) John Wiley & Sons publications

LC no.	LC molecular structures	Direction of λ_{max} shift
(1)	RO-⟨◯⟩-CH=N-⟨◯⟩-R'	Shorter (blue shift)
(2)	R-⟨H⟩-COO-⟨◯⟩-OR'	Shorter (blue shift)
(3)	R-⟨◯⟩-C≡C-⟨◯⟩-OR'	Shorter (blue shift)
(4)	R-⟨H⟩-⟨◯⟩-OR	Shorter (blue shift)
(5)	R-⟨H⟩-⟨◯⟩-CN	Longer (red shift)
(6)	R-⟨◯⟩-⟨◯⟩-CN	Longer (red shift)
(7)	R-⟨◯⟩-⟨◯⟩-⟨◯⟩-CN R-⟨H⟩-⟨H⟩-CN=CH_2	Longer (red shift)
(8)	R-⟨pyrimidine (N, N)⟩-⟨H⟩-CN	Shorter (blue shift)

after the introduction of VOC gases are shown in Figure 19.6. Depending on the kind of gas, the degree of colour shift varied and the direction of colour shift changed with different LC structure. The selective reflection spectrum shifts in the short wavelength direction (defined as blue shift) in LC-1 to LC-4 and LC-8 and in the long wavelength direction (defined as red shift) in LC-5 to LC-7. Also, the behaviour in which the selective reflection spectrum changes varied, depending on the type of VOC gases that enter. Figure 19.7 shows the time variation of the peak wavelength shift of the selective reflection spectrum for different VOCs. From these figures, it can be seen that the direction and magnitude of selective reflection peak wavelength change upon contact with VOC varies depending on the combination of LC structure and VOC gas species. Such a difference in a λ_{max} profile with time is also important information for gas detection.

19.3.2 Substituted functional group

Table 19.6 shows CLC-1 to CLC-6, CLC-10, CLC-11, LC-2 and LC-6 compositions that exhibit selective reflection of green before contact with gas and the shift direction of selective reflection peak wavelengths when these LCs are brought into contact with acetone, toluene, and MEK. CLC-2, LC-11 and LC-2 showed a short wavelength (blue) shift and showed long wavelength (red) shift in CLC-1, CLC-3 to CLC-5, CLC-10 and LC-6. The appearance of colour change before and after contacting toluene with each LC of CLC-2 and CLC-4 was observed with a microscope at magnification of 40 times is shown in Table 19.7.

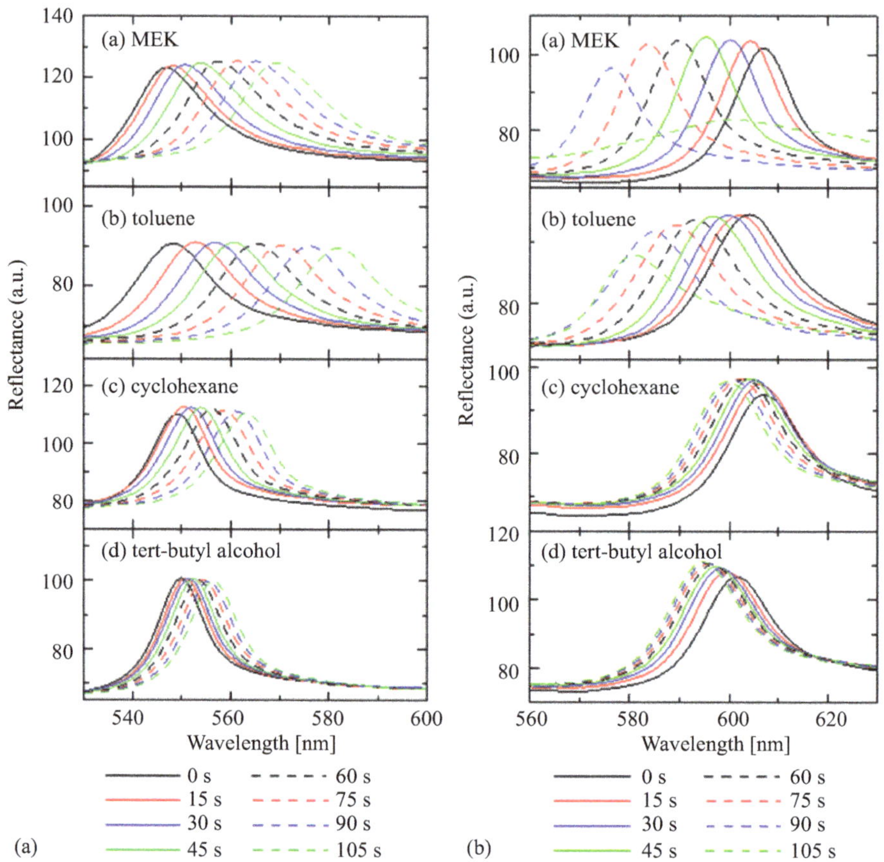

Figure 19.6 (a) Time-lapsed reflection spectra of LC-5 upon contact with 5 μl of (a) MEK; (b) toluene; (c) cyclohexane; and (d) tert-butyl alcohol. (b) Time-lapsed reflection spectra of LC-3 upon contact with 5 μl of (a) MEK; (b) toluene; (c) cyclohexane; and (d) tert-butyl alcohol. Reprinted with permission from Reference [25], Copyright (2017) John Wiley & Sons publications

It is known that an ester compound forms a hydrogen bond and a cholesteric pitch will be shortened. However, although CLC-1,3 and 5 contain a large amount of ester compound, a blue shift did not occur. It is assumed that bulky substituents such as cyclohexyl group, 3,4-difluorophenyl group and 2,3-dicyanophenyl may possibly inhibit blue shift. In order to verify this, CLC-11 was prepared by removing the compound having a 3,4-difluorophenyl group from CLC-5 and contacted with toluene or MEK. Then, CLC-11 showed a blue shift and the hypothesis was demonstrated.

In addition, no blue shift was observed in CLC-10, which was prepared by adding the same proportion of tolane compound as that of CLC-2 to CLC-1.

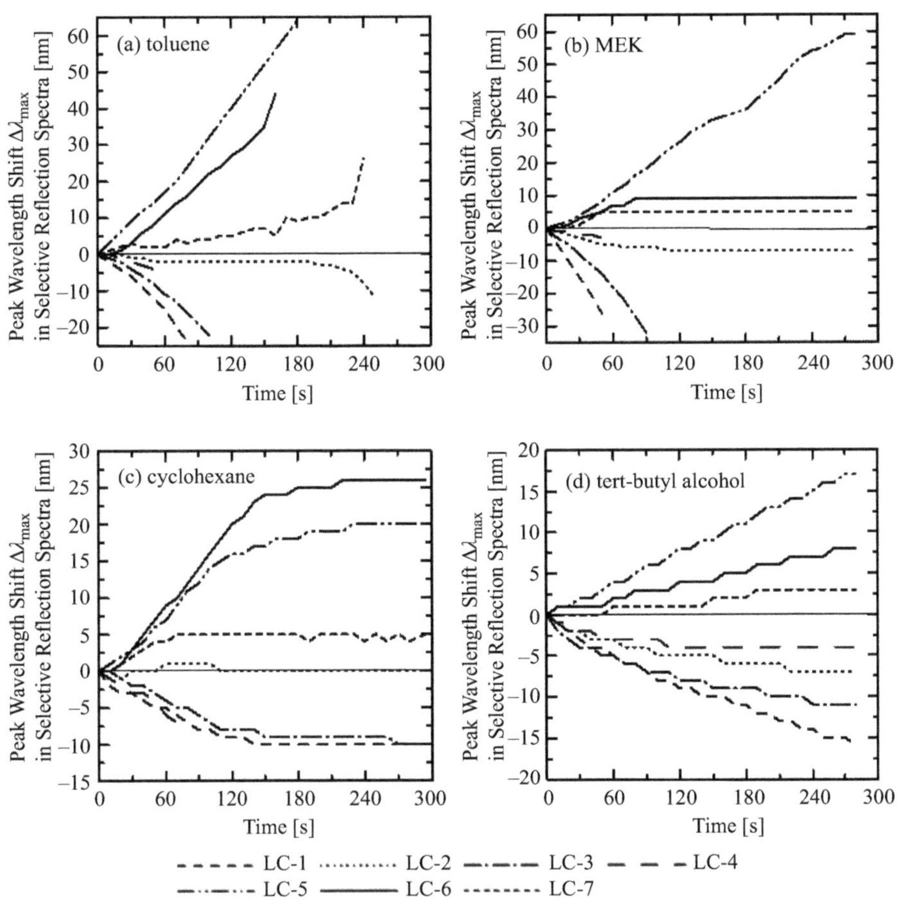

Figure 19.7 Time variation of the peak wavelength shift of the selective reflection spectrum for different VOCs. (VOC: (a) toluene; (b) MEK; (c) cyclohexane; and (d) tert-butyl alcohol.) Reprinted with permission from Reference [48], Copyright (2017) Technical Information Institute Co., Ltd

It seems that this is attributed to the influence of the bulky 2,3-dicyanophenyl compound contained in CLC-1. As described in Table 19.6, the direction in which the selective reflection peak wavelength changes when the Nm*LC contacts the VOC is greatly influenced by the substituent.

19.3.3 Chiral dopant

The performance of the chiral material is expressed by the helical twisting power (HTP). The dopants used in the experiment were purchased from LCC Corporation (Japan). They are DOS-1015L, S-811, NYC-22133L, CN (cholesteryl nonanoate), and their HTP was 25-35, 9-13, 20-25, 6-8 μm^{-1} (LCC catalogue value), respectively.

Table 19.6 Physical properties, molecular structures of the components and mass composition (w/w) of CLC-1 to 5, LC-6, LC-2, CLC-10 and CLC-11 The direction of shift of λ_{max} in a spin-coated CLC when exposed to acetone, toluene, or MEK vapours is shown in bottom row of the table. 'longer' indicates red-shift and 'shorter' blue shift. Reprinted with permission from Reference [25], Copyright (2017) John Wiley & Sons publications

Physical properties of host nematic liquid crystals	**LC no.**	**CLC-1**	**CLC-2**	**CLC-3**	**CLC-4**	**CLC-5**	**LC-6**	**LC-2**	**CLC-10**	**CLC-11**
	T_{NI} /°C	88	75	68	72.8	75.5	60	71.2	83.1	114.3
	n_e	1.564	1.645	1.598	1.634	1.584	1.725	1.567	1.605	1.573
	n_o	1.484	1.496	1.487	1.493	1.490	1.510	1.480	1.491	1.483
	Δn	0.080	0.149	0.111	0.141	0.094	0.215	0.087	0.114	0.090
	$\varepsilon\|\|$	4.27	3.62	9.85	14.89	10.68	18.30	3.2	4.05	2.78
	$\varepsilon\perp$	9.33	3.98	4.75	4.32	5.64	5.60	4.4	8.90	3.82
	$\Delta\varepsilon$	−5.06	−0.36	5.10	10.57	5.04	12.7	−1.00	−4.85	−1.04
R-(H)-COO-(◯)-OR'		50.2 wt.%	44.5 wt.%	59.0 wt.%		38.5 wt.%		100 wt.%	36.2 wt.%	62.6 wt.%
R-(◯)-C≡C-(◯)-OR'			22.2 wt.%						22.2 wt.%	
RO-(◯)-COO-(◯)-R'		4.3 wt.%	29.6 wt.%							
R-(H)-(H)-COO-(◯)-OR'		5.0 wt.%	3.7 wt.%			15.4 wt.%			4.9 wt.%	25.0 wt.%
R-(H)-(H)-COO-(◯)-(H)-R'		2.5 wt.%				7.6 wt.%			2.4 wt.%	12.4 wt.%
R-(H)-COO-(◯)-COO-(◯)-R'		3.7 wt.%								

Table 19.6 (Continued)

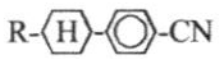			35.0 wt.%	85.0 wt.%					
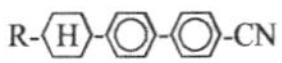			6.0 wt.%	15.0 wt.%					
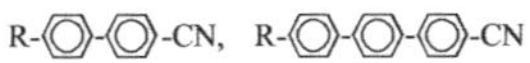						100 wt.%			
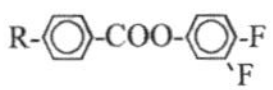					38.5 wt.%				
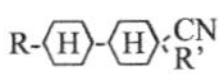	18.3 wt.%							18.4 wt.%	
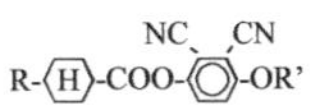	10.0 wt.%							9.8 wt.%	
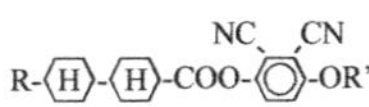	2.4 wt.%							2.4 wt.%	
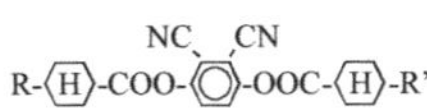	3.6 wt.%							3.7 wt.%	
Direction of λ_{max} shift	Longer	Shorter	Longer	Longer	Longer	Longer	Shorter	Longer	Shorter

Table 19.7 Optical images of CLC-2 and CLC-4 before and after exposure to toluene vapours. All images were captured using the same objective lens. Reprinted with permission from Reference [25], Copyright (2017) John Wiley & Sons publications

CLC type	Direction of λ_{max} shift	Before contact with toluene	After contact with toluene
CLC-2	Shorter (blue shift)		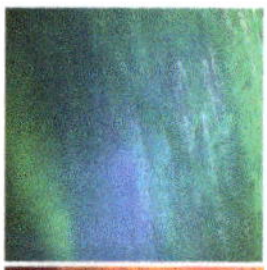
CLC-4	Longer (red shift)		

With the use of a large HTP chiral dopant, its influence on the physical properties of LC materials could be suppressed as we can reduce the amount of the chiral dopant used. Generally, many of them have insufficient solubility in LCs. Therefore, in preparing Nm*LC, a combination of a few chiral dopants are used to realise a desired selective reflection wavelength.

Figure 19.8 shows the peak wavelength shift $\Delta\lambda_{max}$ with time when MEK is introduced in the Nm*LC prepared by adding only each chiral dopant to LC-2. As seen from the figure, the appearance of $\Delta\lambda_{max}$ shift differs depending on the chiral material.

19.3.4 Dielectric anisotropy of LC materials

Figure 19.9 shows the relationship between the peak wavelength shift $\Delta\lambda_{max}$ (1 min value) of the selective reflection and the dielectric anisotropy $\Delta\varepsilon$ of Nm*LC used when MEK is brought into contact with various Nm*LCs. Many Nm*LCs which $\Delta\varepsilon$ is nearly zero tend to show a blue shift. However, no blue shift was observed even when MEK was introduced into a mixture which $\Delta\varepsilon$ was adjusted to be almost zero by mixing CLC-1 and CLC-3 or CLC-1 and CLC-5 (Figure 19.10). It is assumed that the LC component that has bulky 2,3-dicyanophenyl group in CLC-1 inhibits short wavelength shift.

19.3.5 Enhancement of sensitivity and selectivity of gas detection

Table 19.8 shows the shift direction of the selective reflection peak wavelength shift of Nm*LC prepared by mixing the LC material which shows a blue shift and the LC material which show a red shift by changing the mixing ratio. By using this approach, it is possible to distinguish between toluene and MEK, which was difficult by using the LCs in Tables 19.5 and 19.6.

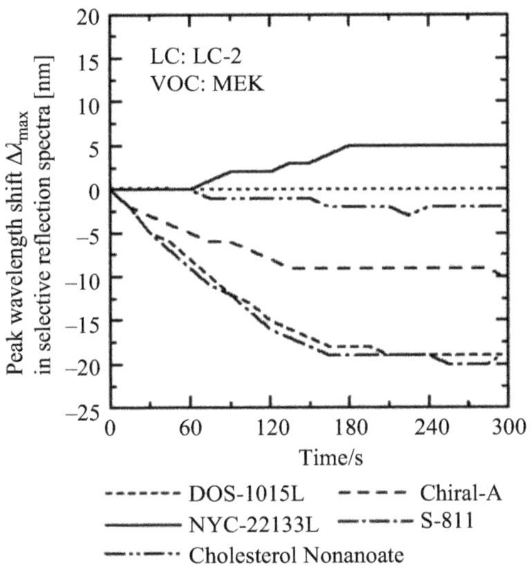

Figure 19.8 Influence of chiral dopants on selective reflection wavelength peak shift $\Delta\lambda_{max}$. Reprinted with permission from Reference [48], Copyright (2017) Technical Information Institute Co., Ltd

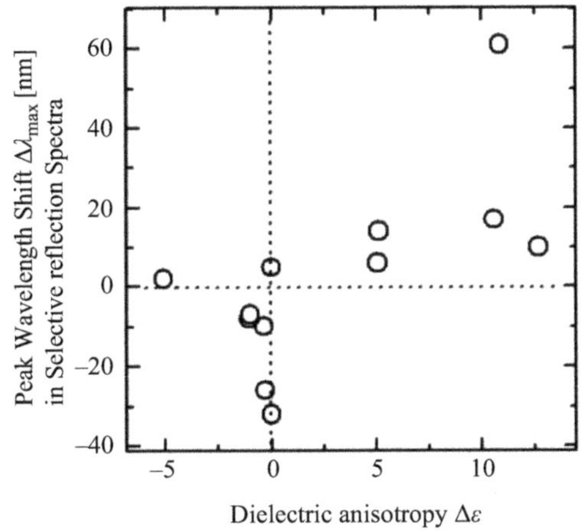

*Figure 19.9 Relationship between the peak wavelength shift $\Delta\lambda_{max}$ (1 min value) of the selective reflection and the dielectric anisotropy $\Delta\varepsilon$ of Nm*LC upon contact with MEK. Reprinted with permission from Reference [48], Copyright (2017) Technical Information Institute Co., Ltd*

Nm*LCs with nematic–isotropic phase transition temperature T_{NI} is 52 °C, 44 °C, or 41 °C were prepared by changing the proportion of 5CB component in LC-6, and the change in the selective reflection peak wavelength upon contact with MEK was measured with time. The results are shown in Figure 19.11. It was found that the

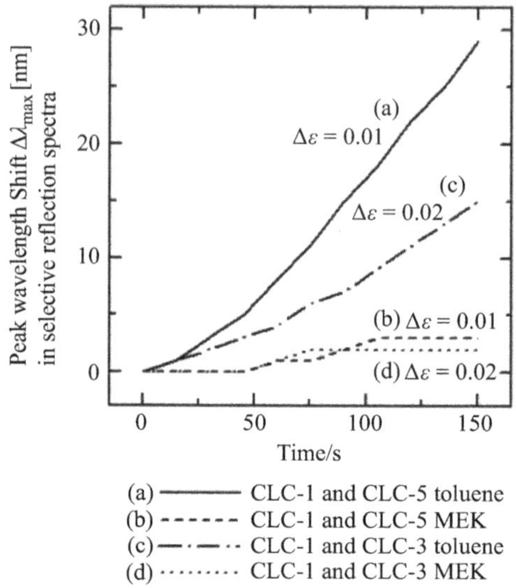

Figure 19.10 *Time variation of the peak wavelength shift $\Delta\lambda_{max}$ (1 min value) of the selective reflection of Nm*LC which has little $\Delta\varepsilon$ and a bulky functional group. Reprinted with permission from Reference [48], Copyright (2017) Technical Information Institute Co., Ltd*

Table 19.8 *Influence of CLC composition on direction of shift of $\Delta\lambda_{max}$ when exposed to MEK, toluene and tert-butyl alcohol (tert-BuOH) vapours. Reprinted with permission from Reference [25], Copyright (2017) John Wiley & Sons publications*

CLCs [wt. %]		**VOC**	
CLC-1	**CLC-2**	**5 μl of MEK**	**5 μl of toluene**
100	0	Red shift	Red shift
77	23	Blue shift	Red shift
52	48	Blue shift	Red shift
0	100	Blue shift	Blue shift
LC-2	**LC-6**	**5 μl of MEK**	**5 μl of tert-BuOH**
35	65	Red shift	Red shift
55	45	Red shift	Red shift
60	40	Blue shift	Red shift
65	35	Blue shift	Blue shift
70	30	Blue shift	Blue shift

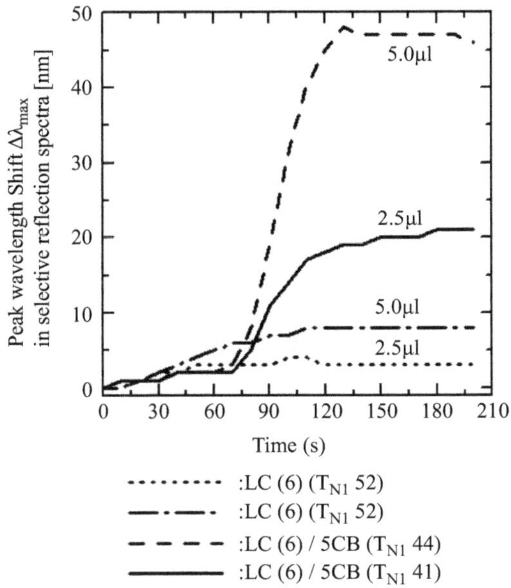

Figure 19.11 *Influence of nematic-isotropic transition temperature T_{NI} on the $\Delta\lambda_{max}$. Reprinted with permission from Reference [25], Copyright (2017) John Wiley & Sons publications*

selective reflection peak wavelength shift when Nm*LC with decreased T_{NI} was used became larger than that of LC-6 alone. This means, the sensitivity of gas detection is enhanced with using Nm*LCs whose order parameter is lower. In other words, the helical structure of Nm*LCs with lower T_{NI} has a tendency to be easily deformed.

19.3.6 Mechanism of selective reflection peak wavelength shift

19.3.6.1 Red shift

The causes of the red shift of the selective reflection peak wavelength λ_{max} upon contact with VOC include (1) the decrease in intermolecular interaction, (2) the decrease in LC order parameter and (3) the decrease in elastic constant. We confirmed that the pre-tilt angle of LC materials increases by several degrees by the dielectric constant measurement and that the elastic constant of the LC material decreases by 20 per cent to 40 per cent due to VOC invasion. The influence of each effect on the selective reflection wavelength was calculated, and it was concluded that the decrease of the LC orientation order was the main cause of the red shift. When VOC penetrates into the helical structure of Nm*LC, molecular tilt occurs in the entire helical structure, and the helical pitch extends. As a result, it is considered that the selective reflection peak wavelength changes in the long wavelength direction.

19.3.6.2 Blue shift

Blue shift of the selective reflection wavelength in Demus ester LC will be explained by the intermolecular hydrogen bonds with VOC gas. However, the reason why selective reflection of diphenylacetylene type LC (tolane-type LC) and pyrimidine type LC shows blue shift upon contact with gas has not been sufficiently elucidated. Among the possible factors contributing to blue shift are (1) hydrogen bonding, as seen in the ester compounds, (2) the shrinkage of chiral pitch due to π-π stacking through VOC gas, (3) the decrease in refractive index and (4) the increase in elastic constant.

19.3.7 Characteristics of chiral nematic LCs

In the development of gas sensor using selective reflection of CLCs, the following guidelines for material design were obtained.

1. Many CLCs shift their selective reflection wavelength in the long wavelength direction (red shift) by contact with VOCs, but some CLCs shift in the short wavelength direction (blue shift).
2. Tolan (diphenylacetylene) derivatives and phenylpyrimidine, one of the few LCs showing blue shift, are extremely effective for the selective detection of VOC gas in the case where no hydrogen bond is formed.
3. This blue shift is inhibited by the presence of LC molecules with a bulky structure.
4. Mixing the blue shift LC and the red shift LC greatly improves the gas selectivity.
5. Selectively detection of specific gas species is achieved based on information on the direction of colour shift, the duration of time change of selective reflection wavelength, the combination of LC and chiral material to be used, and so on.

In these experiments, the LC layer is formed on a glass substrate, but it is suggested that it is practically used on a polymer film or in a form like a polymer network LC. Figure 19.12 shows the colour change of the sensor where Nm*LC layer on a polymer film is covered with a gauze sheet (130 mesh) woven by extremely fine polyester fibres upon contact with acetone and shows the possibility of a flexible gas sensor.

19.4 Conclusion

A gas sensor using LC has a unique feature that it does not require a power supply. Furthermore, the presence and type of a VOC can be easily detected by the change in colour or brightness of the LC. It is expected to be applied to various fields, such as health management, expiry and freshness monitoring of food consumables, monitoring of pesticides, and etc. In addition, this sensor is based on the change in LC alignment and the changes in dielectric properties. Dielectric characteristics can be extracted and inferred as an electric signal. It is also possible to extract the existence of a specific gas as a change in resistance value by adding an anisotropic conductive material. The problem in practical application is the enhancement of

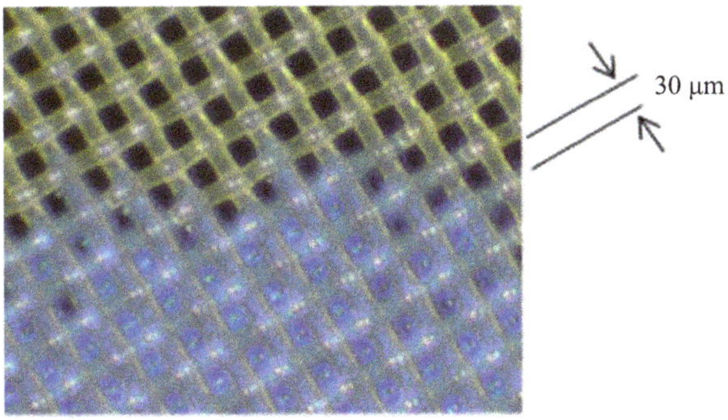

Figure 19.12 *Flexible gas sensor where chiral nematic LCs are sandwiched between a gauze sheet woven with extremely fine polyester fibres and a polymer film. (Manufactured in Yamani Co., Ltd. Sold by Teijin Frontier Co., Ltd.). One side of the void portion surrounded by the fibres is about 30 μm, and the blue portion is a portion where the acetone has penetrated*

selectivity and the detection sensitivity as with other gas sensors. Many of the gas sensors introduced in this paper are reusable, but in detection of microscopic amount of specimens, such as detection of pathogenic bacteria or screening of cancer, utilising irreversible reaction like antigen-antibody reaction and specific chemical reaction is also effective.

References

[1] Ministry of Health, Labour and Welfare, 'The comprehensive reform of social security and tax: Reducing the burden passed on to future generations'. Available from http://www.mhlw.go.jp/english/social_security/kaikaku_3.html (Accessed on 11 April 2018).

[2] Ministry of Health, Labour and Welfare, 'Summary of vital statics'. Available from http://www.mhlw.go.jp/english/database/db-hw/populate/index.html (Accessed on 17 March 2018).

[3] Loizeau F., Lang H.P., Akiyama T., *et al.* 'Piezoresistive membrane-type surface stress sensor arranged in arrays for cancer diagnosis through breath analysis'. *Proc. of MEMS 2013.* Taipei, Taiwan, January, pp. 20–24, 2013, IEEE, pp. 621–24.

[4] Hanai Y., Shimono K., Matsumura K., *et al.* 'Urinary volatile compounds as biomarkers for lung cancer'. *Biosci. Biotechnol. Biochem.* 76(4), pp. 679–84 (2012) doi: 10.1271/bbb.110760.

[5] Adiguzel Y., Kulah H. 'Breath sensors for lung cancer diagnosis'. *Biosensors and Bioelectronics*, 65 (2015) pp. 121–38.

[6] Kim I.-D., Choi S.-J., Kim S.-J. and Jang J.-S. 'Exhaled breath sensors' in Kyung C-M. (ed.) *Smart Sensors for Health and Environment Monitoring*, Springer Science+Business Media Dordrecht 2015, Vol. 323, pp. 19–49, 2015.

[7] Global Information, Inc., Gas sensors market by gas type (oxygen, carbon monoxide, carbon dioxide, ammonia, chlorine, hydrogen sulfide, nitrogen oxide, volatile organic compounds, hydrocarbons), technology, end-use application, geography - global forecast 2023, Marketsandmarkets 2017. Available from https://www.gii.co.jp/report/mama358369-gas-sensors-market-by-technology-electrochemical.html (Accessed on March 17 2018).

[8] de Gennes P.G., Prost J. *The Physics of Liquid Crystals*, Clarendon Press, Oxford, 1993.

[9] Shah R.R. and Abbott N.L. 'Principles for measurement of chemical exposure based on recognition-driven anchoring transitions in liquid crystals'. *Science* 293 (2001) 1296–99. DOI: 10.1126/science.1062293.

[10] Fergason J.L. 'Liquid Crystals'. *Scientific American* 211, pp. 77–85 (1964).

[11] McCamley M. K., Artenstein A.W. and Crawford G.P. 'Chapter 4 Liquid Crystal Biosensors: A New Approach to Medical Diagnostic Devices', in Waltman S. J., Jay G.D., Crawford G.P. (ed.) *LIQUID CRYSTALS Frontiers in Biomedical Applications*, World Scientific Publishing Co. Pte. Ltd., Singapore, pp. 241–96, 2007.

[12] Hag L.V., Gras S.L., Conn C.E. and Drummond C.J. 'Lyotropic liquid crystal engineering moving beyond binary compositional space—Ordered nanostructured amphiphile self-assembly materials by design'. *Chem. Soc. Rev.* 2017, 46, 2705–31.

[13] Zhou S. *Lyotropic Chromonic Liquid Crystals*, Springer International Publishing: Cham, Switzerland, 2017.

[14] Shiyanovskii S.V., Lavrentovich O.D., Schneider T., *et al.* 'Lyotropic chromonic liquid crystals for biological sensing applications'. *Mol. Cryst. Liq. Cryst.* 434, 2005.

[15] de Souza J.F., Pontes K., Alves T., *et al.* 'Spotlight on biomimetic systems based on lyotropic liquid crystal'. *Molecules* 22, 419, 2017.

[16] Stumpf C.H., Bullard1 B., Zhao W., Kuzenko1 S. and Niehaus G.D. 'Rapid detection of listeria monocytogenes using a liquid crystal-based immunoassay'. Crystal Diagnostics LTD. Available from https://www.crystal-diagnostics.com/wp-content/uploads/IAFP-Poster-2016_Listeria-mono_Final.pdf (Accessed on 11 April 2018).

[17] Gupta V.K., Skaife J.J., Dubrovsky T.B. and Abbott N.L. 'Optical amplification of ligand-receptor binding using liquid crystals'. *Science*, 279(5359), pp. 2077–80, 1998. DOI: 10.1126/science.279.5359.2077.

[18] Brake J.M., Daschner M.K., Luk Y.-Y. and Abbott N.L. 'Biomolecular interactions at phospholipid-decorated surfaces of liquid crystals'. *Science* 302, 2094–97 (2003); DOI: 10.1126/science.1091749.

[19] He S., Liang W., Cheng K.-L., Fang J. and Wu S.-T. 'Bile acid–surfactant interactions at the liquid crystal/aqueous interface'. *Soft Matter*, 2014, 10, 4609–14. DOI: 10.1039/c4sm00486h

[20] Adgate J.L., Bartekova A., Raynor P.C., *et al.* 'Detection of organophosphate pesticides using a prototype liquid crystal monitor'. *J. Environ. Monit.*, 2009, 11, 49–55. DOI: 10.1039/b806954a.

[21] Lai Y.-T., Kuo J.-C. and Yang Y.-J. 'A novel gas sensor using polymer-dispersed liquid crystal doped with carbon nanotubes'. *Sensors and Actuators A: Physical* 215 (2014) 83–88. DOI: 10.1016/j.sna.2013.12.021.

[22] Lin I.-H., Miller D.S., Bertics P.J., Murphy C.J., de Pablo J.J. and Abbott N.L. 'Endotoxin-induced structural transformations in liquid crystalline droplets'. *Science* 332, pp.1297–1300. DOI: 10.1126/science.1195639.

[23] Nandia R., Singhb S.K., Singhb H.K., Singhb B. and Singha R.K. 'Fabrication of liquid crystal based sensor for detection of hydrazine vapours'. *Chemical Physics Letters* 614 (2014) 62–66. DOI: 10.1016/j.cplett.2014.09.020.

[24] Ding X. and Yang K.-L. 'Liquid crystal based optical sensor for detection of vaporous butylamine in air'. *Sensors and Actuators B: Chemical* 173 (2012) 607– 13. DOI: 10.1016/j.snb.2012.07.067.

[25] Kek K.J., Lee J.J.Z., Otono Y. and Ishihara S. 'Chemical gas sensors using chiral nematic liquid crystals and its applications'. *Journal of the SID 25/6, 2017*, pp. 366–73, 2017. DOI: 10.1002/jsid.560.

[26] Lai Y.T., Chen Y.M. and Yang Y.J. 'A novel CNT-PDMS-based tactile sensing array with resistivity retaining and recovering by using dielectrophoresis effect'. *Journal of Microelectromechanical Systems* 21 (2012) 217–23.

[27] Sayago I., Fernandez M.J., Fontecha J.L., *et al.* 'New sensitive layers for surface acoustic wave gas sensors based on polymer and carbon nanotube composites'. *Sensors and Actuators B: Chemical* 175 (2012) 67–72. DOI: 10.1016/j.snb.2011.12.031.

[28] Lee D. and Cui T. 'A role of silica nanoparticles in layer-by-layer self-assembled carbon nanotube and In_2O_3 nanoparticle thin-film pH sensors: tunable sensitivity and linearity'. *Sensors and Actuators A: Physical* 188 (2012) 203–11. DOI: 10.1016/j.sna.2012.01.004.

[29] Lee W., Wang C.Y. and Shih Y.C. 'Effects of carbon nanosolids on the electro-optical properties of a twisted nematic liquid-crystal host'. *Applied Physics Letters* 85 (2004) 513–15. DOI: 10.1063/1.1771799.

[30] Basu R. and Iannacchione G.S. 'Nematic anchoring on carbon nanotubes'. *Applied Physics Letters* 95 (2009) 173113. DOI: 10.1063/1.3256013.

[31] Jeong S.J., Park K.A., Jeong S.H., *et al.* 'Electroactive superelongation of carbon nanotube aggregates in liquid crystal medium'. *Nano Letters* 7 (2007) 2178–82. DOI: 10.1021/nl070116u.

[32] Dierking I., Scalia G., Morales P. and LeClere D. 'Aligning and reorienting carbon nanotubes with nematic liquid crystals'. *Advanced Materials* 16 (2004) 865–69. DOI: 10.1002/adma.200306196.

[33] Manna U., Zayas-Gonzalez Y.M., Carlton R.J., Caruso F., Abbott N.L. and Lynn D.M. 'Liquid crystal chemical sensors that cells can wear'. *Angew. Chem. Int. Ed.* 2013, 52, 14011–15. DOI: 10.1002/anie.201306630.

[34] Sato Y., Naito T., Okada K., Kurosawa N., Kimura M. and Shibata T. 'Studies on chiral pitch characteristics of nematic liquid crystals induced by chiral dopants'. *Proceedings of the Domestic Conf. of Jpn. Liq. Cryst. Soc.*, 16, No. 1L313, pp. 114–15, 1990.

[35] Sutarlie L., Qin H. and Yang K.-L. 'Polymer stabilized cholesteric liquid crystal arrays for detecting vaporous amines'. *Analyst*, 2010, 135, 1691–96. DOI: 10.1039/b926674g.
[36] Sutarlie L. and Yang K.-L. 'Monitoring spatial distribution of ethanol in microfluidic channels by using a thin layer of cholesteric liquid crystal'. *Lab Chip*, 2011, 11, 4093–98. DOI: 10.1039/c1lc20460b.
[37] Dickert F.L., Haunschild A. and Hofmann P. 'Cholesteric liquid crystals for solvent vapour detection-elimination of cross sensitivity by band shape analysis and pattern recognition'. *Fresenius J Anal Chem.* (1994) 350, 577–81.
[38] Sutarlie L., Lim J.Y., and Yang K.-L. 'Cholesteric liquid crystals doped with dodecylamine for detecting aldehyde vapors'. *Anal. Chem.* 2011, 83, 5253–58. DOI: 10.1021/ac200589k.
[39] Winterbottom D.A., Narayanaswamy R. and Raimundo I.M., Jr., 'Cholesteric liquid crystals for detection of organic vapours'. *Sensors and Actuators B: Chemical* 90 (2003) 52–57. DOI: 10.1016/S0925-4005(03)00021-2.
[40] Mujahid A., Stathopulos H., Lieberzeit P.A. and Dickert F.L. 'Solvent vapour detection with cholesteric liquid crystals—Optical and mass-sensitive evaluation of the sensor mechanism'. *Sensors* 2010, 10, 4887–97. DOI: 10.3390/s100504887.
[41] Chang C.-K., Bastiaansen C.M.W., Broer D.J. and Kuo H.-L. 'Alcohol-responsive, hydrogen-bonded, cholesteric liquid-crystal networks'. *Adv. Funct. Mater.* 2012, 22, 2855–59. DOI: 10.1002/adfm.201200362
[42] Chang C.-K., Bastiaansen C.W.M., Broer D.J. and Kuo H.-L. 'Discrimination of alcohol molecules using hydrogen-bridged cholesteric polymer networks'. *Macromolecules* 2012, 45, 4550–55. DOI: 10.1021/ma3007152.
[43] Stumpel J.E., Wouters C., Herzer N., *et al.* 'An optical sensor for volatile amines based on an inkjet-printed, hydrogen-bonded, cholesteric liquid crystalline film'. *Adv. Optical Mater.* 2014, 2, 459–64. DOI: 10.1002/adom.201300516.
[44] Han Y., Pacheco K., Bastiaansen C.W.M., Broer D.J. and Sijbesma R.P. 'Optical monitoring of gases with cholesteric liquid crystals'. *J. Am. Chem. Soc.* 2010, 132, 2961–67. DOI: 10.1021/ja907826z.
[45] Chang C.-K., Chiu S.-W., Kuo H.-L. and Tang K.-T. 'Cholesteric liquid crystal-carbon nanotube hybrid architectures for gas detection'. *Applied Physics Letters* 100, 043501 (2012); DOI: 10.1063/1.3679680.
[46] Chang C.-K., Kuo H.-L., Tang K.-T. and Chiu S.-W. 'Optical detection of organic vapors using cholesteric liquid crystals'. *Applied Physics Letters* 99, 073504 (2011); DOI: 10.1063/1.3627162.
[47] Saha A., Tanaka Y., Han Y., Bastiaansen C.M.W., Broer D.J. and Sijbesma R.P. 'Irreversible visual sensing of humidity using a cholesteric liquid crystal'. *Chem. Commun.*, 2012, 48, 4579–81. DOI: 10.1039/c2cc16934g.
[48] Ishihara S. 'Chapter 9, Session 2 Development of Gas Sensors Using Cholesteric Liquid Crystals', in Biometric information sensing and its latest application for healthcare (Tokyo, Technical Information Institute Co., Ltd., 2017), pp. 375–90.

Chapter 20
Smart windows

Hiroshi Kakiuchida[1] *and Akifumi Ogiwara*[2]

Smart windows are a promising technology for use in energy-saving buildings and vehicles. Liquid crystals (LCs) and their composites, for which transmission, reflection and absorption vary in response to external stimuli, have great potential for applications in smart windows. Here, LC materials and their applications are described extensively from fundamental to state-of-the-art technologies. In this chapter, we mainly focus on thermoresponsive polymer network liquid crystals (PNLCs) and introduce other potential composites, classified by environmental stimulus (electric fields, heat, and light) and optical modulation (absorption, reflection, and scattering). Furthermore, the potentiality and possible issues that may arise for use of LC composites for smart windows are discussed.

20.1 Introduction

20.1.1 LCs and other materials for smart windows

Nowadays, LCs, with optical properties that change in response to various external stimuli, are being enthusiastically studied for applications, including not only displays and other information devices, but also smart windows for energy-saving buildings and vehicles. Here, 'smart windows' are often defined as windows which reversibly change their optical transmittance to remain comfort indoors, in response to environmental changes, as shown in Figure 20.1(a). Sunlight, of which intensity distributes in wavelengths ranging from ultraviolet, visible to near-infrared [1,2], as shown in Figure 20.1(b), is transmitted into rooms under control of smart windows. As tabulated in Table 20.1, several materials and devices as well as LCs have been proposed and some have already been commercially developed. Smart windows are classified by two criteria: optical properties of the device (output) and external stimuli to the device (input). The output includes light absorption, reflection (diffraction), and scattering, whereas the input includes electric fields, temperature, and light. Many LC-related materials have been reported for applications in smart windows. Some materials have been investigated for many external stimuli or optical properties,

[1]National Institute of Advanced Industrial Science and Technology, Japan
[2]Department of Electronic Engineering, Kobe City College of Technology, Japan

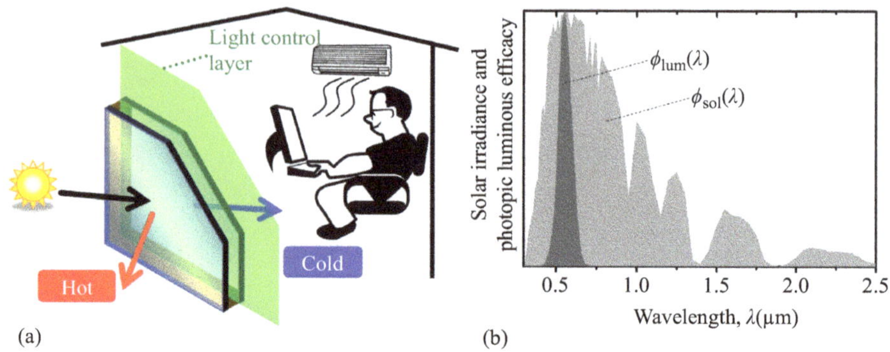

Figure 20.1 Concept of smart windows using reversible light attenuators: (a) light control layers installed or retrofitted on windowpanes autonomously control solar irradiance into rooms of buildings and vehicles, and keep indoors comfortable in all seasons and (b) the normalised spectra of solar irradiance, ϕ_{sol}, and photopic luminous efficacy, ϕ_{lum} [1,2]

whereas others are restricted to only one type of input or output. The wide applicability of LCs is owing to their molecular structure: a rigid anisotropic body, called a mesogen, with terminal groups at the ends of the body. This structure produces the characteristic features of LCs: optical birefringence, anisotropic electric polarisability, and a variety of interesting phases when LC molecules are aggregated.

Generally, an important function of smart windows is to change optical transmittance significantly in response to an external stimulus. Aggregated LC molecules transform their spatial distribution of orientation order in response to electric fields, heat, or light. This transformation produces a change of optical non-uniformity or refractive-index distribution that causes optical refraction and reflection. To have large non-uniformity with using LCs, composites of LCs with other materials have an advantage over simple LCs, because optical refraction and reflection efficiently occur at the boundary surfaces formed by LCs with other materials. This chapter is focused on composites of LCs and polymers, called polymer network LCs (PNLCs). The PNLCs developed for optical applications possess phase separation domains with sizes of an optical wavelength scale or sub- to ten-micron scale, and efficiently control light propagation directly through changes in the distribution of LC molecular orientation in response to external stimuli. Note that the term PNLC is used for composites of LCs and polymers in this chapter, although they are also called polymer-dispersed LCs (PDLCs), polymer-stabilised LCs (PSLCs), and bicontinuous LCs (BCLCs), depending on the compositional ratio of LCs to polymers and consequent phase separation structures. Furthermore, to control the light intensity in selective ranges of wavelengths, angles of incidence, or polarisation states, advanced PNLCs have been developed that have periodic phase separation structures composed of LCs and polymers, such as helical structures formed by cholesteric LCs (CLCs) stabilised with polymer

Table 20.1 The latest materials and devices with potential for smart windows

	Absorption		Reflection		Scattering	
	Materials	**Features**	**Materials**	**Features**	**Materials**	**Features**
Electric field	Nematic LC, RM, DD [3]	Bistability, absorption, haze, homeotropic/focal-conic, electric frequency	PSCLC, negative LC [6]	NIR selective reflection, wide tunability, DC field, negative LC	CLC, ionic liquid [11]	Bistability at zero field, electric-frequency transition
			PSCLC [7]	Band broadening, ionic charge trapping, non-uniform chiral pitch	Positive and negative PDLC [12]	Reverse mode, hybrid alignment, angle-dependent light control
	WO_3 [4]	Many applications for practical use	Mono-, di-acrylate RMs [8]	LC gel, helical pitch, photopolymerisation, switchable reflector	PDLC, polyurethane [13]	PU/PUR microcapsules, LC cores, dye, reduction of electric field
	SPD [5]	Absorption, haze, diffuse transmittance/reflectance	Micro blinds [9]	Magnetron sputtering, laser etching (MEMS), electrostatic force	PDLC, RM, ionic liquid [14]	Self-adjustment response against solar irradiation
			MgNi [10]	WO_3 ion storage, adhesive electrolyte		
Temperature	DD, LC [15]	GH architecture, FDTD calculations	Side-chain LC siloxane [19]	NIR selective reflection, cholesteric↔smectic transition, bar coating	PDLC, PSLC [22]	Hybrid structure for flexible and large plastic substrate
	VWO_2-SiO_2 [16]	Core-shell, nanorod, nanosphere, hydrothermal synthesis	PSCLC, chiral monomer [20]	Co-existing opposite handedness, surface tethered polymer network, tunability	SiO_2 core-shell [23]	Hollow particles, thermo-optic coefficient, organic solvent
	NiO [17]	Splay pyrolysis	PNLC, chiral dopant [21]	Selective reflection, chiral pitch length, chiral nematic acrylate	SiO_2 nano particle [24]	Nano composites, surface modification, silicone matrix
	CoCr [18]	Solar protection			Hydrogel, water [25]	Large size, practical use, cellulose derivative, hydrophobic bonding
Light	LC, DD, gold nano-particles [26]	Electric and optical tunability, LPR, GH, nanorod, nanoplate	CLC, bis(azo) chiral dopant [27]	Photoinduced tunable, VL to NIR range, high helical twisting power dopant	Azo-dye, PSCLC [28]	Azo-dye, UV (strong sunlight), electric fields, homeotropic/focal-conic/isotropic

DD: Dichroic dye, RM: Reactive mesogen, SPD: Suspended particle device, GH: Guest-host, LPR: Localised plasmonic resonance, VL: Visible, NIR: Near-infrared, UV: Ultraviolet, FDTD: Finite-difference time-domain networks, called polymer-stabilised cholesteric LCs (PSCLCs), and holographic structures composed of LC droplets surrounded by polymer networks, called holographic PDLCs (HPDLCs).

20.1.2 Electrically operated or thermoresponsive PNLC

Many LC composites have been studied for applications in electrically operated smart windows, because the optical states between transmittance and non-transmittance can switch quickly and at any time. Another reason that they have been studied is that they have been historically used for applications such as LC displays and other information devices. However, electrically operated windows require the installation of wiring in buildings and electrical power to switch the optical states, although the power is much smaller than that of heating and air conditioning. From the point of view of cost and efficiency, indium tin oxide (ITO) films used as transparent conductive electrodes can potentially increase the product cost and decrease solar irradiance in the infrared range. Recently, several groups have been working on the further reduction of power consumption for transparent/opaque switching. The magnitude of the applied field was lowered by forming an open structure, that is, bicontinuous phase separation of LC and polymer networks, since LC molecules in this open structure can rearrange with few constraints in response to applied fields [29]. Reverse-mode electrically operated PNLCs that are switchable between a transparent off-state and opaque on-state are expected to use less power than normal-mode operation types. Several groups have achieved reverse-mode electrical operation by forming various structures in PNLCs, such as the polymer-wall confinement of negative-dielectric LCs [30] and planar-vertical hybrid alignment of LCs stabilised by polymers [12]. Bistable types of electrically operated PNLCs, which retain both transparent and opaque states after turning off the electric fields, are effective in decreasing power consumption because power is not required to maintain the switched optical states [3,11,31,32]. For example, bistable behaviours were achieved between homeotropic transparent and focal-conic scattering states by employing an electric-hydrodynamic effect [3]. On the other hand, some groups have achieved multiple-stimuli responses with PNLC-related materials for higher-grade smart windows [28]. This multi-responsive light switchability was caused by the reversible phase transitions from a focal-conic state to a homeotropic state by electric fields or to an isotropic state by photo-isomerisation of azo dyes coexistent with LCs. Such reversible optical attenuation controlled by both electric fields and light irradiations can potentially allow for easy and flexible operation.

Thermoresponsive smart windows work autonomously in response to the temperature of ambient air or window surfaces, so that they can be readily installed and do not need ITO films or any complicated controllers. However, the thermoresponsive properties, including optical transmittance and switching temperature, are fixed as a product specification, and thus cannot be changed after installation. Although thermoresponsive PNLCs are not as widely used for smart windows as electrically operated types, they have a potential market because of the aforementioned advantages.

Several types of thermoresponsive PNLCs have been reported. H. Yang, *et al.* have achieved thermoresponsive switchability between SmA* transparent and chiral nematic scattering states [22]. Furthermore, to have practically flexible and large-scale films, they have recently reported a novel coexistent system of PSLC and PDLC structures. Therein, LC molecules were homeotropically stabilised by a polymer network surrounded by micron-scale polymer walls.

We have studied thermoresponsive PNLCs and their related materials. This chapter is focused on composites of LCs with polymers and their related materials for applications in thermoresponsive smart windows, including PNLCs for haze switchability, PSCLCs for wavelength-selective reflection, and HPDLCs for angular-selective diffraction.

20.2 Thermoresponsive PNLCs

20.2.1 Transparency/haze switchability

PNLCs are a promising candidate for thermoresponsive switchable windows, which transmit sunlight into rooms in low winter temperatures and block it in high summer temperatures to provide comfort in buildings and vehicles year-round, as shown in Figure 20.1(a). The LC molecules in the PNLCs transform their phase and change their orientation distribution in response to ambient temperature. This transition changes the relationship of the refractive indices between the LC and polymer phases. If the refractive indices of these two phases match, the structure in the PNLC is optically uniform and the incident light travels straight through the PNLCs without any refraction and reflection, which creates transparency. If the refractive indices are different between these phases, the incident light experiences a non-uniformity and consequently scatters, creating haze.

Nematic-to-isotropic (NI) phase transition is a fundamental property of LCs. We employed the change in refractive index due to NI phase transition to achieve thermoresponsive switchability between the transparent state at lower temperatures and opaque state at higher temperatures, which we call 'reverse-mode thermoresponse'. The structure formed in PNLCs requires three features to exhibit a reverse-mode thermoresponse: (i) switchability between optical uniformity at lower temperatures and non-uniformity at higher temperatures, (ii) a monodomain structure in optical anisotropy, and (iii) phase separation domains appropriately sized to scatter light efficiently.

Our thermoresponsive PNLCs possess meso- (submicron- to micron-) scale phase separation with orientation-ordered LCs and anisotropically polymerised reactive mesogens (RMs). Reversible thermoresponsive light attenuation was achieved by changing the deviation in refractive indices between the LC and polymer phases due to the NI phase transition. The molecular orientation of the LC phase is ordered to form the nematic phase below the temperature of NI phase transition, T_{NI}, whereas it is disordered to form the isotropic phase above T_{NI}. As shown in Figure 20.2(a), the aggregated LCs in the nematic state below T_{NI} show optical anisotropy or birefringence, where they have two different refractive indices

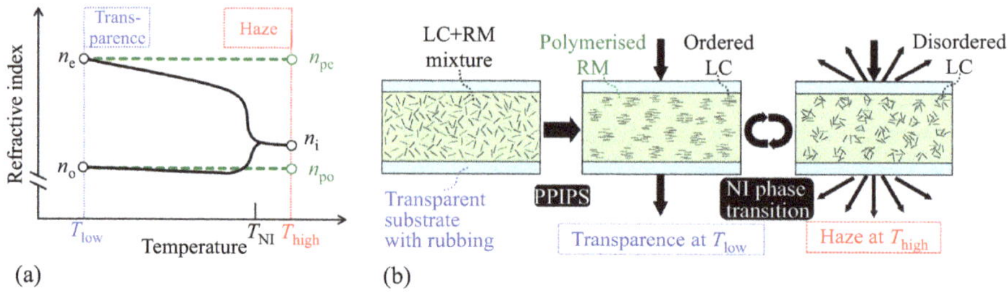

Figure 20.2 Reverse-mode thermoresponsive behaviour explained by (a) temperature dependence of the refractive index and (b) schemes of structural formation in PPIPS and thermoresponse due to NI phase transition. Solid curves in (a) show refractive indices of the LC phase in the nematic state for (n_o) ordinary and (n_e) extraordinary rays at temperatures below the NI phase transition temperature, T_{NI}, and in (n_i) the isotropic state at temperatures above T_{NI}. Dotted lines in (a) show the refractive indices (n_{po} and n_{pe}) of the anisotropic polymer phase

for ordinary and extraordinary rays, denoted by n_o and n_e, respectively. When the aggregated LCs are in the isotropic state at temperatures above T_{NI}, n_o and n_e converge to $n_i = (n_e + 2n_o)/3$. To produce a thermoresponsive change between a transparent state below T_{NI} and opaque state above T_{NI}, the polymer phase must be optically anisotropic and have refractive indices equal to n_o and n_e. The refractive indices of the polymer phase, which are denoted by n_{po} and n_{pe} for ordinary and extraordinary rays, respectively, are independent of temperature. By adjusting the refractive indices between the LC and polymer phases for two orthogonal polarisation azimuths, that is, $n_{po} = n_o$ and $n_{pe} = n_e$, the LC and polymer phases have optical uniformity, and consequently, the PNLC is transparent at a temperature below T_{NI}. This structure becomes optically non-uniform, and the PNLC changes to an opaque state when the refractive indices of the LC phase deviate from those of the polymer phase, that is, $n_{pe} \neq n_i$ and $n_{po} \neq n_i$, when the temperature rises above T_{NI}.

The use of a monodomain anisotropic structure is a convenient way to practically achieve reverse-mode thermoresponse, as schematically shown in Figure 20.2(b). LC and RM molecules are uniaxially oriented by a rubbing treatment on the substrate surfaces. We employed bifunctional main-chain RMs as a raw material, which have a molecular structure with a rigid anisotropic body like LCs and two functional groups at the ends of the body, for the anisotropic polymers formed in PNLCs. This structure has the advantage of allowing polymerisation while maintaining uniaxial order in its molecular orientation. The uniaxial orientation ordering of RMs was stabilised by a rubbing treatment on the substrate surfaces during and after polymerisation, and then an anisotropic polymer phase was formed with refractive indices that matched those of the LC phase in the nematic state.

Our PNLC structures were fabricated using a method based on photopolymerisation-induced phase separation (PPIPS). In PPIPS, a non-uniform distribution was formed, consisting of LC and polymer phase domains. As the photo-exposure proceeds, RM molecules in the mixture polymerise and extrude LC molecules from the polymerised domains. Consequently, the LC molecules locally aggregate to form LC phase domains. The size of LC phase domains depends not only on polymerisation temperature (a larger size at higher temperatures) but also on monomer functionality (a smaller size from lower functionality) [33–35]. Thus, the monomer functionality is an essential factor for controlling the size of LC domains. RMs possessing two functional groups at the ends of the body, as used in this study, can undergo stable polymerisation with uniaxial orientation ordering and form monodomain anisotropic structures. However, PPIPS produces LC domains with a ten-nanometre scale or smaller using this small functionality monomer [33], which is not an appropriate size to scatter visible and near-infrared light. In this study, we chose a non-uniform irradiation technique using a laser speckling method. Then, we fabricated reverse-mode thermoresponsive PNLCs that possessed appropriately sized phase separation domains composed of orientation-ordered LCs and anisotropically polymerised RMs [36].

The PNLCs were fabricated from the raw mixtures of LCs, RMs and a trace amount of photoinitiator and co-initiator. A nematic LC, 4-cyano-4’-hexylbiphenyl, called 6CB (Tokyo Chemical Industry Co., Ltd.), with physical properties $n_e = 1.692$, $n_o = 1.537$, and $T_{NI} = 29$ °C was used. The 6CB was mixed with a main-chain type of diacrylate RM: 1,4 bis-[4-(3-acryloyloxypropyloxy) benzoyloxy]-2-methylbenzene, commercially called RM257 (Merck Corp.), with physical properties $n_{pe} = 1.687$ and $n_{po} = 1.508$. A co-initiator, N-phenylglycine (NPG), and a photoinitiator, dibromofluorescein (DBF), (Tokyo Chemical Industry Co., Ltd.) were added to the mixtures. The mixture ratio of LC and RM was adjusted to 64 and 36 weight per cent, respectively. The added concentrations of DBF and NPG were both 0.1 weight per cent. The mixtures were stirred at a temperature of 60 °C until they became uniform transparent liquids, and then filled in the 30 μm clearance between a pair of $25\times20\times0.7$ mm transparent glass substrates. To have LC and RM molecules uniaxially oriented, the substrate surfaces were coated with polyimide and rubbing-treated along antiparallel directions beforehand. Next, the sample cells were mounted on a thermoregulator to maintain a curing temperature of 20 °C ($\approx T_{NI}-10$ °C). The cells were exposed for 5 min to a Nd:YVO_4 laser (J150GS, Showa Optronics Co., Ltd.), whose wavelength was 532 nm and intensity was 200 mW/cm^2 on the cells. The cells were irradiated with uniform intensity by direct irradiation or with non-uniform intensity through a light diffuser. The light diffuser (DFSQ1-50C02-1500, SIGMAKOKI Co., Ltd., Tokyo, Japan) was placed 20 mm in front of the cell to project laser speckle patterns onto the cells. The speckle patterns on the sample cell result from optical interference of the laser rays randomly refracted by the coarse surface of the light diffuser. After the exposure, possible residual monomers were completely polymerised by 5 min irradiation from an ultraviolet lamp with an intensity of 0.8 mW/cm^2.

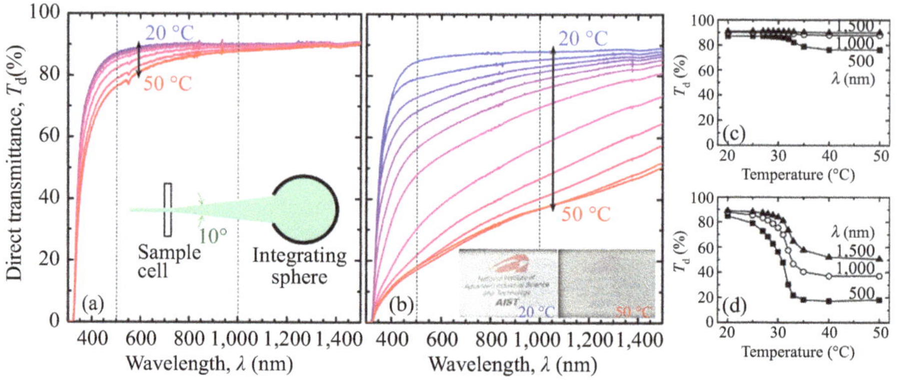

Figure 20.3 Direct transmittance spectra, T_d, measured at different temperatures between 20 °C and 50 °C for samples obtained through PPIPS by (a) uniform and (b) non-uniform irradiations. (c) and (d) show the temperature dependence of T_d at wavelengths of 500, 1,000 and 1,500 nm. The inset in (a) is a set-up to measure T_d. The inset photographs in (b) are the snapshots of sample cell taken at 20 °C and 50 °C

The direct transmittance, T_d, was measured at different temperatures between 20 °C and 50 °C, as a function of wavelength, shown in Figure 20.3, where T_d was measured by the detector within a divergence angle of 10°. Figure 20.3(a) shows the spectral T_d for the sample prepared by uniform irradiation and the temperature dependence is shown in Figure 20.3(c). T_d changes as a function of temperature, although the magnitude of the change is not large enough for smart windows. Figure 20.3(b) shows the spectral T_d for the sample prepared by non-uniform irradiation. T_d widely changes in response to temperature and exhibits values above 80 per cent at 20 °C and below 20 per cent at 50 °C around a 500 nm wavelength, as shown in Figure 20.3(d). The inset snapshots in Figure 20.3(b) show the letters and logo directly placed 20 mm behind the sample cell with transparency and haze.

These two samples were observed by scanning electron microscopy (SEM) and polarising optical microscopy (POM), as shown in Figure 20.4. According to the observation at 20 °C by POM with crossed nicols, as shown in III of Figures 20.4(a) and 20.4(b), the brightness appears throughout the area observed in both the samples prepared by non-uniform and uniform irradiations, which suggests the existence of monodomain optical anisotropy. As shown in the POM images observed at 50 °C in IV of Figures 20.4(a) and 20.4(b), the colour contrasts change, although complete darkness does not appear, when temperature rises. This indicates that LC molecular orientation transforms between uniaxially ordered and disordered states, whereas the polymer network composed of RMs remains uniaxially ordered along the rubbing directions. Phase separation domains with a few-micrometre size were observed in the case of non-uniform irradiation, as shown in IV of Figure 20.4(b),

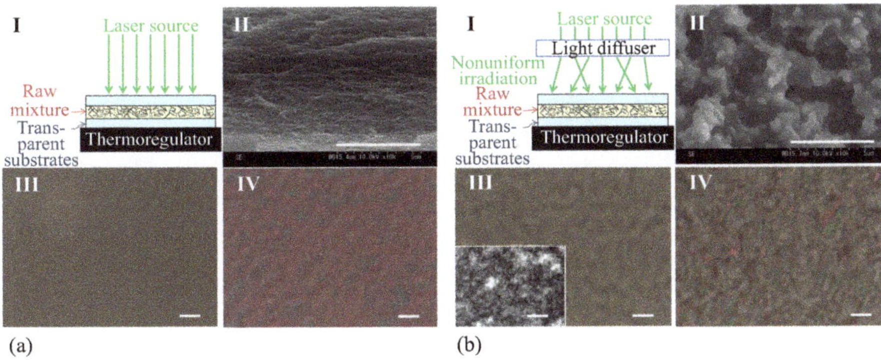

Figure 20.4 Laser irradiation schemes for PPIPS (I), and the samples' cross-sectional views by SEM (II) and top views by POM with crossed nicols (III and IV). POM images were taken at temperatures around (III) 20 °C and (IV) 50 °C. The samples were prepared by (a) uniform and (b) non-uniform irradiations. The inset image in III of (b) shows intensity distribution of the speckle patterns produced by non-uniform irradiation. The bar length in the photographs corresponds to 5 μm

whereas there are no obvious domains in the case of uniform irradiation, as shown in IV of Figure 20.4(a). Cross-sections observed by SEM clearly show the difference of phase separation morphology between these two samples, as shown in II of Figures 20.4(a) and 20.4(b), where the images are of the polymer network after rinsing the LC molecules away with methanol. According to the microscopic observation, the formation of small domains by uniform irradiation causes insufficient light scattering and consequently an insufficient magnitude of change in T_d.

20.2.2 Wavelength or angular selectivity

Smart windows ideally allow the control of view and lighting for users' comfort, although a large change in the magnitude of the transmittance and a simple fabrication process are indispensable for thermoresponsive PNLCs to be practical for wide use. Sunlight, which is intense in the ultraviolet, visible, and near-infrared ranges [1,2], as shown in Figure 20.1(b), has an intensity of approximately 1 kW/m^2 in Japan, for example. If smart windows change the near-infrared transmittance of sunlight but retain visible light transmittance, approximately half of the solar intensity passing through the windows into the room is controllable. Such wavelength-selective controls have been achieved by electrically tuning the transmittance wavelength range of PSCLCs [6–8]. In another case, light transmittance at particular angles of incidence can be controlled, while leaving other angles unaffected [12]. The wavelength- or angular-selective control allows the users to control the solar intensity and keep outdoor visibility. We describe here our

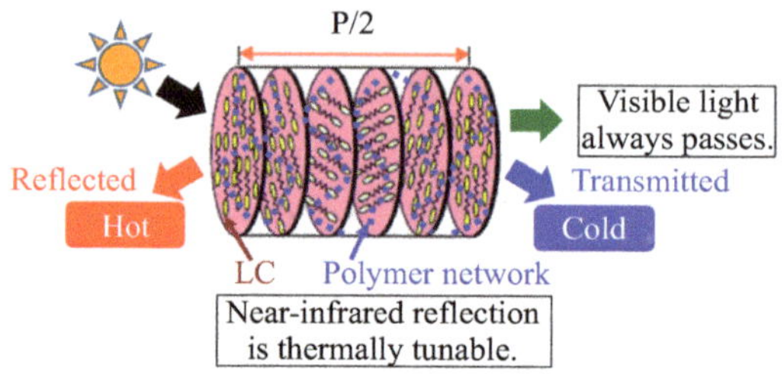

Figure 20.5 Conceptual illustration of thermoresponsive wavelength-selective PSCLCs for smart windows, that is, sunlight blocked in summer and passed in winter with retaining visibility

wavelength-selective PSCLCs and angular-selective HPDLCs, which are applicable in thermoresponsive smart windows.

Figure 20.5 shows our thermoresponsive wavelength-selective PSCLCs possessing CLCs with temperature-dependent chiral pitches. A nematic LC phase is transformed to chiral nematic phase, which possesses a helical structure, by containing chiral dopants. The pitch of the helical period is tunable, as a function of type and concentration of the chiral dopants [37,38]. The pitch length of the helix, corresponding to a molecular rotation, is determined by the macroscopic helical twisting power (HTP) based on the concentration and temperature dependence of the chiral dopant. Moreover, the bandwidth and magnitude of the selective reflection spectra are modulated by polymer stabilisation using different types and contents of diacrylate RMs in the PPIPS [7,39–42]. The wavelength and bandwidth of the selective reflection spectrum in PSCLCs is controlled by the chiral dopant molecules and polymer network structures formed by polymer stabilisation in the PPIPS process. We describe the effects of thermal modulation on the selective transmittance spectra obtained by superposing CLC cells [41]. Several types of CLC cells are formed by using chiral dopants with different magnitudes and signs of the HTP. The effect of the superposition of such different CLC cells on selective wavelength is investigated by the spectroscopic analysis under thermal modulation.

Figure 20.6 shows the transmittance spectra composed of CLCs with different temperature dependencies. The CLCs were prepared with combinations of various concentrations of chiral dopants by using nematic LCs (Merck E8) and various chiral dopants such as CB15 (Merck), CNL617L, and CNL611R (Daily Polymer Corp.), which possess different magnitudes and signs of the HTP. The signs of CB15 and CNL611R are defined as positive for right-handed helices, while the sign of CNL617L is defined as negative for left-handed helices. The chiral dopants also show different temperature dependencies. The selective transmittance spectrum of CB15 shifts to the longer wavelength range as temperature increases, while those of CNL617L and CNL611R shift to the shorter wavelength as temperature increases. PSCLCs were also fabricated using 15 weight per cent of the chiral dopant (CNL617L) and diacrylate RM (RM257, Merck) with concentrations ranging from

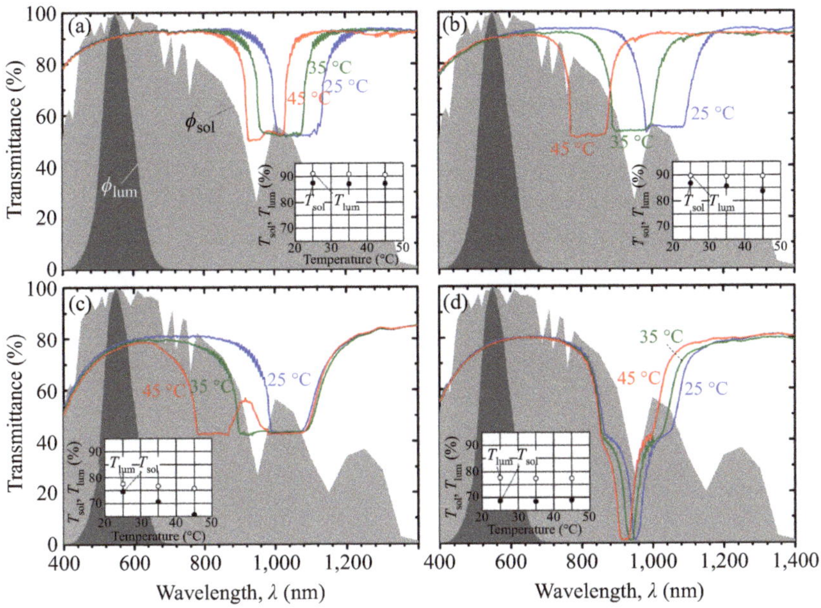

Figure 20.6 Transmittance spectra of CLCs and PSCLCs measured at different temperatures, where the preparation recipes of (a) to (d) are described in the text. Light and dark grey regions are the spectral solar irradiance, ϕ_{sol}, and the photopic luminous efficacy, ϕ_{lum}, which are the same as shown in Figure 20.1(b). Temperature dependencies of solar and luminous transmittances, T_{sol} and T_{lum}, estimated using (20.1) and (20.2) are shown in the insets

1 to 10 weight per cent based on the fraction of all ingredients, including nematic LC. NPG and DBF were added with both 0.1 weight per cent in content ratio. Then, the CLC composites were injected into the 25 μm air gap between a pair of 25×20×1 mm glass plates with polyimide layers for a rubbing treatment in anti-parallel directions. The PPIPS process for the PSCLCs was performed by UV irradiation for 3 min in the UV curing chamber (120 mW/cm^2 at 365 nm).

The CLCs in Figures 20.6(a) and 20.6(b) were composed of nematic LC (E8) and chiral dopants with various concentrations, such as (a) E8 (85 weight per cent) and CNL617L (15 weight per cent), and (b) E8 (44 weight per cent), CB15 (24 weight per cent), and CNL617L (32 weight per cent). The mixture ratios of chiral dopants were adjusted to control the temperature dependence of the selective transmittance spectra of the CLCs, using the previous experimental HTP results for different chiral dopants [41]. Figures 20.6(c) and 20.6(d) shows the transmittance spectra measured by superposing two different temperature-dependent cells. Figure 20.6(c) shows the result of the superposed cells: the one is a CLC cell prepared from the mixture of E8 (44 weight per cent), CB15 (24 weight per cent), and CNL617L (32 weight per cent) shown in Figure 20.6(b), and the other is a PSCLC cell prepared from the mixture of RM257 (3 weight per cent), E8 (82 weight per cent), and CNL617L (15 weight per cent). Figure 20.6(d) shows the result of the superposed cells containing different chiral dopants: the one was formed from the mixture of E8 (83 weight per cent) and a left-handed helical chiral dopant CNL617L (17 weight per cent), and the other was formed from the mixture of E8 (88 weight per cent) and a right-handed helical chiral dopant CNL611R (12 weight per cent). The results in Figure 20.6 show that the location and width of the reflection band can be widely modulated by the temperature. The spectral solar irradiance, $\phi_{sol}(\lambda)$, and the photopic luminous efficacy, $\phi_{lum}(\lambda)$, are shown as light and dark grey regions, respectively, in Figure 20.6, which are the same as in Figure 20.1(b). The solar and luminous transmittances, T_{sol} and T_{lum}, respectively, which are performance indicators for a solar-ray controller, are found by the following expressions [1,2]:

$$T_{sol} = \int_0^{\infty} \phi_{sol}(\lambda)T(\lambda)d\lambda / \int_0^{\infty} \phi_{sol}(\lambda)d\lambda \tag{20.1}$$

$$T_{lum} = \int_0^{\infty} \phi_{lum}(\lambda)T(\lambda)d\lambda / \int_0^{\infty} \phi_{lum}(\lambda)d\lambda \tag{20.2}$$

Note that $T(\lambda)$ denotes the transmittance as a function of wavelength. In this study, T_{sol} in (20.1) was calculated using $\phi_{sol}(\lambda)$ of the wavelength range from 300 to 2,500 nm, which approximately covers the distribution of solar-ray intensity. T_{lum} in (20.2) was calculated by $\phi_{lum}(\lambda)$ in the specific wavelength region corresponding to visible light (400 to 700 nm).

The insets of Figures 20.6(a) to 20.6(d) show the temperature dependence of the solar and luminous transmittances, T_{sol} and T_{lum}, calculated using (20.1) and (20.2) from the transmittance spectra, revealing the control effects of the near-infrared spectra in solar rays. T_{lum}, as shown in the insets of Figures 20.6(a) and 20.6(b), remains constant around 90 per cent despite temperature changes, while T_{sol}, as shown in the inset of Figure 20.6(b), decreases with an increase of temperature from 25 °C to 45 °C. The insets of Figures 20.6(c) and 20.6(d) show the temperature dependence of T_{sol} and T_{lum} for the superposed CLC cells that have different temperature-dependent reflection band. As the temperature changes from 25 °C to 45 °C, T_{lum} is larger than 75 per cent, while T_{sol} decreases by10 per cent (75 per cent to 65 per cent) in Figure 20.6(c). The difference between T_{lum} and T_{sol} is 10 per cent or more at a temperature of 45 °C. Figure 20.6(d) shows the temperature dependence of T_{sol} and T_{lum}, obtained from the narrow selective wavelength band, where the measurements were carried out for superposed CLC cells with different chiral helices. T_{lum} is approximately 10 per cent higher than T_{sol} in Figure 20.6(d). The superposition of CLC cells with different chiral helices demonstrates that the light filtering of solar transmittance in specific wavelength regions is tunable as a function of temperature. This thermally tunable filtering can be designed by the formation process of PSCLCs using chiral dopants and diacrylate RMs. Thus, the PSCLCs with thermal tunability in reflection bandwidth show potential for use as high transparent thermoresponsive smart windows.

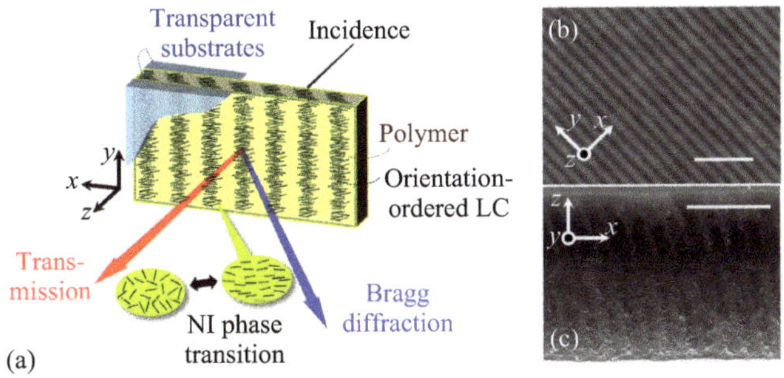

Figure 20.7 Scheme and micrographs of a slant-type Bragg-diffractive HPDLC including LC molecules orientation-ordered through a kind of self-assembly process in PPIPS: (a) diffraction/transmission states reversibly change as a function of temperature, through nematic-to-isotropic (NI) phase transition; (b) top view observed by POM with crossed nicols; and (c) cross-sectional view observed by SEM. The scale bars correspond to 5 μm

HPDLCs are a promising candidate for use in angular-selective thermo-responsive switchable windows, where 'angle' here means the angle of incidence to the windows. HPDLCs have been enthusiastically proposed for various optoelectronic devices [43]. The structure of HPDLCs possesses a meso-scale periodic composite of orientation-ordered LC molecules and polymer networks, as shown in Figure 20.7. They can thermoresponsively switch states between diffraction and transmission, which is caused by the refractive index change due to NI phase transition of the LC phase [44,45].

The HPDLCs were fabricated through PPIPS by holographic exposure from a raw mixture of an LC material, a pre-polymer blend, and photo-initiators. The LC material was 4-cyano-4′-pentylbiphenyl, called 5CB (K15, Merck Ltd.), with properties $n_e = 1.729$, $n_o = 1.533$, and $T_{NI} = 35$ °C [46]. The blend of commercial pre-polymers was as follows. A pre-polymer (AH600), which possesses urethane groups formed by reaction of 2-hydroxy 3-phenoxy propyl acrylate (refractive index: $n = 1.5258$) with hexamethylene diisocyanate ($n = 1.453$), was blended with the other pre-polymers, dimethylol tricyclodecane diacrylate (DCPA, $n = 1.5034$) and 2-hydroxyethyl methacrylate (HO, $n = 1.453$). All the pre-polymers were commercial products of Kyoeisha Chemical Co., Ltd. The blended proportions of AH600, DCPA and HO were 80, 13 and 7 weight per cent, respectively. The LC and the pre-polymer blend were mixed in a ratio of 45 and 55 weight per cent, respectively, together with additional 0.1 weight per cent co-initiator (NPG) and 0.1 weight per cent photoinitiator (DBF). This raw mixture was stirred at room temperature until it formed a transparent solution.

The raw mixture was injected to uniformly fill the 10 μm clearance between a pair of 25×20×1 mm transparent glass substrates. This sample cell was mounted on a holographic exposure setup with a thermoregulator to control the curing

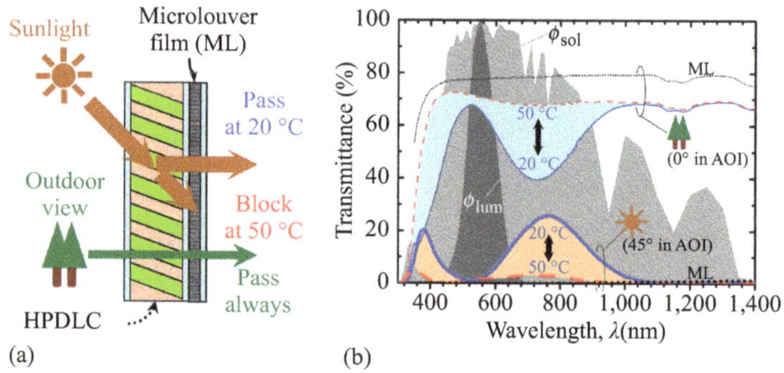

Figure 20.8 Transmittance spectra of the slanted HPDLC combined with a microlouver (ML) film. The transmittances were measured at 0° and 45° angles of incidence (AOI), to represent an outdoor view and sunlight, respectively. The measurement temperatures were 20 °C and 50 °C. Solar irradiance, ϕ_{sol}, and photopic luminous efficacy, ϕ_{lum}, after those in Figure 20.1(b) and (20.1) and (20.2)

temperature. The LC molecules in the mixture became uniaxially oriented self-organisationally during PPIPS by the holographic exposure. The light source for holographic exposure, a Nd:YVO_4 laser oscillated in single mode at a wavelength of 532 nm, was collimated, expanded, linearly polarised, and divided into two beams for interferential irradiation onto the sample cell. The sample cell was thermoregulated at 60 °C and irradiated by two beams superposed at 5° and 35° angles of incidence at 20 mW/cm^2 intensity. This holographic exposure produced a 13°-slanted periodic phase separation with a pitch of 1 μm in the sample cell, as shown in Figures 20.7(b) and 20.7(c). After 5 min of holographic exposure, the possible residual monomer was cured by uniform irradiation using an ultraviolet lamp with an intensity of 0.8 mW/cm^2.

To apply the HPDLC to practical sunlight control windows, the light must be appropriately guided after being switched between the transmittance and diffraction states. This HPDLC sample was combined with a commercial film whose transmittance depends on the angle of incidence (microlouver film, 3M), as schematically shown in Figure 20.8(a). The transparency of this microlouver film at 0° angle of incidence decreases with an increase in the angle and becomes nearly zero at angles larger than 35°. The microlouver film attached to the backside of the HPDLC efficiently controls the intensities of the transmitted and diffracted sunlight. The sunlight entering the windows is classified by diffuse and direct components. The former represents an outdoor view from a low altitude at around a 0° angle of incidence and is considered to have a low intensity, whereas the latter represents light entering a window from a high altitude with large intensity. As shown in Figure 20.8(b), the optical transmittances of the HPDLC attached to the microlouver (ML) film were measured at 20 °C and 50 °C. The transmittances at 0° and 45° angles of incidence, as shown by the solid and dashed curves, respectively, correspond to outdoor view and obliquely incident sunlight. The ML film used singly is transparent at a 0° angle of incidence (thin dotted curve), whereas the transmittance is nearly zero at a 45° angle of incidence (thick dotted curve). The thin solid and broken curves show the transmittances at 20 °C and 50 °C, respectively, at an angle of incidence of 0° to the HPDLC attached with the ML. The thermoresponsive change in the transmittance occurs mainly in the near-infrared range. The 0°-incident transmittance, which corresponds to an outdoor view, remains large in the visible range at any temperature, although the near-infrared component changes (solid curves). On the other hand, 45°-incident transmittance, which represents direct sunlight coming from a high altitude, reversibly decreases with a rise in temperature (thick solid and broken curves), particularly in near-infrared range. This angular-selective thermoresponse is a promising solution for achieving efficient control of near-infrared components in sunlight without losing the outdoor view.

20.3 Summary and perspectives

We described recent technology developments in smart windows and provided a detailed explanation of our work on haze switchable PNLCs, wavelength-selective reflection PSCLCs, and angular-incidence selective diffraction HPDLCs for applications in thermoresponsive smart windows. PNLCs and their related devices are mainly composed of only LCs and polymers with molecular-orientation ordering controllable by temperature. These simple structures and operation principles, which are less complicated than those of electrically operated smart windows, have advantages for both users and manufacturers.

To succeed commercially, smart windows must be able to meet fulfil requirements such as specific properties, cost, and durability, and they undoubtedly should be improved in switchable performance. The properties required for smart windows will vary depending on the location, lifestyle, individual preferences, safety, etc. A promising solution to meet such complicated demands is to make use of the multiple stimuli-responsiveness of LCs, that is, responsiveness to temperature, electrical fields, and light. On the other hand, issues of cost and durability are unavoidable for smart windows to be put to practical use. Unlike LC displays, direct sunlight, intense heat, and severe cold may degrade the switchable performance of the windows. Recently, since fundamental performance has improved, smart window durability has attracted attention. There are research groups examining the long-term performance of switchable PNLCs exposed to harsh environments, and some of the groups have proposed ways to improve durability [47–50].

Smart windows are a promising candidate for a new energy-saving tool to be used in buildings and vehicles, and have potential to become a major industrial market. The technology and knowledge derived from the historical success of LC displays can be applied to accelerate the development of LC smart windows.

References

[1] ISO 9845-1: 'Solar energy—reference solar spectral irradiance at the ground at different receiving conditions', 1992, pp. 1–14.

[2] 'Standard practice for calculation of photometric transmittance and reflectance of materials to solar radiation', *Annual Book of ASTM Standards*, vol. **12.02**, Easton, 1987, E971-83, pp. 829–837.

[3] Kim J.-H., Huh J.-W., Oh. S.-W., *et al.* 'Bistable switching between homeotropic and focal-conic states in an ion-doped chiral nematic liquid crystal cell'. *Opt. Express* 2017, **25**, pp. 29180–88.

[4] Granqvist C. G. 'Electrochromics for smart windows: Oxide-based thin films and devices'. *Thin Solid Films* 2014, **564**, pp. 1–38.

[5] Barrios D., Vergaz R, Sanchez-Pena J. M., Granqvist C. G. and Niklasson G. A. 'Toward a quantitative model for suspended particle devices: Optical scattering and absorption coefficients'. *Sol. Energy Mater. Sol. Cells* 2013, **111**, pp. 115–22.

[6] Khandelwal H., Debije M. G., White T. J. and Schenning A. P. H. J. 'Electrically tunable infrared reflector with adjustable bandwidth broadening up to 1100 nm'. *J. Mater. Chem. A* 2016, **4**, pp. 6064–69.

[7] Tondiglia V. P., Natarajan L. V., Bailey C. A., *et al.* 'Bandwidth broadening induced by ionic interactions in polymer stabilized cholesteric liquid crystals'. *Opt. Mater. Express* 2014, **4**, pp. 1465–72.

[8] Hikmet R. A. M. and Kemperman H. 'Electrically switchable mirrors and optical components made from liquid-crystal gels'. *Nature* 1998, **392**, pp. 476–79.

[9] Lamontagne B., Barrios P., Py C. and Nikumb S. 'The next generation of switchable glass: the micro-blinds'. *Proceedings of the Conference Glass Performance Days*; Tampere, Finland, Jun 2009. Tampere: Glass Performance Days; 2009, pp. 637–39.

[10] Tajima K., Hotta H., Yamada Y., Okada M. and Yoshimura K. 'Electrochromic switchable mirror glass fabricated using adhesive electrolyte layer' *Appl. Phys. Lett.* 2012, **101**, pp. 251907–1–3.

[11] Lan Z., Li Y., Dai H. and Luo D. 'Bistable smart window based on ionic liquid doped cholesteric liquid crystal'. *IEEE Photon. J.* 2017, **9**, pp. 2200307–1–7.

[12] Yamaguchi R., Inoue K., Oikawa Y. and Takasu T. 'Electro-optical property in hybrid aligned reverse mode cell using liquid crystals with positive and negative dielectric constant anisotropies'. *J. Photopolym. Sci. Technol.* 2017, **30**, pp. 463–66.

[13] Kim M., Park K. J., Seok S., *et al.* 'Fabrication of microcapsules for dye-doped polymer-dispersed liquid crystal-based smart windows'. *ACS Appl. Mater. Interfaces* 2015, **7**, pp. 17904–90.

[14] Cupelli D., Nicoletta F. P., Manfredi S., *et al.* 'Self-adjusting smart windows based on polymer-dispersed liquid crystals'. *Sol. Energy Mater. Sol. Cells* 2009, **93**, pp. 2008–12.

[15] Powers R. M. and McCarthy W. 'Thermally switched optical filter incorporating a guest-host architecture'. *US Pat.* 2013-0033738.

[16] Zhou Y., Ji S., Li Y., Gao Y., Luo H. and Jin P. 'Microemulsion-based synthesis of $V_{1-x}W_xO_2$@SiO_2 core–shell structures for smart window applications'. *J. Mater. Chem. C*, 2014, **2**, pp. 3812–19.

[17] Kamal H., Elmaghraby E. K., Ali S. A. and Abdel-Hady K. 'The electrochromic behavior of nickel oxide films sprayed at different preparative conditions'. *Thin Solid Films*, 2005, **483**, pp. 330–39.

[18] Gadžurić S., Vraneš M. and Dožic S. 'Thermochromic cobalt(II) chlorocomplexes in different media: Possible application for auto-regulated solar protection'. *Sol. Energy Mater. Sol. Cells*, 2012, **105**, pp. 309–16.

[19] Zhang W., Kragt S., Schenning A. P. H. J., de Haan L. T. and Zhou G. 'Easily processable temperature-responsive infrared-reflective polymer coatings'. *ACS Omega*, 2017, **2**, pp. 3475–82.

[20] McConney M. E., Tondiglia V. P., Hurtubise J. M., Natarajan L. V., White T. J. and Bunning T. J. 'Thermally induced, multicolored hyper-reflective cholesteric liquid crystals'. *Adv. Mater.*, 2011, **23**, pp. 1453–57.

[21] Yang H., Mishima K., Matsuyama K., Hayashi K., Kikuchi H. and Kajiyama T. 'Thermally bandwidth-controllable reflective polarizers from (polymer network/liquid crystal/chiral dopant) composites'. *App. Phys. Lett.*, 2003, **82**, pp. 2407–09.

[22] Guo S.-M., Liang X., Zhang C.-H., *et al.* 'Preparation of a thermally light-transmittance-controllable film from a coexistent system of polymer-dispersed and polymer-stabilized liquid crystals'. *ACS Appl. Mater. Interfaces*, 2017, **9**, pp. 2942–47.

[23] Fujiwara M., Shiokawa K., Monobe H. and Shimizu Y. 'Temperature control of light transmission using mixed system of silica hollow particles with nanoparticle shell and organic components'. *ACS Appl. Mater. Interfaces*, 2015, **7**, pp. 1107–13.

[24] Cheng D. F. and Hozumi A. 'Highly loaded silicone nanocomposite exhibiting quick thermoresponsive optical behavior'. *ACS Appl. Mater. Interfaces*, 2011, **3**, pp. 2219–23.

[25] Watanabe H. 'Intelligent window using a hydrogel layer for energy efficiency'. *Sol. Energy Mater. Sol. Cells*, 1998, **54**, pp. 203–11.

[26] Liu Q., Yuan Y. and Smalyukh I. I. 'Electrically and optically tunable plasmonic guest−host liquid crystals with long-range ordered nanoparticles'. *Nano Lett.*, 2014, **14**, pp. 4071–77.

[27] White T. J., Freer A. S., Tabiryan N. V. and Bunning T. J. 'Photoinduced broadening of cholesteric liquid crystal reflectors'. *J. Appl. Phys.*, 2010, **107**, pp. 073110–1–6.

[28] Oh S.-W., Baek J.-M., Kim S.-H. and Yoon T.-H. 'Optical and electrical switching of cholesteric liquid crystals containing azo dye'. *RSC Adv.*, 2017, **7**, pp. 19497–501.

[29] Ouskova E., Sio L. De, Vergara R., White T. J., Tabiryan N. and Bunning T. J. 'Ultra-fast solid state electro-optical modulator based on liquid crystal

polymer and liquid crystal composites'. *Appl. Phys. Lett.*, 2014, **105**, pp. 231122–1–4.

[30] Wang H., Gong H., Song P., *et al.* 'Reverse-mode polymer dispersed liquid crystal films prepared by patterned polymer walls'. *Liq. Cryst.*, 2015, **42**, pp. 1320–28.

[31] Bao R., Liu C.-M. and Yang D.-K. 'Smart bistable polymer stabilized cholesteric texture light shutter'. *Appl. Phys. Express*, 2009, **2**, pp. 110421–1–3.

[32] Lee K. M., Tondiglia V. P. and White T. J. 'Bistable switching of polymer stabilized cholesteric liquid crystals between transparent and scattering modes'. *MRS Commun.*, 2015, **5**, pp. 223–27.

[33] Pogue R. T., Natarajan L. V., Siwecki S. A., Tondiglia V. P., Sutherland R. L. and Bunning T. J. 'Monomer functionality effects in the anisotropic phase separation of liquid crystals'. *Polymer*, 2000, **41**, pp. 733–41.

[34] White T. J., Natarajan L. V., Bunning T. J. and Guymon C. A. 'Contribution of monomer functionality and additives to polymerization kinetics and liquid crystal phase separation in acrylate-based polymer-dispersed liquid crystals (PDLCs)'. *Liq. Cryst.*, 2007, **34**, pp. 1377–85.

[35] Kakiuchida H., Ogiwara A. and Matsuyama A. 'Multiple Bragg diffractions with different wavelengths and polarizations composed of liquid crystal/polymer periodic phases'. *ACS Omega*, 2017, **2**, pp. 6081–90.

[36] Kakiuchida H. and Ogiwara A. 'Reverse-mode thermoresponsive light attenuators produced by optical anisotropic composites of nematic liquid crystals and reactive mesogens'. *Opt. Mater.*, 2018, **78**, pp. 273–78.

[37] Kang S. W., Sprunt S. and Chien L. C. 'Structure and morphology of polymer-stabilized cholesteric diffraction gratings'. *Appl. Phys. Lett.*, 2000, **76**, pp. 3516–18.

[38] Wilson M. R. and Earl D. J. 'Calculating the helical twisting power of chiral dopants'. *J. Mater. Chem.*, 2001, **11**, pp. 2672–77.

[39] Guo R., Li K., Cao H., *et al.* 'Chiral polymer networks with a broad reflection band achieved with varying temperature'. *Polymer*, 2010, **51**, pp. 5990–96.

[40] Gao Y., Yao W., Sun J., *et al.* 'A novel soft matter composite material for energy saving smart windows: from preparation to device application'. *J. Mater. Chem. A*, 2015, **3**, pp. 10738–46.

[41] Ogiwara A., and Kakiuchida H. 'Thermally tunable light filter composed of cholesteric liquid crystals with different temperature dependence'. *Sol. Energy Mater. Sol. Cells*, 2016, **157**, pp. 250–58.

[42] Khandelwal H., Schenning A. P. H. J. and Debije M. G. 'Infrared Regulating Smart Window Based on Organic Materials'. *Adv. Energy Mater.*, 2017, **7**, pp. 1602209–1–18.

[43] Peng H., Ni M., Bi S., Liao Y. and Xie X. 'Highly diffractive, reversibly fast responsive gratings formulated through holography'. *RSC Adv.*, 2014, **4**, pp. 4420–26.

[44] Kakiuchida H., Tazawa M., Yoshimura K. and Ogiwara A. 'Thermal control of transmittance/diffraction states of holographic structures composed of

polymer and liquid crystal phases'. *Sol. Energy Mater. Sol. Cells*, 2010, **94**, pp. 1747–52.

[45] Ogiwara A., Kakiuchida H., Yoshimura K., Tazawa M., Emoto A. and Ono H. 'Effects of thermal modulation on diffraction in liquid crystal composite gratings'. *Appl. Opt.*, 2010, **49**, pp. 4633–40.

[46] Chirtoc I., Chirtoc M., Glorieux C. and Thoen J. 'Determination of the order parameter and its critical exponent for *n*CB ($n = 5-8$) liquid crystals from refractive index data'. *Liq. Cryst.*, 2004, **31**, pp. 229–40.

[47] Park S. and J. Hong W. 'Polymer dispersed liquid crystal film for variable-transparency glazing'. *Thin Solid Films*, 2009, **517**, pp. 3183–86.

[48] Kotsopoulos S. D., Graybill W. and Casalegno F. 'Designing a connected sustainable living environment'. *Int. J. Architect. Comput.* 2014, **2**, pp. 183–204.

[49] Lee J. W., Kim J. K., Ahmada F., Jamil M. and Jeon Y. J. 'Properties of thiol-vinyl PDLC films without additional photoinitiator'. *Liq. Cryst.*, 2014, **41**, pp. 1109–15.

[50] Ghosh A. and Mallick T. K. 'Evaluation of optical properties and protection factors of a PDLC switchable glazing for low energy building integration'. *Sol. Energy Mater. Sol. Cells*, 2018, **176**, pp. 391–96.

Chapter 21
Antenna technology for satellite communications

Hiroshi Moritake[1]

Liquid crystals (LCs) have a large anisotropy of refractive index in the visible light region and the microwave region. With the increasing use of mobile communication systems, electrically controllable phase shifters that use LCs for the dielectric layer have gained considerable attention. In this chapter, we first introduce the fundamental characteristics of LCs in the microwave region. In this study, we describe a complex dielectric constant in the microwave region. Next, we describe the microwave phase shifter loaded with LCs. Furthermore, we explain the new antenna technology for satellite communications, which uses LCs as a dielectric material of the variable capacitor for microwave metamaterials.

21.1 Introduction

With the phenomenal increase in the use of mobile phones, communications satellites are being used for mobile communications besides mobile network. The frequency range from 3 to 30 GHz is allotted for satellite communications, and the frequency of approximately 12 GHz is mainly used by a broadcast satellite. A parabolic satellite antenna is normally used for receiving electromagnetic waves from the broadcast satellite functions. The parabolic antenna has a parabolic reflector. To receive electromagnetic waves, it is necessary to orient the reflector exactly in the direction of the satellite. Therefore, the parabolic antenna is mainly used at stationary locations such as the roofs of buildings or houses. In other words, it is difficult to use the parabolic antenna from the roof of a car or from any other moving location. A phased array antenna can be used for satellite communications. The phased array antenna, which is an array of antennas with phase shifters, can be electrically steered to point in different directions without moving the antennas. Several types of electrically controllable phase shifters have been proposed; phase shifters that use an LC material as a dielectric layer has received considerable attention [1–13].

[1]National Defense Academy, Japan

LCs have large anisotropy of refractive index not only in the visible light region but also in the microwave and milliwave regions. In this chapter, the fundamental characteristics of LCs in the microwave region are introduced first; subsequently, the LC-based microwave phase shifter is discussed. Furthermore, a new antenna technology using LC for satellite communications is explained.

21.2 Fundamental characteristics of LCs in the microwave region

LCs have large anisotropy of dielectric constant between the parallel and perpendicular to the director. The dielectric constants are also frequency dependent. In the low-frequency region, the dielectric constants are large values because of the rotation of the dipole moments along the molecular long and short axes, as shown in Figure 21.1(a) and (b), respectively. The relaxation frequency of the dielectric constant parallel to the director (denoted by $\varepsilon_{//}$) is lower than that of the dielectric constant perpendicular to the director (denoted by $\varepsilon_{\perp}$). Figure 21.2 shows the frequency dependencies of the measured complex dielectric constants in a nematic LC [4]. In this measurement, the dielectric constant in the lower frequency part

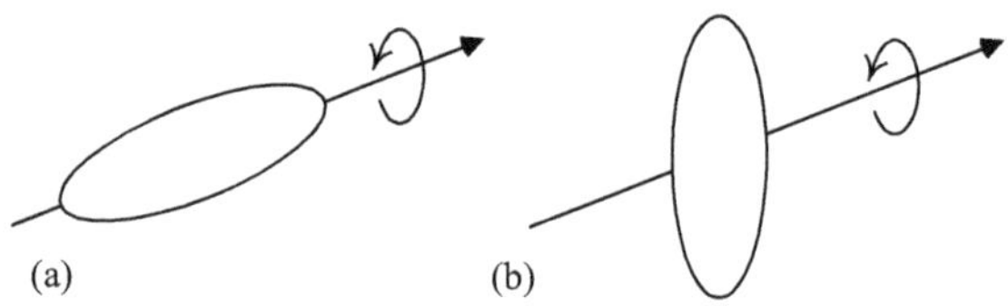

Figure 21.1 Schematics of rotation of dipole moments along (a) molecular long axis and (b) molecular short axis

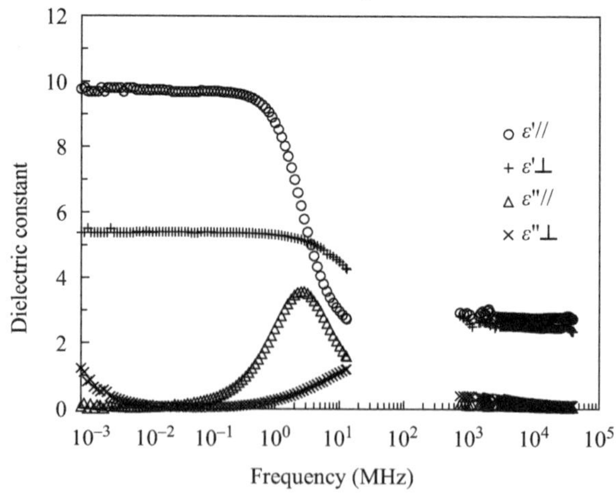

Figure 21.2 Frequency dependence of measured complex dielectric constant

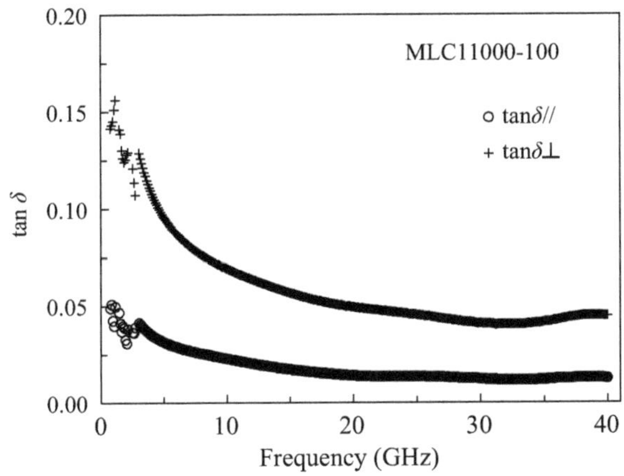

Figure 21.3 Structure of the MSL loaded with NLC

(<13 MHz) was measured by using an impedance analyser. However, the dielectric constant in the high frequency region (>700 MHz) was measured by the cutback method, in which two coaxial tubes with different lengths were filled with the LC. Then, the dielectric properties were measured on the basis of their respective group delay and transmission losses by using a network analyser. The relaxation frequencies $\varepsilon_{//}$ and $\varepsilon_{\perp}$ were approximately 1 and 100 MHz, respectively. The relaxation frequency $\varepsilon_{\perp}$ was approximately a hundred times higher than that of $\varepsilon_{//}$.

In the microwave application, the parameter tanδ is very important because of the transmission loss; this parameter is calculated by diving the imaginary part of the dielectric constant by the real part of the constant. The typical tanδ in the dielectric substrate for a high-frequency device is below 0.01. Figure 21.3 shows the frequency dependence of tanδ in the high-frequency region (>700 MHz). Form this figure, we can see that tanδ decreases with increasing frequency and the tanδ parallel to the director ($\varepsilon_{//}$) is smaller than that perpendicular to the director ($\varepsilon_{\perp}$). This difference was caused by the difference between the relaxation frequencies of $\varepsilon_{//}$ and $\varepsilon_{\perp}$. This result shows that when the LC material is selected, it is important to maintain a low relaxation frequency for the molecular rotation, especially $\varepsilon_{\perp}$.

21.3 Microwave phase shifter that uses LCs

21.3.1 Structure of the microstrip line loaded with LC

A microstrip line (MSL) is one of the most popular transmission lines in the microwave region. Figure 21.4(a) shows a schematic structure of MSL. The MSL structure consists of a central conductor (that acts as a microstrip transmission line), a dielectric layer, and a ground plane. In the MSL structure, the electromagnetic wave propagating in the MSL exists mainly in the dielectric substrate. However, the MSL is filled with nematic LCs, as shows in Figure 21.4(b). It is composed of

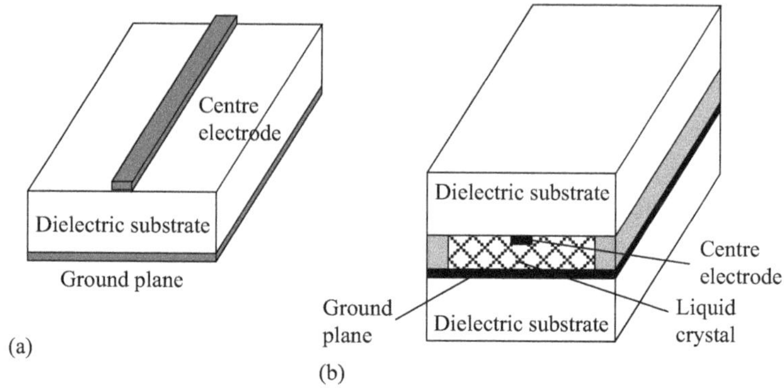

Figure 21.4 Schematic structures of the MSL and MSL phase shifter loaded nematic liquid crystal

three layers: a lower dielectric substrate layer with a grounded conductor, a nematic LC layer, and an upper dielectric substrate layer having a central conductor of the MSL. The upper dielectric substrate is set parallel to the lower dielectric substrate via two spacers along two sides of the transmission line. The LC layer is sandwiched between two dielectric substrate layers and acts as a dielectric layer in an ordinary MSL structure.

To assign the initial alignment of LC molecules, the surfaces of both the dielectric substrates, that contact the LC layer, were coated with a polyimide alignment layer and rubbed for unidirectional alignment. In this case, there are two possible rubbing directions; one direction is parallel and the other is perpendicular to the propagation direction of the microwave in the MSL. To control the director orientation of the LC molecules, a low frequency bias voltage of 1 kHz was applied between the central conductor and the grounded plane of the MSL. Figure 21.5 shows the orientation distributions of LC molecules in a cross section of the phase shifter in the 'ON' and 'OFF' states for both (a) parallel and (b) perpendicular rubbing directions. The OFF state is when no bias voltage is applied; in this state, the LC molecules remain in the initial distribution because of the anchoring force from the aligned layers. The ON state is when a bias voltage is applied; in this state, the LC molecules reorient along the direction of the electric field.

In the OFF state, the director under the condition shown in Figure 21.5(a) was exactly perpendicular and under the conditions shown in Figure 21.5(b), the director was almost perpendicular to the direction of the microwave electric field in the MSL structure. Thus, the microwave electric field was subjected to $\varepsilon_{\perp}$. However, in the ON state, the LC molecules were reoriented in the direction of the bias electric field and the microwave electric field was subjected to $\varepsilon_{//}$. This change in the dielectric constant of the LC molecules caused a change in the effective dielectric constant of the MSL structure; this caused a phase shift in the microwaves. The change from the OFF state to the ON state is called the 'rise' and the reverse process is called the 'decay'.

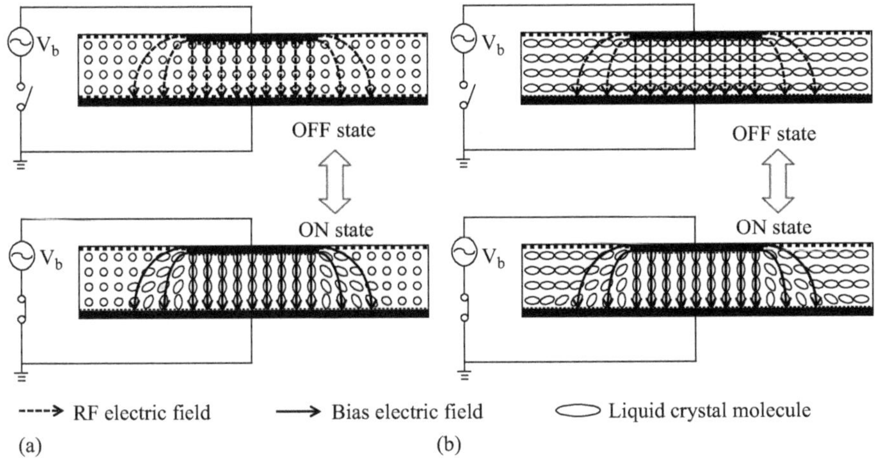

Figure 21.5 Orientation distributions of liquid crystal molecules in a cross section of phase shifter in on and off states for both (a) parallel and (b) perpendicular rubbing directions

21.3.2 Calculation of the effective dielectric constant and phase shift

In the MSL loaded with nematic LCs, we can assume that a microwave only propagates in the transverse electromagnetic wave mode, which means that the microwave electric field denoted by $\boldsymbol{E}_\mathrm{h}$ is perpendicular to the direction of propagation. When the microwave propagates along the z-axis, there is no component parallel to the z-axis of $\boldsymbol{E}_\mathrm{h}$, and $\boldsymbol{E}_h = \boldsymbol{E}_h(x,y)$. The orientation distribution of the LC molecules under the application of the bias electric field can be calculated by solving the elastic and electric free energy equation and the director momentum conservation law [14–17]. By using the orientation distribution of the LC molecules $\boldsymbol{n} = \boldsymbol{n}(x,y,t)$ and the distribution of the microwave electric field $\boldsymbol{E}_h = \boldsymbol{E}_h(x,y)$, the dielectric constant distribution of the LC layer and the effective dielectric constant of the MSL structure were calculated.

The dielectric constant distribution $\varepsilon_\mathrm{lc}(x,y,t)$ of the LC layer in the microwave electric field $\boldsymbol{E}_h(x,y)$ of the MSL is calculated as

$$\varepsilon_{lc}(x,y,t) = \frac{\varepsilon_{lc\perp}\boldsymbol{E}_h{}^2(x,y) + \Delta\varepsilon_{lc}\{\boldsymbol{E}_h(x,y)\cdot\boldsymbol{n}(x,y,t)\}}{\boldsymbol{E}_h{}^2(x,y)}, \tag{21.1}$$

where $\Delta\varepsilon_\mathrm{lc} = \varepsilon_{\mathrm{lc}//} - \varepsilon_\mathrm{lc}$ is the dielectric constant anisotropy of the NLC at high frequencies. The effective dielectric constant $\varepsilon_\mathrm{eff}(t)$ of the MSL structure is calculated from the dielectric constant distribution $\varepsilon_\mathrm{lc}(x,y,t)$ of the LC layer and the distribution of the microwave electric field $\boldsymbol{E}_h(x,y)$ as follows:

$$\varepsilon_\mathrm{eff}(t) = \frac{\int_{S_1}\boldsymbol{E}_h{}^2(x,y)dS + \varepsilon_r\int_{S_2}\boldsymbol{E}_h{}^2(x,y)dS + \int_{S_3}\varepsilon_{lc}(x,y,t)\boldsymbol{E}_h{}^2(x,y)dS}{\int_{S_1+S_2+S_3}\boldsymbol{E}_h{}^2(x,y)dS}. \tag{21.2}$$

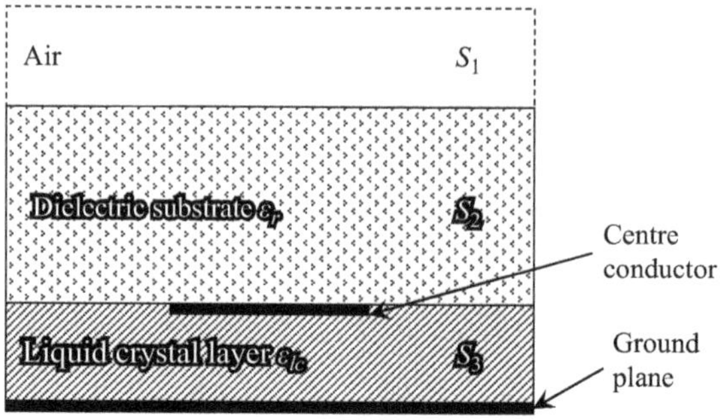

Figure 21.6 Layer structure of phase shifter loaded with LC for calculation

Here, S_3 and S_2 are the areas of the LC and dielectric substrate layers, respectively, as shown in Figure 21.6. S_1 is the area of the air layer on the top of the dielectric substrate layer (as shown in Figure 21.6), where the high-frequency electric field energy is assumed to exist.

The phase shift $\Delta\Phi(t)$ of a microwave signal with the frequency f is calculated by using the effective dielectric constant of the MSL $\varepsilon_{\text{eff}}(t)$ as follows:

$$\Delta\Phi(t) = \Phi(t) - \Phi(0) = \frac{2\pi fl}{c}\left(\sqrt{\varepsilon_{\text{eff}}(t)} - \sqrt{\varepsilon_{\text{eff}}(0)}\right). \tag{21.3}$$

Here, l is the length of the MSL loaded with an NLC. In this study, the response times, including the rise time and the decay time, of the MSL loaded with an NLC are calculated from $\Delta\Phi(t)$. In the rise process, the rise time is defined as the time required for $\Delta\Phi(t)$ to reach 90 per cent of its saturation value in the ON state after applying the bias voltage. In the same way, in the decay process, the decay time is defined as the time required for $\Delta\Phi(t)$ to reach 90 per cent of its saturation value in the OFF state after removing the bias voltage.

21.3.3 Response characteristics of phase shifter

Figure 21.7 shows the bias voltage dependencies of the measured phase shift of the microwave phase shifters for both the parallel and perpendicular rubbing direction. In this measurement, the frequency of the microwave is 20 GHz. The phase shift in the parallel rubbing device is 23° when 100 V is applied. However, the phase shift in the perpendicular rubbing device is 20° under the same voltage. This difference is caused by the initial alignment of the LC molecules. As mentioned in Section 21.3.2, the initial orientation of the LC molecules is exactly perpendicular to the electric field, as shown in the OFF state of Figure 21.5(a). However, the initial orientation of the LC molecules is almost but not exactly perpendicular to the electric field, as shown in the OFF state of Figure 21.5(b). Consequently, the $\Delta\varepsilon_{\text{eff}}$ in the parallel rubbing device is slightly higher than that in the perpendicular direction. Therefore, the phase shift in the parallel rubbing device is larger than that in the perpendicular device, as shown in Figure 21.7

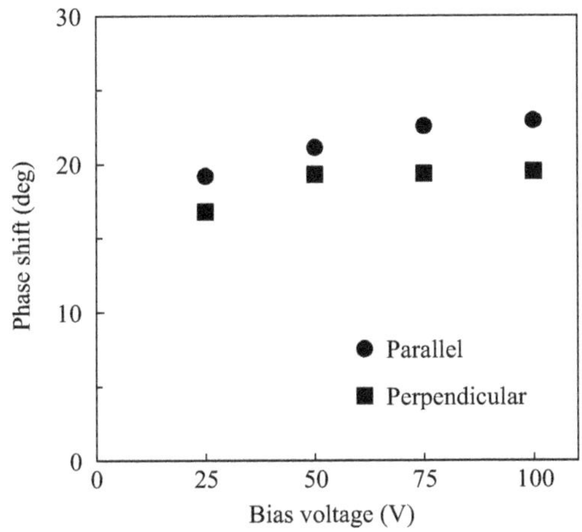

Figure 21.7 Experimental results for the bias voltage dependences of phase shift at frequency of 20 GHz for both the parallel and perpendicular rubbing directions

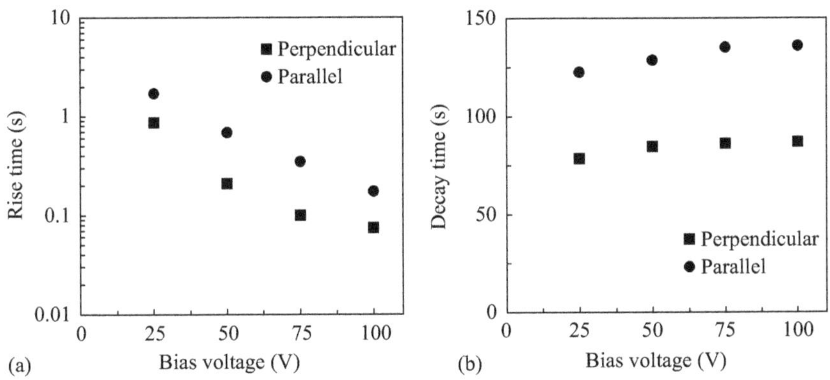

Figure 21.8 Experimental results for the bias voltage dependences of the (a) rise time and (b) decay time for both the parallel and perpendicular rubbing directions

The bias voltage dependencies of the rise and decay times are shown in Figure 21.8. In the both parallel and perpendicular directions, the rise time decreases with increasing bias voltage, and the decay time was almost independent of the bias voltage except when a low voltage was applied. These response time results indicate the same results in the electro-optic effect. The rise and decay times in the parallel rubbing device are longer than those in the perpendicular device. This difference was caused by the differences among the motions of the LC molecules; these results were confirmed by the simulation results.

The parallel and perpendicular devices showed long decay times of 136 and 87 s, respectively. The long decay time is caused by the thick LC layer of 100 μm. The long decay time is disadvantageous in practical applications. To reduce the decay time of the LCs in these devices, polymer-stabilised LCs have been employed. Polymer-stabilised LCs can be obtained by doping a polymerisable monomer and a photoinitiator into an LC and photopolymerising the mixture in the LC phase to form a polymer network. In the polymer-LC composite system, the decay time is shortened because the thickness of the LC was effectively reduced. However, many studies on the polymer stabilised LCs reported an increase in the driving voltage because of the strong elastic interactions between the LCs and the polymer network. Using this technique, a decay below 1 s has been achieved. A further reduction of the decay time can be performed by increasing the concentration of polymerisable monomer, but the threshold voltage increases strongly with the concentration.

21.3.4 Phased array antenna

In this section, a phased array antenna that uses an LC loaded phase shifter is explained. Figure 21.9 shows a schematic diagram of the phased array antenna. An element of the phased array antenna consists of a phase shifter and an antenna. When the elements are arrayed in a row, the focusing direction of the microwave can be steered in a direction perpendicular to the row of elements. If the microwave needs to focus on any direction, the elements need to be arranged in a matrix in a plane. In this case, the phased shift amount of the phase shifter must be able to vary

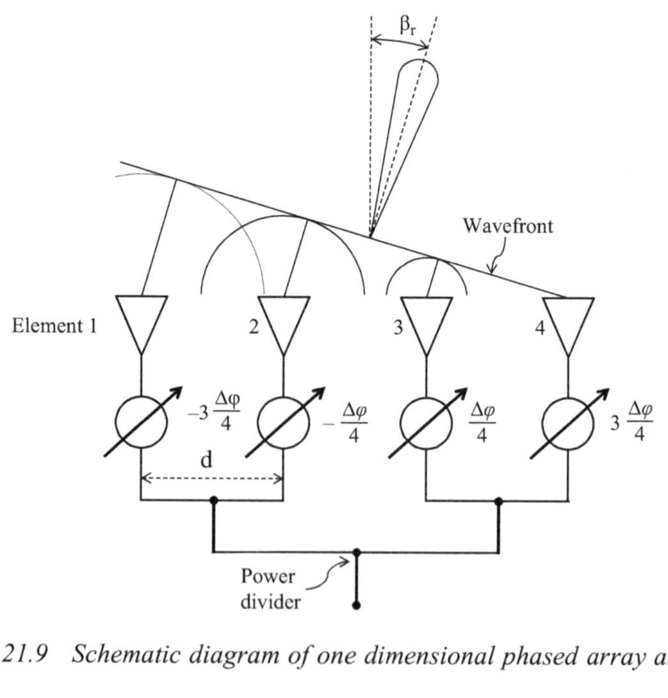

Figure 21.9 Schematic diagram of one dimensional phased array antenna

up to 360°. Typically, the number of the elements in a row range from 10 to 100, and the number of elements in the entire matrix ranges from 100 to 10,000. The element of the phased array antenna often uses a patch antenna because the patch antenna is a non-directional antenna and can be easily connected to the MSL.

21.4 New antenna technology using LCs for satellite communications

New antenna technologies that do not use phased array antennas have been proposed. In these technologies, a metamaterial is used [18–20]. The metamaterial is an artificial material and possesses a negative refractive index. In the microwave region, the metamaterial is obtained by using a composite right- and left-handed (CRLH) transmission line [21]. Figure 21.10 shows an equivalent circuit model of the CRLH transmission line. In the CRLH transmission line, a negative refractive index is achieved for a specified frequency range. The negative and positive refractive indexes can be changed when the capacitance of the capacitor in the CRLH transmission line changes by the reorientation of the LC molecules. The CRLH leaky wave (CRLH LW) antenna consists of a CRLH transmission line and periodical metal slots [22]. When the refractive index is positive, the transmission line is right-handed and the radiation direction is forward. However, when the refractive index is negative, the transmission line is left-handed and the radiation direction is backward. Therefore, the radiation direction can be steered, as shown in Figure 21.11. Therefore, this antenna has potential applications in milliwave radar antennas of automotives.

However, another microwave application has been proposed for satellite communications. This technique consists of metamaterials and a holographic beam form [23]. In this application, a number of antenna cells are arranged compared with the phased array antenna. The antenna cell does not possess the phase shifter, therefore, each antenna cell has a certain specific phase shift. A metamaterial has an electromagnetic band gap (EBG) between the right- and left-handed regions. In the EBG, the electromagnetic wave cannot exist and the microwave in the frequency range of the EBG cannot propagate. In this application, the coupling between the antenna and the transmission line is controlled by the EBG. In other words, each antenna can switch on or turn off by controlling the orientation of the LC molecules by applying a bias voltage. Therefore, only a few selected antennas

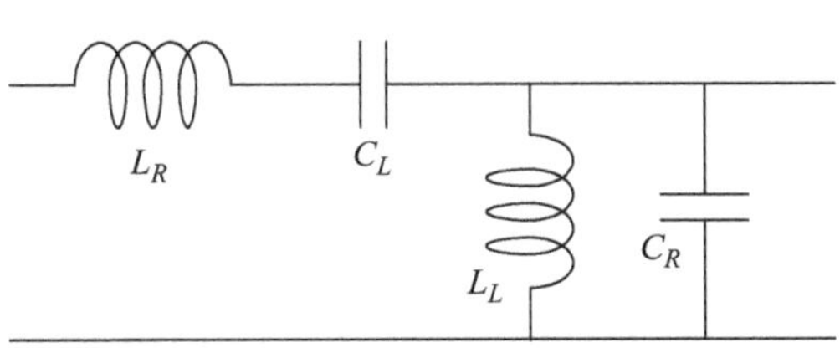

Figure 21.10 Equivalent circuit model of composite right- and left-handed transmission line

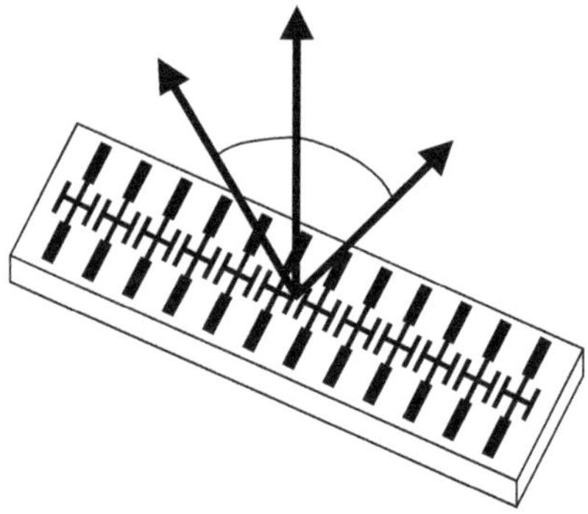

Figure 21.11 Schematic structure of CRLH LW antenna

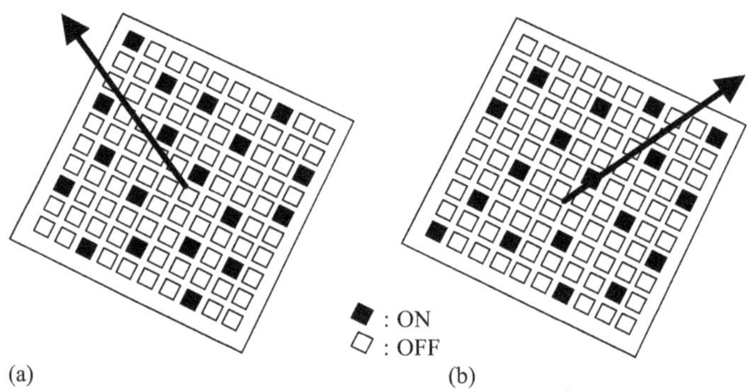

Figure 21.12 Schematic structure of new antenna

cells are used in this system. In this application, the cells are preliminarily selected by using the computing holographic theory, as shown in Figure 21.12. This technique is beneficial for the LC layer. In the phased array antenna using LCs, the LC is used as the dielectric layer in the transmission line, and the thickness cannot decrease because of the increase in the dielectric loss. However, in the proposed application, the LC layer is used as the dielectrics of the capacitor in the CRLH transmission line and the thickness can be the same as that of the LC display. Consequently, the response time in the new application is faster than that in the ordinary LC loaded phase shifter. Therefore, this technique is promising for the microwave applications that use LCs.

21.5 Summary

In this chapter, the fundamental properties of LCs in the microwave region were described first. In the microwave region, the real parts of the complex dielectric constants in both the parallel and perpendicular directions to the director of the LC molecules are almost independent of frequency. However, in the imaginary part of the dielectric constants, tanδ, which is an index of the transmission loss, still cannot be neglected in the low frequency region of the microwave; this is because of the relaxation owing to the rotation of the dipole moments along the molecular short axis. Next, the microwave phase shifter loaded with nematic LCs was described. The molecular reorientation under the application of the bias voltage in the LC loaded MSL phase shifter was explained, and the effective dielectric constant in the phase shifter was calculated. The measured phase shift, rise time and decay time were explained, and the problem of a long decay time was pointed out. Finally, new antenna techniques using metamaterials were described. The use of LCs in the microwave region is promising. In the near future, there will be milliwave and terahertz wave applications that use LCs.

References

[1] Lim K. C., Margerum J. D. and Lackner A. M. 'Liquid crystal millimeter wave electronic phase shifter', *Appl. Phys. Lett.* 1993, 62, 1065–67.

[2] Dolfi D., Labeyrie M., Joffre P. and Huignard J. P. 'Liquid crystal microwave phase shifter', *Electron. Lett.* 1993, 29, 926–28.

[3] Guerin F., Chappe J. M., Joffre P. and Dolfi, D. 'Modeling, synthesis and characterization of a millimeter-wave multilayer microstrip liquid crystal phase shifter', *Jpn. J. Appl. Phys.* 1997, 36, 4409–13.

[4] Kamei T., Utsumi Y., Moritake H., Toda K. and Suzuki Y. 'Measurements of the dielectric properties of nematic liquid crystals at 10 kHz to 40 GHz and application to a variable delay line', *Electron. and Commun. in Japan*, Part 2, 2003, 86, 49–60.

[5] Kuki T., Fujikake H., Kamoda H. and Nomoto T. 'Microwave variable delay line using a membrane impregnated with liquid crystal', *IEICE Trans. Electron.* 2003, E86-C, 1699–1703.

[6] Morita S., Ozaki R., Moritake H., Kamei T. and Utsumi, Y. 'Microwave variable phase shifter of coplanar waveguide using ferroelectric liquid crystal', *Mol. Cryst. Liq. Cryst.* 2007, 476, 105–13.

[7] Moritake H., Morita S., Ozaki R. Kamei T. and Utsumi Y. 'Microwave variable phase shifter of coplanar waveguide using ferroelectric liquid crystal', *Jpn. J. Appl. Phys.* 2008, 47, 1367–70.

[8] Kuki T., Fujikake H. and Nomoto T. 'Microwave variable delay line using dual-frequency switching-mode liquid crystal', *IEEE Trans. Microwave Theory Tech.* 2002, 50, 2604–09.

[9] Utsumi Y., Kamei T., Saito K. and Moritake H. 'Increasing the speed of microstrip-line-type polymer-dispersed liquid-crystal loaded variable phase shifter', *IEEE Trans. Microwave Theory Tech.*, 2005, 53, 3345–53.

[10] Moritake H., Morita S., Ozaki R., Kamei T. and Utsumi Y. 'Fast-switching microwave phase shifter of coplanar waveguide using ferroelectric liquid crystal', *Jpn. J. Appl. Phys.* 2007, 46, L519–L521.

[11] Utsumi Y., Nguyen T. B., Kamei T., Ozaki R. and Moritake H. 'Comparison of microwave measurements and theoretical calculations of dielectric birefringence for a liquid crystal loaded CPW-FE phase shifter', *Mol. Cryst. Liq. Cryst.*, 2009, 510, 197–213.

[12] Kamei T., Moritake H. and Utsumi Y. 'Microwave and light-wave measurements for nematic-liquid-crystal-loaded phase shifter using coplanar waveguide with floating electrode', *Jpn. J. Appl. Phys.*, 2010, 49 01AF03-1–01AF03-6.

[13] Moritake H., Ozaki R., Kamei T. and Utsumi, Y. 'Effective dielectric permittivity of coplanar waveguide type microwave phase shifter using ferroelectric liquid crystal', *Mol. Cryst. Liq. Cryst.* 2011, 542, 158–66.

[14] Nguyen T., Nguyen B. T., Ozaki R. and Moritake H. 'Analyses of nematic liquid crystal molecule motion and response characteristics in microstripline-type microwave phase shifter', *Jpn. J. Appl. Phys.* 2013, 53, 121701-1–121701-10.

[15] Jakeman E. and Raynes E. P. 'Electro-optic response times in liquid crystals', *Phys. Lett. A* 1972, 39A, 69–70.

[16] Yokoyama H. and van Sprang H. A. 'A novel method for determining the anchoring energy function at a nematic liquid crystal-wall interface from director distortions at high fields', *J. Appl. Phys.* 1985, 57, 4520–26.

[17] Panasyuk G. and Allender D. W. 'Approximate description of the three dimensional director and electric field in a liquid crystal display at a high voltage', *J. Appl. Phys.* 2000, 87, 649–57.

[18] Veselago V. G. 'The electrodynamics of substances with simultaneously negative values of a ε and μ', *Sov. Phys. –Usp.* 1968, 10, 509–14.

[19] Smith D. R., Padilla W. J., Vier D. C., Nemat-Nasser S. C. and Schultz S. 'Composite medium with simultaneously negative permeability and permittivity', *Phys. Rev. Lett.* 2000, 84, 4184–87.

[20] Caloz C. and Itoh T. 'Application of the transmission line theory of left-handed (LH) materials to the realization of a microstrip "LH line', *IEEE AP-S Int. Symp.* 2002, 2, 412–15.

[21] Sato K. and Matsuzawa S. 'Electronically scanned lefthanded leaky wave antenna for millimeter-wave automotive applications', *Proc. Int. Workshop Antenna Tech*; New York, USA, 2005, 420–24.

[22] Matsuzawa S., Sato K., Inoue Y. and Nomura, T. 'Nonmembers W-band steerable composite right/left-handed leaky wave antenna for automotive applications', *IEICE Trans. Electron.* 2006, E89-C, 1337–44.

[23] Stevenson R. A., Bily A. H., Cure D., Sazegar M. and Kundtz N. 'Rethinking wireless communications: Advanced antenna design using LCD technology', *Digest of Technical Papers - SID International Symposium* 2015, 46, 827–30.

Index

www.ingramcontent.com/pod-product-compliance
Lightning Source LLC
LaVergne TN
LVHW051048210926
848486LV00001B/50